Full Scale Plant Optimization in Chemical Engineering

Full Scale Plant Optimization in Chemical Engineering

A Practical Guide

Živorad R. Lazić

WILEY-VCH

Author

Dr. Živorad R. Lazić
Milutina Milankovica 74/14
11070 Novi Beograd
Serbia

Library of Congress Card No.: applied for

British Library Cataloguing-in-Publication Data
A catalogue record for this book is available from the British Library.

Bibliographic information published by the Deutsche Nationalbibliothek
The Deutsche Nationalbibliothek lists this publication in the Deutsche Nationalbibliografie; detailed bibliographic data are available on the Internet at <http://dnb.d-nb.de>.

Print ISBN: 978-3-527-35038-4
ePDF ISBN: 978-3-527-83729-8
ePub ISBN: 978-3-527-83728-1
oBook ISBN: 978-3-527-83727-4

Cover Design: SCHULZ Grafik-Design
Typesetting Straive, Chennai, India
Printing and Binding CPI Group (UK) Ltd, Croydon, CR0 4YY

Printed on acid-free paper

C119652_230522

Contents

Preface

"Let your systems learn the wisdom of age and experience"

It is well known that chemical engineers are involved in each step of the process development (lab-scale and pilot-scale) and later in the scale-up and full-scale plant process improvement. So far, there are three powerful methodologies/tools for chemical engineers with wide application in industry:

- Design of Experiments – DOE
- Evolutionary Operation – EVOP
- Data Mining using Neural Networks – DM

The objective of this new book is to compare all three methods in full-scale plant optimization applications and to focus on a simple but powerful technique which provides the experimental tools for full-scale plant optimization. It is called Evolutionary Operation or EVOP.

Chapter I provides the basic principles necessary for a simple understanding of the difference between these three methodologies.

Chapter II covers only two- and three-factor full factorial designs from the Design of Experiments because of the subject of this book. DOE experiments are very efficient in the lab or pilot plant but NOT in the full-scale process. There are two main reasons for that:

- An owner of the large-scale process would hardly accept DOE experiment. To detect significant factor's effect in industrial process, a large change in control factors is required, and that on the other hand will produce off-spec product.
- Also replications required by DOE to reduce noise will produce even more off-spec products.

Chapter III describes theory and application of Data Mining using the Neural Networks modeling tool. In the last twenty or more years, with more data historians installed in process industry, Data Mining, using neural network software, is widely used for industrial process optimization. The main problem with this approach without experiment is static operation of the plant, which means running the plant with almost no change in process factors. Standard Operating Procedures keep factors constant for a very long time. A greater change in process factors is not allowed. If it happens, it will produce off-spec product. This means, all collected data are in very narrow experimental space range, and usually local or global optimum of the process is out of that range.

An analysis of historical data typically has only 10% to 20% chance of success, despite the fact that hundreds or thousands of data points may be available.

Chapter IV, DOE and Data Mining disadvantages, strongly suggests reusing Evolutionary Operation – EVOP experimental technique. Box G.E.P proposed a special Operation as a technique for improving industrial productivity. Its basic philosophy is that it is nearly always inefficient to run an industrial process to produce a product alone; a process should be run so as to generate a product plus information on how to improve that product. The procedure consists of carefully planned cycle of minor variants on the standard work process. The routine procedure consists of running each of the variants continually repeating the cycle. Usually the effects of these deliberate changes in the factors-variables are masked by the large errors inherent in large-scale production units. However, since production will continue anyway, a cycle of variants, which do not affect production significantly, can be run almost indefinitely. Because of constant repetition the effect of small changes can be detected. The wisdom of the age and long experience suggests that EVOP tools help chemical engineers dealing with two different objectives:

- Scale-up from lab and pilot-plant to full-scale operation and "fine tuning" of large-scale production.
- Once the optimum is achieved at certain point of time, it is going to be a moving target and it is very hard or impossible to use tools like DOE and DM due to variability of raw materials quality, energy supply fluctuation, tooling wear, gradual equipment misalignment, shifts in ambient temperature and humidity, corrosion, decreasing heat transfer, aging equipment and instrumentation, aging catalyst and scale build-up, etc.

Chapter V describes theory and applications of five different EVOP methods including Box EVOP, with numerous examples and simulations to compare efficiency:

- Box Evolutionary Operation – BEVOP
- Rotating Square Evolutionary Operation – ROVOP
- Random Evolutionary Operation – REVOP
- Simplex Evolutionary Operation – SEVOP
- Quick Start EVOP – QSEVOP.

These simulators are the perfect tools for training operators in different EVOP techniques. Also they could be used to run full-scale plant optimization. In meantime, a special software "EVOP Engine" is under construction and will be released after the book is published.

Research and development scientists and engineers may find the book helpful in optimizing various processes to be more efficient, less energy intensive, less time consuming, more reliable, and less wasteful of raw materials.

This book has come into being a product of many years of research activities at Military Technical Institute, Belgrade, and long experience in process and product development with large-scale production, including Lenzing Fibers Corporation, Austria; MacDermid, TN, US; and BASF Catalysts LLC., GA, US. The author is especially pleased to offer his gratitude to Helena Smuckler and Anica Lazic for editing of the manuscript. I express my special gratitude to Milica Pojič. for helping in a search for the literature.

Petrovac, Montenegro *Živorad R. Lazič*
June 2021

Biography

Živorad R. Lazič is the author of *Design of Experiments in Chemical Engineering: A Practical Guide*, published by J. Wiley in January 2004. He has produced a unique, "how to do it," a practical guide for the statistical design of experiments. It is the ideal book for the industrial scientist or engineer who wants to take advantage of DOE techniques without becoming a statistician. Basic statistical ideas are presented clearly and simply with numerous examples. This is one of the few books that are practically suited for self-study by a busy technologist, engineers, and scientists. He is a Certified Six-Sigma Black Belt professional with interests in advanced statistical tools, Design of Experiments (DOE), Statistical Process Control (SPC), Evolutionary Operation (EVOP), and process modeling via application of neural networks.

Živorad received his B.S., M.S., and Ph.D. in chemical engineering from the University of Belgrade, in Belgrade, Serbia. He began his career with Viscosa-Loznica Corporation (1975–1979). After that he worked for Military Technical Institute (VTI), Belgrade, where he took a position as Head of R&D Department for composite rocket propellant. He was trained in Hercules, McGregor, TX, in 1982 and 1986. Lazić moved abroad in 1994 due to the war which took place in former Yugoslavia.

He spent more than six years as Vice President for process and product development in P.T. South Pacific Viscose, Indonesia, subsidiary of Lenzing Group, Austria. After he moved to USA in 2001, he worked as Quality Assurance manager at Lenzing Fibers Corporation, Lowland, TN, and Black Belt Six Sigma engineer at MacDermid Graphics Solutions Morristown, TN.

From 2008 Živorad worked as a Senior Research Scientist at BASF Catalysts LLC., Gordon, GA., USA. He retired from this position in 2014.

1

The Basic Ideas

1.1 Introduction

A typical industrial process passes through many stages of development. Thus, in the evolution of a chemical process to produce a certain product, first comes the idea for a promising manufacturing step, followed by long laboratory work to approve concept. The laboratory results provide a preliminary estimate of feasibility and may lead to the possible industrial process. Those results may then be used to build a pilot plant. The pilot plant is a next step between lab and full scale and will have sufficient flexibility to test more extreme conditions. Engineers will use data from pilot plant to design a full-scale plant. Assuming that the plant is built then, ideally, plant will run with operating conditions generated from lab and pilot plant. The small-scale work will have provided estimates of process parameters like concentrations, temperatures, pressures, and flow rates. These estimates of process conditions represent only good first guesses in a continuing process adjustment. This is the reason for special attention to plant startup. A special technical team is usually assigned during the startup, since it is realized that major adjustments may be necessary before the process can be made to perform reasonably well (or sometimes to perform at all).

With startup phase completed, a further stage of iteration has been gone through and, hopefully, the process will produce a stable product. The most likely is that product is being produced at lower production rate, at lower yield, and at lower quality than the plant is capable of. The process of "fine tuning" still remains to be done. This "tuning" or optimization process is the main subject of this book.

It is common to find chemical process 10 years after startup with production rates two or three times those originally thought possible, with major yield increases, and with a product of greatly improved quality [1]. Before optimization can be conducted, precise representations of the process, or models, have to be built.

There are two methods, available for chemical engineer, to build a model:

- DM – Data mining, using Historical Data (Data Historian) and neural network modeling technique to build model.
- DOE – Design of experiment to build a regression model based on experimental data.

Full Scale Plant Optimization in Chemical Engineering: A Practical Guide, First Edition. Živorad R. Lazić.

Whenever feasible, the alternative of taking new data to build a model is preferable. This approach is supported by long practical experience. A rule of thumb is that an experimental design – DOE, executed according to the principles, has a 90% or better chance of obtaining the information required for meeting the objective.

An analysis of historical data typically has only a 10–20% chance of success, even though hundreds/thousands of data points may be available.

Typical defects in historical data are:

- The independent variables-factors are often highly confounded, making determination of their effects difficult and ambiguous.
- Independent variables will often vary over narrow ranges, since the process has been operating with the intention of making uniform product, not for the purpose of demonstrating the effects of the factors.
- The lack of the randomization leads to large bias errors.
- The data often contain many erroneous bad data that must be corrected or deleted before further analysis.
- Proportion of data where one or more variables have missing data is often large. One missing factor disqualifies an entire row in regression analysis.

Historical Databases–Data Historians are more widely available with the proliferation of on-line process computers and database management programs. In many cases, the abundance of numbers greatly exceeds the useful information contained in the data.

Frequently, more than half of the effort in the analysis of historical data is required to clean up the files to a state that will yield preliminary estimates of descriptive statistics. Furthermore, the successful analysis of historical data requires the skills of a statistician experienced in the analysis of "messy data." The latest breakthrough in process modeling and optimization is based on historical data and neural network with fuzzy logic is used for nonlinear modeling. The "Process Insights"[1] (Pavilion Technologies Co.) is one of the best neural network model-based applications for chemical industry.

These remarks should not be accepted as a reason against all analyses of historical data. Such analyses are necessary in fields of economics and social science, where direct experimentation is difficult if not impossible. But in the fields of engineering and technology, where direct experimentation is possible, the chances of success are greatly enhanced by using the DOE strategy [3]. As a preliminary step for designed experimentation, simple plotting of historical data is often useful to obtain a clue for potential problem cause and suggestion for the next steps. On the other hand, the design, execution, and analysis of well-planned experiment will often be less costly in both time and money than an elaborate and massive analysis of historical data. The special DOE software packages available are "Design Expert,"[2] "Echip,"[3] "MiniTab,"[4] etc.

1 Process Insights, Pavilion Technologies Co., Austin, TX, USA.
2 Design-Expert, Stat-Ease, Inc., Minneapolis, MN, USA.
3 ECHIP, Incorporated, 724 Yorklyn Road, Hockessin, DE, USA.
4 Minitab LLC, State College, Pennsylvania, USA.

2

Design of Experiments – DOE

Design of experiments (DOE) is a discipline that applies statistics to the experimental process. It enables researchers and engineers to systematically vary a number of independent (input) variables-factors to evaluate their effect on dependent (output) variables or responses. As the factors are changed in the experimental process, their relationships, effects, and interactions are measured, analyzed, and mapped. Statistical models of the system play a key role in this strategy.

DOE gives the unparalleled understanding of, and power over, the experimental process. There is simply no other approach that yields the same level of understanding or insight into the way a given system of factors interact to produce responses.

Moreover, the approach is also far more economical and efficient than alternatives. Other experimental techniques are typically ad hoc. They involve testing while holding the others constant. This approach is dependent for its success upon informed hunches, guesswork, experience, luck, and intuition. Moreover, since each experimental run only looks at one factor, many runs are necessary even to get a limited amount of information about system.

DOE, by contrast, is a planned approach to experimentation. The design is a plan for deliberately and systematically changing factors to evaluate their effect upon responses. The design tells the experimenter ahead of time how much experimentation is necessary to reach the required objective. Moreover, because a design is based on statistics, the greatest amount of meaningful data possible is collected from the fewest number of runs. As a result, costs in both time and resources can be anticipated, controlled, and reduced.

DOE takes the art out of experimentation and substitutes science in its place. But perhaps even more important, DOE provides a range of information about the process or product being studied that is simply not available through classical one-factor-at-a-time methods [2].

Generally speaking, there are the three most common objectives of any experimental research:

- Screening potential factors
- Mathematical modeling, and
- Optimization.

Full Scale Plant Optimization in Chemical Engineering: A Practical Guide, First Edition. Živorad R. Lazić.

Dealing with three different types of factors:

- Process factors
- Mixture factors
- Process and mixture factors combined together.

My previous book *Design of Experiments in Chemical Engineering* covers the basic theory of DOE and provides numerous practical examples of its application in the chemical process industries, including all three experimental objectives and types of factors [3].

Because of the subject of this book, only full factorial designs are covered. The process conditions, which we may vary, are defined by process factors of two kinds, quantitative and qualitative. Quantitative factors can be varied on a continuous scale. Examples are temperature, concentration, pressure, and agitation speed. Qualitative factors are concerned with attributes and cannot be varied on a continuous scale. Examples are type of catalyst, type of raw materials, shifts, and operators.

2.1 The 2^2 Factorial Designs

Factorial designs are particularly useful arrangements, which may be used to investigate either quantitative or qualitative factors. Suppose we are interested at a particular stage of development of a certain batch process, and two-process factors are being studied. These are the percentage concentration of a certain feed material and temperature at which the reaction is conducted [1].

The four sets of process conditions forming the design are by convention shown in the vertices of a square as in Figure 2.1.

The average for the response (y) fluidity after four complete cycles/replicates are as follows (Table 2.1):

Table 2.1 2^2 Factorial design.

Condition number	**Concentration (%)**	**Temperature (°C)**	**Fluidity average of four cycles**
1	13.5	124	$\bar{y}_1 = 60.2$
2	14.5	124	$\bar{y}_2 = 67.6$
3	13.5	128	$\bar{y}_3 = 73.2$
4	14.5	128	$\bar{y}_4 = 76.2$

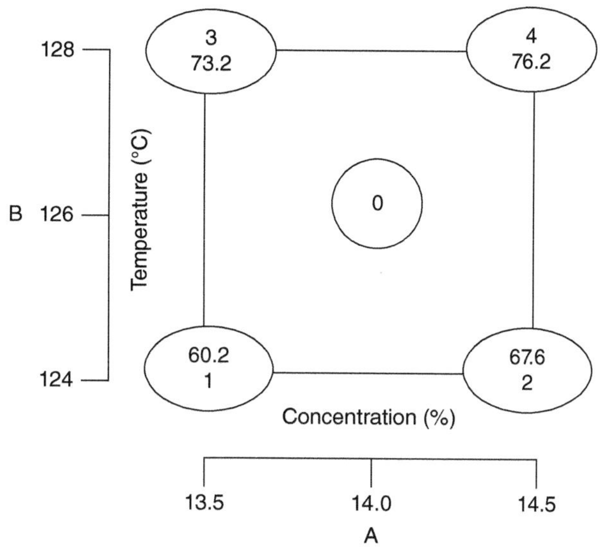

Figure 2.1 Fluidity averages after four cycles.

The four sets of conditions for the factors A (X_1) and B (X_2) can be written down using the symbol minus (−) to indicate the low level and the symbol plus (+) to indicate the high level. At the same time, the symbol zero (0) indicates the center of the experiment.

Mathematically speaking, the change of factor from the center to the higher or lower level corresponds to factor's interval of variation (Δx). It means the change of each factor in units of the factor variation interval is the coded value of the factor (X_i).

$$X_i = \frac{x_i - x_{i0}}{\Delta x} \tag{2.1}$$

where

X_i is coded value of ith factor (nondimensional magnitude)
x_i, x_{i0} are natural or real factor values, their current and center values, respectively
Δx is the natural value of factor variation interval. In a design matrix when we vary factors on two levels (+1; −1), only signs (+; −) exist.

The four sets of process conditions forming the design are shown in Table 2.2.

Table 2.2 2^2 Factorial design in coded values.

Condition number	Concentration $A(X_1)$	Temperature $B(X_2)$	Interaction $AB(X_1X_2)$	Fluidity average of four cycles
1	−	−	+	$\bar{y}_1 = 60.2$
2	+	−	−	$\bar{y}_2 = 67.6$
3	−	+	−	$\bar{y}_3 = 73.2$
4	+	+	+	$\bar{y}_4 = 76.2$

2.2 Effects for 2^2 Factorial Designs

The estimate of the effect on fluidity by changing concentration from 13.5 to 14.5 when the temperature is held at its high level of 128 (°) is $\overline{y}_4 - \overline{y}_3 = 76.2 - 73.2 = 3.0$. This estimate is the simple effect of concentration at the high level of temperature. Similarly, the simple effect of concentration when temperature is at its lower level is: $\overline{y}_2 - \overline{y}_1 = 67.6 - 60.2 = 7.4$. The average of these two simple effects is called the main effect of concentration. It is usually indicated by a single capital letter symbol, in this case A.

$$\begin{aligned} \text{Average} &= \text{main effect } A = 1/2(\overline{y}_4 + \overline{y}_2) - 1/2(\overline{y}_3 + \overline{y}_1) \\ &= 1/2(\overline{y}_4 + \overline{y}_2 - \overline{y}_3 - \overline{y}_1) = 5.2 \end{aligned} \tag{2.2}$$

Equivalently, as before, the main effect of temperature is defined as the average response at high temperature less the average response at low temperature.

$$B = 1/2(\overline{y}_4 + \overline{y}_3) - 1/2(\overline{y}_2 + \overline{y}_1) = 1/2(\overline{y}_4 + \overline{y}_3 - \overline{y}_2 - \overline{y}_1) = 10.8 \tag{2.3}$$

It means the main effects are:

Concentration main effect: $A = 5.2$
Temperature main effect: $B = 10.8$.

Each of these effects is calculated by using all data. It is important to notice that there are four repetitions at each of four sets of conditions, generating 16 observations in all, and each of the main effects is the difference between two averages of eight results.

2.3 Interactions Between Factors

Those two estimates for main effects are considerably different. Some difference would, of course, be expected because of experimental error. Even if there were no such error, however, we should expect that one factor would frequently have a different effect depending on the level of a second factor: for example, the effect produced by change in temperature would usually depend on the setting of other factors such as concentration. Let us consider how we might measure interaction with the data of the present example.

Using Table 2.2, AB interaction signs are product of corresponding signs of factors. The difference of averages indicated by plus and minus signs is estimate for two-factor interaction.

$$\text{AB} = 1/2(\overline{y}_4 + \overline{y}_1) - 1/2(\overline{y}_3 + \overline{y}_2) = 1/2(\overline{y}_4 + \overline{y}_1 - \overline{y}_3 - \overline{y}_2) = -2.2 \tag{2.4}$$

The linear regression model based on 2^2 full factorial design is shown below:

$$\widehat{y} = b_0 + b_1X_1 + b_2X_2 + b_{12}X_1X_2 \tag{2.5}$$

The regression coefficients are calculated as follows:

$$b_0 = \frac{\sum_1^N \bar{y}_u}{N} \tag{2.6}$$

$$b_i = \frac{\sum_1^N X_{iu} \cdot \bar{y}_u}{\sum_1^N X_{iu}^2} = \frac{\sum_1^N X_{iu} \cdot \bar{y}_u}{N} \tag{2.7}$$

$$b_{ij} = \frac{\sum_1^N X_{iu} X_{ju} \bar{y}_u}{\sum_1^N X_{iu}^2} = \frac{\sum_1^N X_{iu} X_{ju} \bar{y}_u}{N} \tag{2.8}$$

where

X_{iu} is coded X_i factor value in u-th design point-trial
$\bar{y}_u$ is response average in u-th design point-trial
u is current number of design points-trials
N is total number of design points-trials in design matrix.

Since the effects are doubled in comparison with regression coefficients, it means:

$b_1 = A/2 = 5.2/2 = 2.6$ $\quad b_2 = B/2 = 10.8/2 = 5.4$ $\quad b_{12} = \text{AB}/2 = -2.2/2 = -1.1$ $\quad b_0 = 69.3$

The regression model in coded factors looks like:

$$\hat{y} = 69.3 + 2.6X_1 + 5.4X_2 - 1.1X_1X_2 \tag{2.9}$$

2.4 Standard Error for the Effects

We have shown how the data coming from the 2^2 full factorial design may be summarized in terms of main effects and interaction. These values are, of course, subject to sampling error. We can easily calculate standard errors for these effects.

$$V(A) = V\left(\frac{1}{2}\bar{y}_4 + \frac{1}{2}\bar{y}_2 - \frac{1}{2}\bar{y}_3 - \frac{1}{2}\bar{y}_1\right) = \left(\frac{1}{4} + \frac{1}{4} + \frac{1}{4} + \frac{1}{4}\right)\frac{\sigma^2}{n} = \frac{\sigma^2}{n} \tag{2.10}$$

And, if we have an estimate s of σ, then standard error (S.E.) for the main effect is:

$$\text{S.E.}(A) = \frac{s}{\sqrt{n}} \tag{2.11}$$

This is also standard error for the main effect B and the interaction AB.

2.5 2^3 Factorial Design

As before the factors involved may be quantitative, such as temperature, concentration, pressure, or qualitative, such as type of catalyst or type of raw material. For illustration, we consider a 2^3 factorial design used to study the effects of temperature A, concentration B, and pressure C on the response: yield of by-product [1]. The factors were run at the following levels (Table 2.3):

Table 2.3 Factors settings.

	Low	High
Temperature *A* (°C)	170	180
Concentration *B* (%)	22	28
Pressure *C* (psi)	30	40

The resulting 2^3 factorial design may then be represented by the eight vertices of a cube, numbered 1–8, as shown in Figure 2.2. The eight sets of conditions, which include all combinations of the pairs of levels, are shown in Table 2.4. Table 2.4 shows the average yields of by-product after three cycles/repetitions. In Figure 2.2, the average by-product yields are shown next to the appropriate corner of the cube.

Table 2.4 The eight sets of conditions of a 2^3 factorial design.

Standard order	Temperature *A*	Concentration *B*	Pressure *C*	Average by-product yields (three cycles)
1	170	22	30	3.9
2	180	22	30	4.8
3	170	28	30	2.9
4	180	28	30	3.6
5	170	22	40	3.5
6	180	22	40	5.1
7	170	28	40	2.8
8	180	28	40	4.4

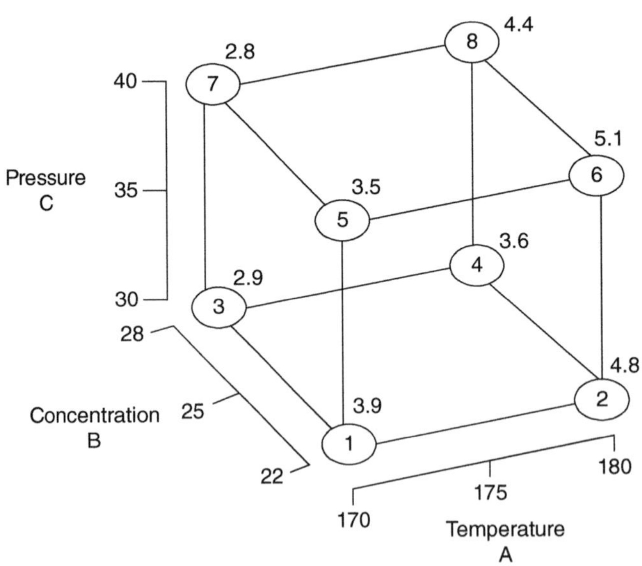

Figure 2.2 2^3 factorial design with average by-product yields after three cycles.

The corresponding table with coded factors and interactions are shown in Table 2.5.

Table 2.5 2^3 Factorial design in coded values.

Standard order	Temperature $A(X_1)$	Concentration $B(X_2)$	Pressure $C(X_3)$	Interactions AB (X_1X_2)	AC (X_1X_3)	BC (X_2X_3)	ABC $(X_1X_2X_3)$	$\bar{y}_i$ (%)
1	−	−	−	+	+	+	−	3.9
2	+	−	−	−	−	+	+	4.8
3	−	+	−	−	+	−	+	2.9
4	+	+	−	+	−	−	−	3.6
5	−	−	+	+	−	−	+	3.5
6	+	−	+	−	+	−	−	5.1
7	−	+	+	−	−	+	−	2.8
8	+	+	+	+	+	+	+	4.4

2.6 Effects for the 2^3 Factorial Designs

Using the same approach like in 2^2 factorial design, main effects and interactions are as follows:

$$A = \frac{1}{4}(\bar{y}_2 + \bar{y}_4 + \bar{y}_6 + \bar{y}_8) - \frac{1}{4}(\bar{y}_1 + \bar{y}_3 + \bar{y}_5 + \bar{y}_7) = 1.20 \tag{2.12}$$

$$B = \frac{1}{4}(\bar{y}_3 + \bar{y}_4 + \bar{y}_7 + \bar{y}_8) - \frac{1}{4}(\bar{y}_1 + \bar{y}_2 + \bar{y}_5 + \bar{y}_6) = -0.90 \tag{2.13}$$

$$C = \frac{1}{4}(\bar{y}_5 + \bar{y}_6 + \bar{y}_7 + \bar{y}_8) - \frac{1}{4}(\bar{y}_1 + \bar{y}_2 + \bar{y}_3 + \bar{y}_4) = 0.15 \tag{2.14}$$

$$AB = \frac{1}{4}(\bar{y}_1 + \bar{y}_4 + \bar{y}_5 + \bar{y}_8) - \frac{1}{4}(\bar{y}_2 + \bar{y}_3 + \bar{y}_6 + \bar{y}_7) = -0.05 \tag{2.15}$$

$$AC = \frac{1}{4}(\bar{y}_1 + \bar{y}_3 + \bar{y}_6 + \bar{y}_8) - \frac{1}{4}(\bar{y}_2 + \bar{y}_4 + \bar{y}_5 + \bar{y}_7) = 0.40 \tag{2.16}$$

$$BC = \frac{1}{4}(\bar{y}_1 + \bar{y}_2 + \bar{y}_7 + \bar{y}_8) - \frac{1}{4}(\bar{y}_3 + \bar{y}_4 + \bar{y}_5 + \bar{y}_6) = 0.20 \tag{2.17}$$

$$ABC = \frac{1}{4}(\bar{y}_2 + \bar{y}_3 + \bar{y}_5 + \bar{y}_8) - \frac{1}{4}(\bar{y}_1 + \bar{y}_4 + \bar{y}_6 + \bar{y}_7) = 0.05 \tag{2.18}$$

Interpretation of the effects for the 2^3 factorial design is available in Ref. [3]. The regression model with coded factors looks like:

$$\hat{y} = b_0 + b_1X_1 + b_2X_2 + b_3X_3 + b_{12}X_1X_2 + b_{13}X_1X_3 + b_{23}X_2X_3 + b_{123}X_1X_2X_3 \tag{2.19}$$

$$\hat{y} = 3.875 + 0.6X_1 - 0.45X_2 + 0.075X_3 - 0.025X_1X_2 + 0.20X_1X_3 + 0.1X_2X_3 + 0.025X_1X_2X_3 \tag{2.20}$$

2.7 Standard Errors of Effects for Two- and Three-Level Factorial Designs

We have discussed in some detail the 2^2 and the 2^3 factorial design. If we perform n cycles-replicates of a 2^p design in p factors, each main effect and interaction will be a contrast between the average of one half of the observations and the average of the other half.

If σ^2 is the variance of the individual observations supposed independently distributed, then the variance of an average of $\frac{1}{2}(n \times 2^p)$ observations is $2\sigma^2/n2^p$. Each main effect and interaction is a difference of two such independent differences and, therefore, has variance [1]:

$$\frac{2\sigma^2}{n2^p} + \frac{2\sigma^2}{n2^p} = \frac{4\sigma^2}{n2^p} \quad (2.21)$$

The standard error for each effect will be obtained by taking the square root of this quantity and substituting the estimate s for the standard deviation σ, as shown in Table 2.6.

Table 2.6 Standard Errors (S.E.) for effects.

Design	2^P	2^2	2^3
Variance (Effect)	$\frac{4\sigma^2}{n2^p}$	$\frac{\sigma^2}{n}$	$\frac{\sigma^2}{2n}$
S.E. (Effect)	$\frac{2s}{\sqrt{n2^p}}$	$\frac{s}{\sqrt{n}}$	$\frac{s}{\sqrt{2n}}$

For illustration, we return to by-product example. From the single data, the estimated standard deviation s after three cycles of a 2^3 design is $s = 0.38$. The 2 S.E. limits for the effects are therefore calculated by:

$$\text{Effect} \pm 2\left(\frac{s}{\sqrt{2n}}\right) \quad (2.22)$$

where $n = 3$, $s = 0.38$. Thus, the 2 S.E. limits for each main effect and interaction are calculated by:

$$\text{Effect} \pm 0.31$$

It was mentioned earlier the basic theory of DOE and numerous practical examples of its application in the chemical process industries, including all three experimental objectives and type of factors, are provided in practical guide [3].

For the purpose of this book, a few examples of DOE application in full scale plant will show what kind of obstacles engineers will face to run DOE.

Example 2.1 Continuous industrial kaolin ozonation and bleaching processing with huge capacity 40 (t/h) are shown in Figure 2.3 [4]. The objective was the experimental cost optimization. After long discussion with plant management, it was accepted to run response surface DOE [3], divided into four blocks to minimize one-time off-spec production. Two factors, ozone dosage and compressed air dosage, were investigated. The responses were final ozone product GEB (General Electric Brightness) brightness and final bleach product GEB brightness. Each block had been completed in separate month and other factors, which were not included in DOE, were calculated after that as a block effect (crude quality changes, float plant performance changes, cold weather effect, rate changes, pH changes, etc.). Ozone dosage range was 0.5–1.0 (lb/t) and compressed air dosage 0.0–40.0 (lb/t).

The experimental design, analysis, and optimization were conducted using "Design Expert" software.

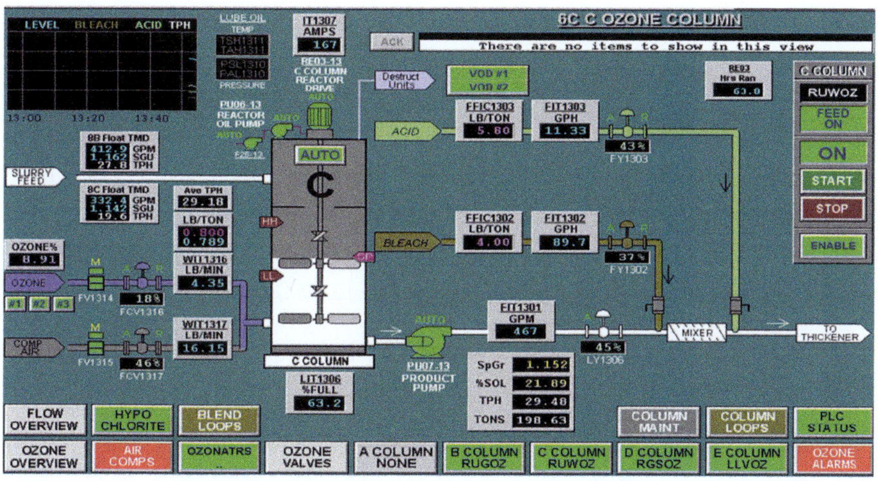

Figure 2.3 Ozone column with bleaching.

The response surface experimental design is presented in Table 2.7.

Table 2.7 Response surface design ($n = 2$; $b = 4$; $N = 20$).

		Responses												
		Ozone feed				Ozone product				Bleach product				
Ozone (lb/t)	Air (lb/t)	GEB	GEB	GEB	Aver.	GEB	GEB	GEB	Aver.	GEB	GEB	GEB	Aver.	BP-OP
1	20	88.5	88.2	88.4	88.37	89.5	89.4	89.6	89.5	90.1	90.1	90.1	90.1	0.6
0.5	40	88.3	88.3	88.3	88.3	88.8	88.9	89	88.9	89.6	89.9	89.7	89.73	0.83
1	20	88.3	88.4	88.4	88.37	89.4	89.4	89.4	89.4	90	89.9	90	89.97	0.57
0.75	30	88.3	88.4	88.4	88.37	88.7	88.8	88.8	88.77	90	90.1	90.1	90.07	1.3
0.5	0	88.2	88.2	87.9	88.1	88.4	88.3	88.4	88.37	88.7	88.7	88.6	88.67	0.3
0.75	0	88.1	88.1	88.1	88.1	88.7	88.9	88.7	88.77	89.6	89.4	89	89.33	0.57
1	0	88.1	88.1	87.9	88.03	89	88.9	88.6	88.83	89.8	89.7	89.5	89.67	0.83
0.5	20	88.1	88.1	88.2	88.13	88.7	88.7	88.7	88.7	89.7	89.8	89.3	89.6	0.9
1	40	88.3	88.3	88.2	88.27	89.1	89	89	89.03	90.2	90.3	90.1	90.2	1.17
1	40	88.1	88	88	88.03	88.7	88.8	88.9	88.8	90.2	90.2	90.1	90.17	1.37
0.5	40	88.5	88.6	88.6	88.57	89.1	89	88.9	89	89.8	89.6	89.7	89.7	0.7
1	0	88.5	88.5	88.5	88.5	88.7	88.6	88.7	88.67	88.9	88.7	89	88.87	0.2
1	0	88.3	88.3	88.4	88.33	88.8	88.7	88.8	88.77	89	89	89.2	89.07	0.3
0.625		88.4	88.3	88.4	88.37	88.9	88.8	88.9	88.87	89.7	89.6	89.7	89.67	0.8
0.875	20	88.4	88.4	88.5	88.43	89	89.2	89.2	89.13	90.1	90.2	90.2	90.17	1.03
0.5	0	88.4	88.3	88.3	88.33	88.8	88.7	88.6	88.7	89.4	89.4	89	89.27	0.57
0.5	40	88.2	88.3	88.4	88.3	88.6	88.7	88.9	88.73	90.3	90.2	90.3	90.27	1.53
0.5	40	88.3	88.3	88.3	88.3	89	88.9	88.9	88.93	90.4	90.3	90.3	90.33	1.4
0.875	10	88	87.9	88.1	88	89	89.2	89.28913		90.1	90	90	90.03	0.9
0.5	0	88	87.9	88	87.97	88.8	88.9	88.8	88.83	89.2	89.3	89.3	89.27	0.43

The ozone product GEB brightness calculated by software is shown in Figure 2.4.

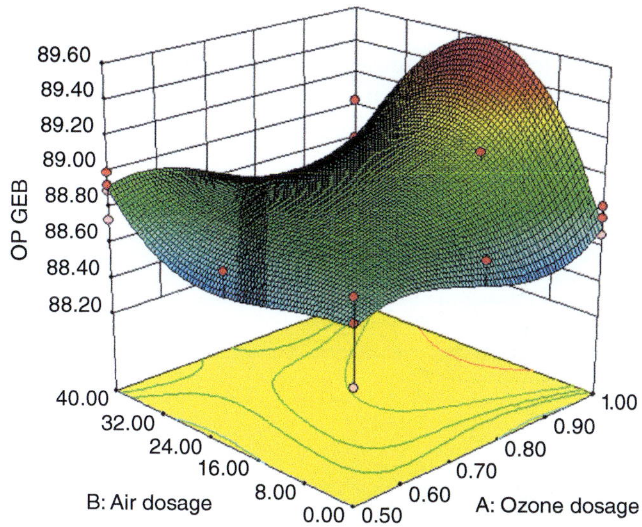

Figure 2.4 Ozone product GEB response surface.

The contour diagram for the same response is seen in Figure 2.5.

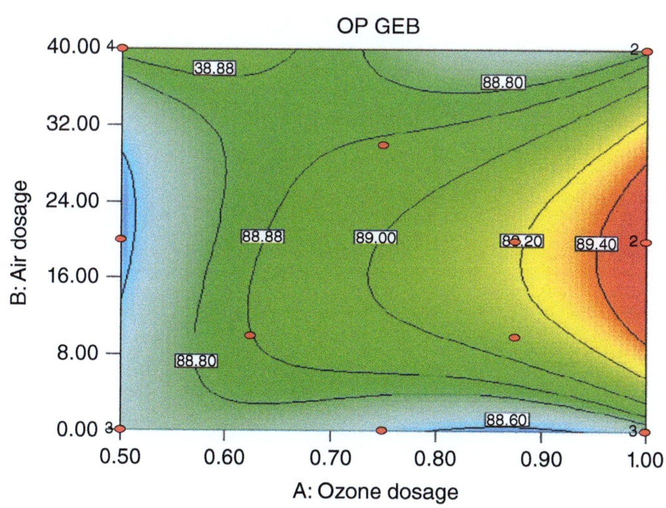

Figure 2.5 Contour diagram for ozone product GEB response surface.

Based on the response surface and contour diagram, conclusions are as follows:

- The right amount of compressed air with higher ozone dosage will increase ozone product brightness (OP-GEB).
- With too less or too much air dosage, ozone dosage will have no effect on OP-GEB.
- Possible explanation is: too low air dosage means poor contact between ozone and clay and with more air, ozone dilution will have negative effect on ozone product brightness.

The bleach product GEB brightness calculated by software is shown in Figure 2.6.

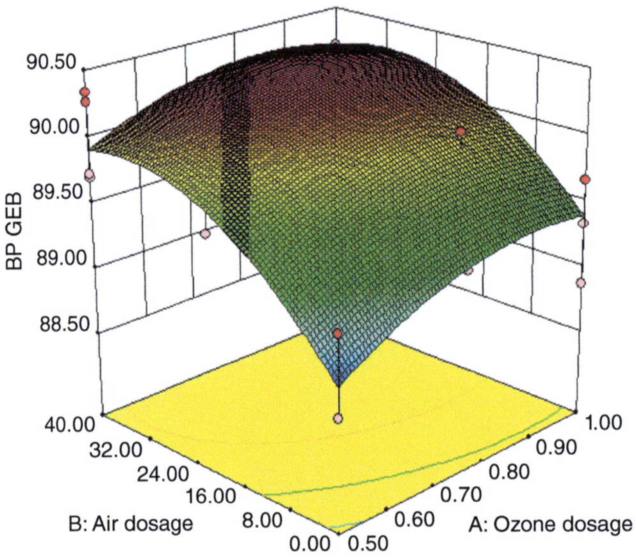

Figure 2.6 Bleach product GEB response surface.

The contour diagram for the same response is seen in Figure 2.7.

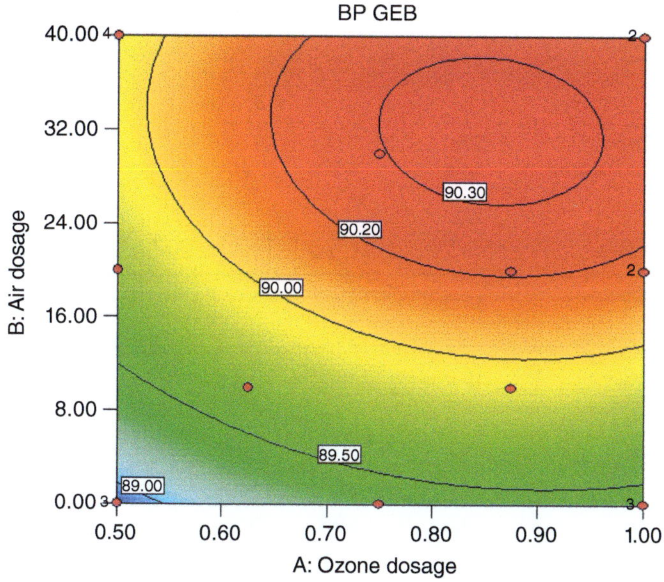

Figure 2.7 Contour diagram for bleach product GEB response surface.

Based on those two diagrams for bleach product brightness (BP-GEB), it is very easy to conclude:

The interaction between ozonation and bleaching converted OP-GEB saddle response surface into parabolic BP-GEB surface. The bleach product brightness optimum or "sweet spot" is between 0.65 to 1.0 (lb/t) of ozone dosage and 20 to 40 (lb/t) of compressed air dosage.

The optimization tool from software suggested optimum settings for ozone column, Ozone Dosage 0.8 (lb/t) and Compressed air Dosage 33.0 (lb/t).

It was demonstrated by verification trial that reduction in ozone dosage from 1.0 (lb/t) to 0.8 (lb/t) resulted in significant savings in plant $960,000 per year based on the current production level. It is also important to mention off-quality product requiring additional beneficiation and costs from runs with lower brightness.

The off-spec product cost was significant but not included in savings calculation. That was the main reason why it was so difficult for plant management to accept DOE experiment on process with huge production rate.

Example 2.2 The Shockwave Power Reactor (SPR) unit with controlled cavitation was used in kaolin production experiment to improve ozonation process [5]. Plant management accepted to run D-Optimal designed experiment for a limited time to minimize time off-quality production. Three factors, ozone dosage, ozone concentration, and slurry flow, are investigated. The responses were final ozone product GEB brightness and final product brightness bleached with 4, 7, and 9 lb of bleach per ton of product. The experimental design and analysis were conducted by using software "Design Expert" (Table 2.8).

Table 2.8 D-Optimal design ($n = 3$; $N = 19$).

	Factors			PropertiesSPR product			
	Ozone dosage	Ozone concentration	Slurry flow	GEB AS IS	GEB+4 (lb/t)	GEB+7 (lb/t)	GEB+9 (lb/t)
Run order	(lb/t)	(%)	$1/3 \times 10^2$ (gpm)	(%)	(%)	(%)	(%)
2	0.5	10	15.0	89.2	90.1	90.4	90.5
19	1.5	8	5.0	89.2	90.2	90.4	90.4
18	0.5	6	10.0	89.3	90.2	90.3	90.3
16	1.0	6	5.0	89.4	90.1	90.4	90.5
15	1.5	6	5.0	89.3	90.1	90.4	90.3
1	1.0	10	10.0	89.3	90.1	90.4	90.4
17	1.5	10	5.0	89.4	90.1	90.3	90.3
13	0.5	8	15.0	89.2	90.0	90.4	90.2
9	1.0	8	10.0	89.1	90.1	90.4	90.5
4	1.5	8	5.0	89.2	90.1	90.3	90.4
7	1.0	8	10.0	89.3	90.1	90.3	90.3
10	0.5	8	15.0	89.2	90.1	90.2	90.3
14	0.5	6	10.0	89.3	90.1	90.5	90.4
5	1.0	6	5.0	89.1	90.0	90.3	90.4
11	1.0	10	10.0	89.2	90.0	90.5	90.5
12	0.5	10	15.0	89.2	90.2	90.5	90.5
8	1.5	10	5.0	89.3	90.1	90.4	90.4
3	1.0	6	5.0	89.3	90.1	90.4	90.4
6	1.5	8	5.0	89.3	90.2	90.5	90.4
Control	**1.5**	**10.0**		**89.3**	**90.1**	**90.5**	**90.5**

Based on the raw data, it was easy to see there is no significant difference between runs compared with control sample. "Design Expert" software also confirmed that the model is not significant relative to the noise. There is a 47.18% chance that a "Model F-value" this large could occur due to noise. As mentioned in introduction, there were two reasons for this unwanted outcome: Too small factors variations and/or too high noise in full-scale plant. This can be fixed by using greater change in factors and/or with more replicates. Those steps are possible on lab and pilot plant scale but very rare in full-scale plant because of the management concern about off-quality (waste) product.

3

Neural Network Modeling – Data Mining

This is known as predictive modeling, a term used to describe the process of mathematically or mentally representing a phenomenon or occurrence with a series of equations or relationships. These models are composed of inputs and outputs, or manufacturing data for factors and responses. Also called data mining, or knowledge discovery, this often computer-assisted process can uncover hidden patterns and relationships in data. Meaning is extracted from enormous databases through advanced statistical analysis and modeling techniques. Data mining has emerged as a key technology for companies that want to improve the quality of their decision making and gain competitive advantage by exploiting the data available to them.

The goal of predictive modeling technology is to accurately predict outcomes and uncover relationships in data. A model relates a set of inputs such as process factors to outputs-responses such as production yields. Once built, the model can take new manufacturing inputs and produce a prediction of outputs. The closeness with which these predicted outputs match actual performance is a quality measure of the model.

Optimization provides a strategic view of the environment, allowing the user to make a complex playing field understandable through advanced optimization software. The decision maker can control the process by specifying constraints and comparing competing strategies, monitoring the effects, and making decisions.

Optimization allows decision makers to run a simulation environment on the desktop in which they can change controllable factors to see the marginal differences such changes make to the output being optimized. Such changes are extremely complex, and multiple changes happening at the same time make prediction even more difficult. Not only can optimization predict how much an output will be affected by changing a manufacturing factor, but it can also predict how it will change when multiple other factors under the decision-maker's control are adjusted at the same time.

These tools are useful when you have good wealth of data and not much physical knowledge. With neural networks you don't have to worry about the form of the empirical model. Neural network modeling technique is mathematical approach to mimic neuro-biological processes. The technique came out of a psychological research, research originated by McCollock and Pitts (1942) [6], Hebb in early 1950's [7] and revolutionized by Rummelhart (backpropagation method) in early 1980's [8].

Full Scale Plant Optimization in Chemical Engineering: A Practical Guide, First Edition. Živorad R. Lazić.

Example 3.1 Rayon is a manufactured regenerated cellulosic fiber produced from naturally occurring polymers [9]. Therefore it is not a truly synthetic nor a natural fiber. Most commercial rayon manufacturers today utilize viscose process. This process is very complex, classic technology, which includes a few steps: steeping, pressing, shredding, aging, xanthation, dissolving, ripening, filtering, degassing, spinning, cutting, washing, and drying. The historical data were collected from production line for sixteen process factors and eight properties in the last five years (15082 data records/rows). Process factors include Cl_2 concentration in bleach, lye content in viscose, cellulose content in viscose, viscose maturity, sulfur content in viscose, hemicellulose content in viscose, berol concentration in spinbath, spinneret type, sodium sulfate concentration in spinbath, viscose throughput per day, zinc sulfate concentration in spinbath, sulfuric acid concentration in spinbath, spinning speed, viscose viscosity, stretch, and 2nd "Kolben-Korb-Filter" (KKF) filterability (measured by Coulter counter).

The measured properties-responses include Conditioned Fiber Strength (CONSTR), Fiber Yellowness, Fiber Whiteness, Dyeability, Sliver Cohesion, and three different splinters counts: Spl_1, Spl_2&3, and Spl_4.

The most important fiber property is Condition Fiber Strength (CONSTR), and objective is to model and to find setting for optimal fiber strength, which means maximum fiber strength with minimum cost. The second objective is to use model to predict certain properties because certain critical properties cannot be measured on-line or quickly enough to be useful in controlling the process. Lab results are often too late, infrequent, or inaccurate to be useful for control purposes.

The complete process modeling and optimization will include a few steps:

1. Format and load data files
2. Data preprocessing
3. Create and train a model
4. Use sensitivity analysis to determine which input factors have the most significant effect on the fiber strength
5. Use neural network model for the most important process factors to find optimum.

3.1 Data Preprocessing

Raw data usually come from different systems with different types of formats, and data may be sampled at different time frequencies. To extract data from raw data files and use them to produce a dataset, suitable for building and training a model, is called formatting procedure. Synchronizing the sampling intervals will be done in the preprocessing step.

The data preprocessing includes Time Merge routine, which will put all data (process and QC data) on a uniform sampling interval. The Time Merge tool will interpolate, extrapolate, or compress variables as necessary to place all variables on the sampling interval specified in the interval box. As a general rule, the value for

Table 3.1 Five years, process data and fiber strength.

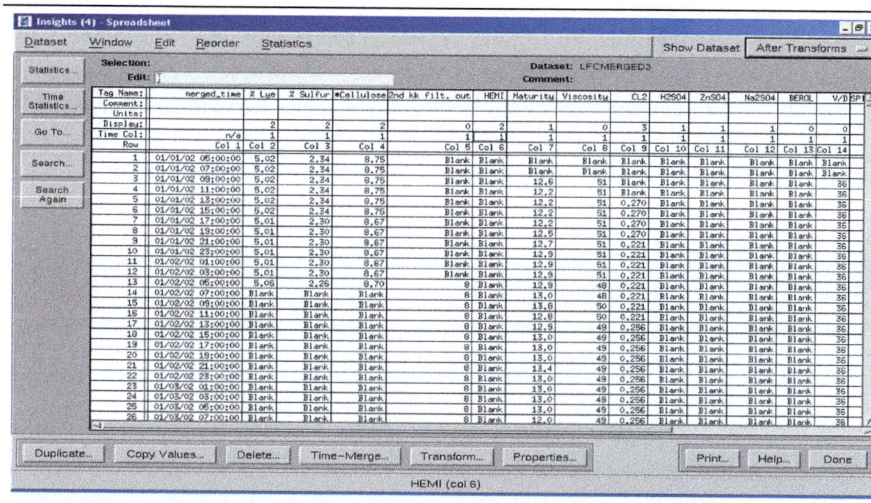

Tag Name:	merged_time	% Lye	% Sulfur	*Cellulose	2nd kk filt. out	HEMI	Maturity	Viscosity	CL2	H2SO4	ZnSO4	Na2SO4	BEROL	V/D	SP
Comment:															
Units:															
Display:		2	2	2	0	2	1	0	3	1	1	1	0	0	
Time Col:	n/a	1	1	1	1	1	1	1	1	1	1	1	1	1	
Row	Col 1	Col 2	Col 3	Col 4	Col 5	Col 6	Col 7	Col 8	Col 9	Col 10	Col 11	Col 12	Col 13	Col 14	
1	01/01/02 05:00:00	5.02	2.34	8.75	Blank	Blank	Blank	Blank	Blank	Blank	Blank	Blank	Blank	Blank	
2	01/01/02 07:00:00	5.02	2.34	8.75	Blank	Blank	Blank	Blank	Blank	Blank	Blank	Blank	Blank	Blank	
3	01/01/02 09:00:00	5.02	2.34	8.75	Blank	Blank	12.6	51	Blank	Blank	Blank	Blank	Blank	36	
4	01/01/02 11:00:00	5.02	2.34	8.75	Blank	Blank	12.2	51	Blank	Blank	Blank	Blank	Blank	36	
5	01/01/02 13:00:00	5.02	2.34	8.75	Blank	Blank	12.2	51	0.270	Blank	Blank	Blank	Blank	36	
6	01/01/02 15:00:00	5.02	2.34	8.75	Blank	Blank	12.2	51	0.270	Blank	Blank	Blank	Blank	36	
7	01/01/02 17:00:00	5.01	2.30	8.67	Blank	Blank	12.2	51	0.270	Blank	Blank	Blank	Blank	36	
8	01/01/02 19:00:00	5.01	2.30	8.67	Blank	Blank	12.5	51	0.270	Blank	Blank	Blank	Blank	36	
9	01/01/02 21:00:00	5.01	2.30	8.67	Blank	Blank	12.7	51	0.221	Blank	Blank	Blank	Blank	36	
10	01/01/02 23:00:00	5.01	2.30	8.67	Blank	Blank	12.9	51	0.221	Blank	Blank	Blank	Blank	36	
11	01/02/02 01:00:00	5.01	2.30	8.67	Blank	Blank	12.9	51	0.221	Blank	Blank	Blank	Blank	36	
12	01/02/02 03:00:00	5.01	2.30	8.67	Blank	Blank	12.9	51	0.221	Blank	Blank	Blank	Blank	36	
13	01/02/02 05:00:00	5.06	2.26	8.70	8	Blank	12.9	48	0.221	Blank	Blank	Blank	Blank	36	
14	01/02/02 07:00:00	Blank	Blank	Blank	8	Blank	13.0	48	0.221	Blank	Blank	Blank	Blank	36	
15	01/02/02 09:00:00	Blank	Blank	Blank	8	Blank	13.8	50	0.221	Blank	Blank	Blank	Blank	36	
16	01/02/02 11:00:00	Blank	Blank	Blank	8	Blank	12.8	50	0.221	Blank	Blank	Blank	Blank	36	
17	01/02/02 13:00:00	Blank	Blank	Blank	8	Blank	12.9	49	0.256	Blank	Blank	Blank	Blank	36	
18	01/02/02 15:00:00	Blank	Blank	Blank	8	Blank	13.0	49	0.256	Blank	Blank	Blank	Blank	36	
19	01/02/02 17:00:00	Blank	Blank	Blank	8	Blank	13.0	49	0.256	Blank	Blank	Blank	Blank	36	
20	01/02/02 19:00:00	Blank	Blank	Blank	8	Blank	13.0	49	0.256	Blank	Blank	Blank	Blank	36	
21	01/02/02 21:00:00	Blank	Blank	Blank	8	Blank	13.4	49	0.256	Blank	Blank	Blank	Blank	36	
22	01/02/02 23:00:00	Blank	Blank	Blank	8	Blank	13.0	49	0.256	Blank	Blank	Blank	Blank	36	
23	01/03/02 01:00:00	Blank	Blank	Blank	8	Blank	13.0	49	0.256	Blank	Blank	Blank	Blank	36	
24	01/03/02 03:00:00	Blank	Blank	Blank	8	Blank	13.0	49	0.256	Blank	Blank	Blank	Blank	36	
25	01/03/02 05:00:00	Blank	Blank	Blank	8	Blank	13.0	49	0.256	Blank	Blank	Blank	Blank	36	
26	01/03/02 07:00:00	Blank	Blank	Blank	8	Blank	12.0	49	0.256	Blank	Blank	Blank	Blank	36	

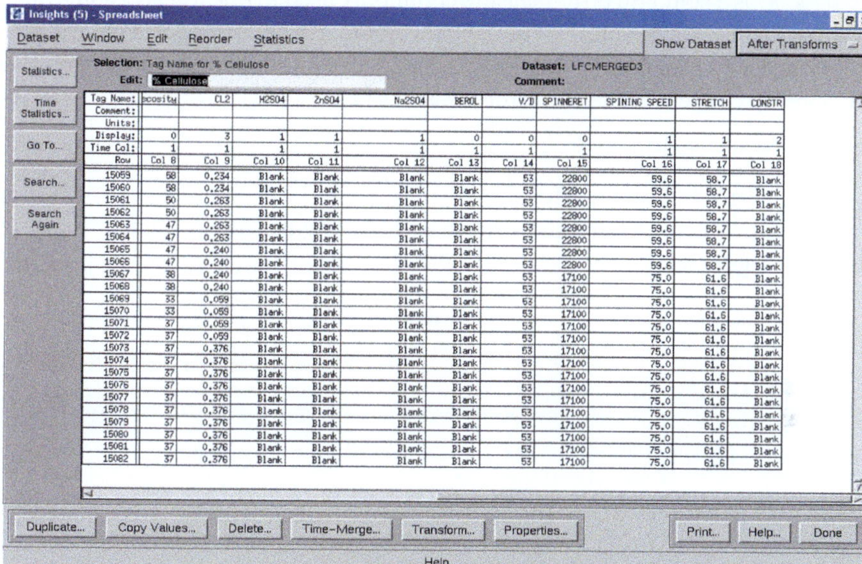

Tag Name:	scosity	CL2	H2SO4	ZnSO4	Na2SO4	BEROL	V/D	SPINNERET	SPINING SPEED	STRETCH	CONSTR
Comment:											
Units:											
Display:	0	3	1	1	1	0	0	0	1	1	2
Time Col:	1	1	1	1	1	1	1	1	1	1	1
Row	Col 8	Col 9	Col 10	Col 11	Col 12	Col 13	Col 14	Col 15	Col 16	Col 17	Col 18
15059	58	0.234	Blank	Blank	Blank	Blank	53	22800	59.6	58.7	Blank
15060	58	0.234	Blank	Blank	Blank	Blank	53	22800	59.6	58.7	Blank
15061	50	0.263	Blank	Blank	Blank	Blank	53	22800	59.6	58.7	Blank
15062	50	0.263	Blank	Blank	Blank	Blank	53	22800	59.6	58.7	Blank
15063	47	0.263	Blank	Blank	Blank	Blank	53	22800	59.6	58.7	Blank
15064	47	0.263	Blank	Blank	Blank	Blank	53	22800	59.6	58.7	Blank
15065	47	0.240	Blank	Blank	Blank	Blank	53	22800	59.6	58.7	Blank
15066	47	0.240	Blank	Blank	Blank	Blank	53	22800	59.6	58.7	Blank
15067	38	0.240	Blank	Blank	Blank	Blank	53	17100	75.0	61.6	Blank
15068	38	0.240	Blank	Blank	Blank	Blank	53	17100	75.0	61.6	Blank
15069	33	0.059	Blank	Blank	Blank	Blank	53	17100	75.0	61.6	Blank
15070	33	0.059	Blank	Blank	Blank	Blank	53	17100	75.0	61.6	Blank
15071	37	0.059	Blank	Blank	Blank	Blank	53	17100	75.0	61.6	Blank
15072	37	0.059	Blank	Blank	Blank	Blank	53	17100	75.0	61.6	Blank
15073	37	0.376	Blank	Blank	Blank	Blank	53	17100	75.0	61.6	Blank
15074	37	0.376	Blank	Blank	Blank	Blank	53	17100	75.0	61.6	Blank
15075	37	0.376	Blank	Blank	Blank	Blank	53	17100	75.0	61.6	Blank
15076	37	0.376	Blank	Blank	Blank	Blank	53	17100	75.0	61.6	Blank
15077	37	0.376	Blank	Blank	Blank	Blank	53	17100	75.0	61.6	Blank
15078	37	0.376	Blank	Blank	Blank	Blank	53	17100	75.0	61.6	Blank
15079	37	0.376	Blank	Blank	Blank	Blank	53	17100	75.0	61.6	Blank
15080	37	0.376	Blank	Blank	Blank	Blank	53	17100	75.0	61.6	Blank
15081	37	0.376	Blank	Blank	Blank	Blank	53	17100	75.0	61.6	Blank
15082	37	0.376	Blank	Blank	Blank	Blank	53	17100	75.0	61.6	Blank

interval should be the interval at which you wish to predict the outputs. However, it should be no smaller than the smallest sampling interval of the inputs. The formatted data are presented in Table 3.1.

3.2 Building, Training, and Verifying Model

This third steps includes building, training, and verifying of the model. When model was built, "Insights" software will start to train it. It means software will begin tuning the model by passing alternately through a set of training data and a set of testing data. Each combined training/testing pass is called "epoch." During the training pass of each epoch, software modifies the internal structure of the model based on the error between the original output value (from the actual data) and the predicted output value (predicted value from the model). An additional set of data is reserved for validating the model after it is trained and is never seen by the model during training. When you train a model, you can stop it manually or you can let the software stop it based on various criteria. The stopping criteria are:

- Auto stop, training stops when the model no longer improves on the test set
- Final Epoch, training stops when the software has completed the specified number of epoch
- Train Relative Error, training stops when the relative error for the training set is equal to or less than the specified error.

The lower these Relative Errors are, the better the model is. A Relative Error of 0.0 indicates that the model predicts data perfectly. Relative Error greater than 1.0 indicates that the model is worse than a model that constantly predicts the mean of the data. R^2 is a standard error measure commonly used in linear regression. Unlike Relative Error, a higher number for R^2 indicates better performance. Relative Error will decrease, while R^2 will increase as the model improves during training. The relationship between these two error measures is:

$$R^2 = 1 - (\text{Rel-Error})^2$$

Relative Error is always positive.

The Relative Error R^2 and the strip chart with original and predicted values are shown in Figure 3.1.

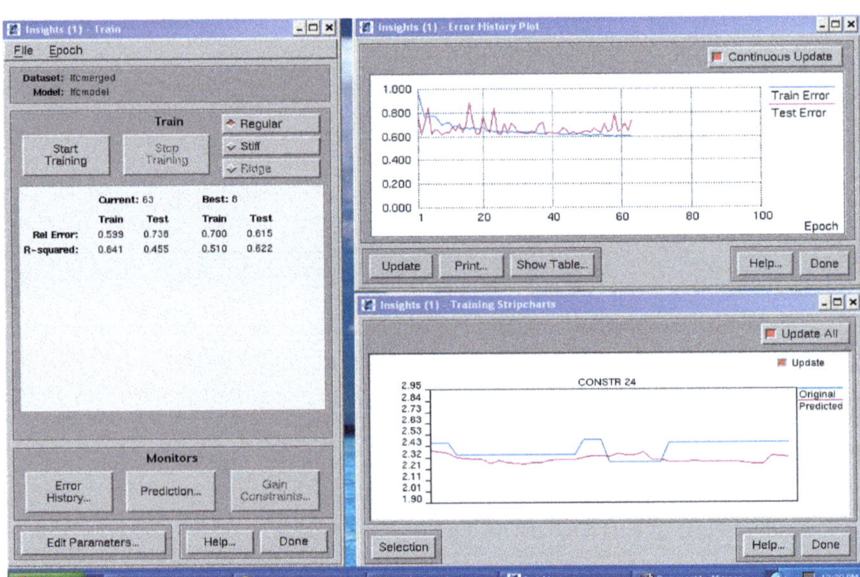

Figure 3.1 The Relative Error R^2 and the strip chart with original and predicted values.

It is easy to see when training begins, the discrepancy between the predicted and actual output values is relatively high. At this point, the model can't generalize very well from data; it cannot yet accurately predict output values from input values. As training progress, the model continues to modify its internal structure to better represent the relationship between input and output variables. The best prediction was reached after only eight epochs. As a general rule, R^2 greater than 0.35 or so (or equivalently, Relative Error of 0.80 or less) indicates that the model is useful for prediction and set point recommendation.

3.3 Model Analyzing

A few tools are available to analyze predictive model:

- Predicted vs. Actual
- Sensitivity vs. Rank
- Sensitivity Report
- Output vs. Percent
- Sensitivity vs. Percent.

A Predicted vs. Actual analysis is used to run a dataset through a trained model and compare the model's Predicted output values with the Actual output values recorded in the dataset. The principal reasons for doing this are:

- To validate a model by running it on data that it never saw during training
- To identify points that a trained model did not learn well
- To view the training/testing Relative Error and R^2 for individual variables in a model with multiple output variables
- To generate predicted model values and include them in the Dataset.

The Predicted vs. Actual plot is a "scatter plot" of the output Conditioned Fiber Strength-CONSTR (Figure 3.2) value predicted by the model against the actual output as recorded in the dataset.

This "perfect model" line, and a line displaying the best fit to the plotted points, is drawn on the plot. This plot visually displays the quality of the modeling fit, which is seen to be quite good. The Sensitivity vs. Rank tool shows which input variables have the greatest effect on the output variable. The influence of a change in input on an output is called the sensitivity or gain. The strategy for process optimization is to identify the most significant factors and optimize them first. This provides a good global (or static) optimization of the process.

The three types of sensitivity analysis are:

- Sensitivity vs. Rank analysis plots the average sensitivities, displayed in descending order, Figure 3.3
- Output vs. Percent, analysis creates plots of input–output relationships, Figure 3.4
- Sensitivity vs. Percent analysis creates plots that are the derivatives of the Output vs. Percent plots, Figure 3.5.

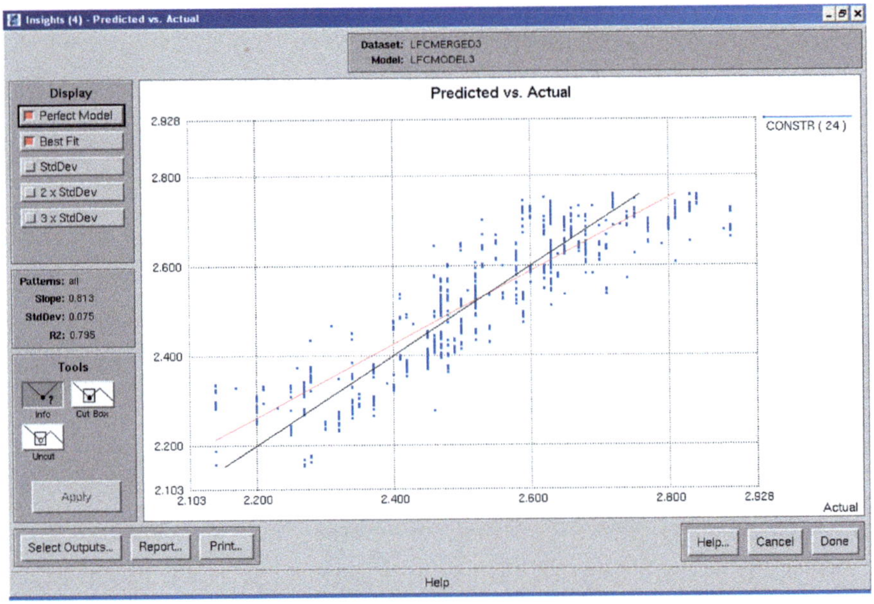

Figure 3.2 The Predicted vs. Actual-Scatter plot.

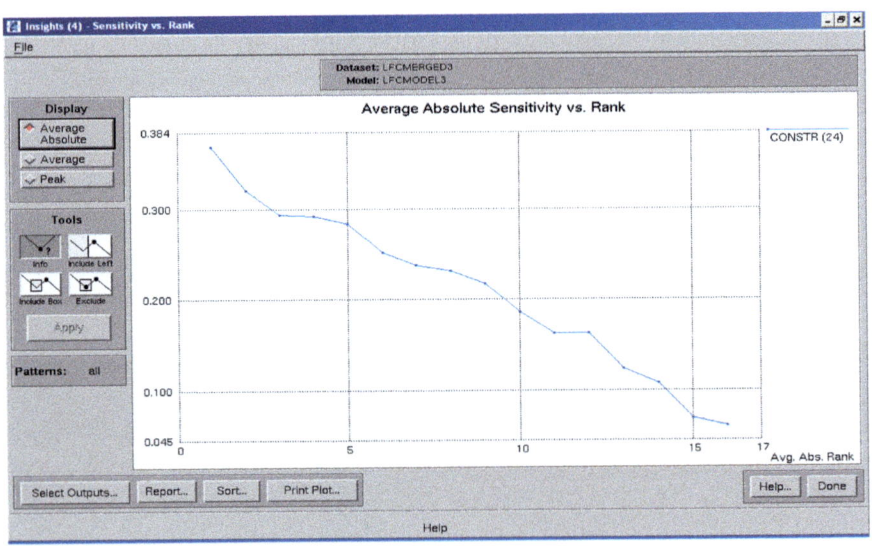

Figure 3.3 Sensitivity vs. Rank analysis plots.

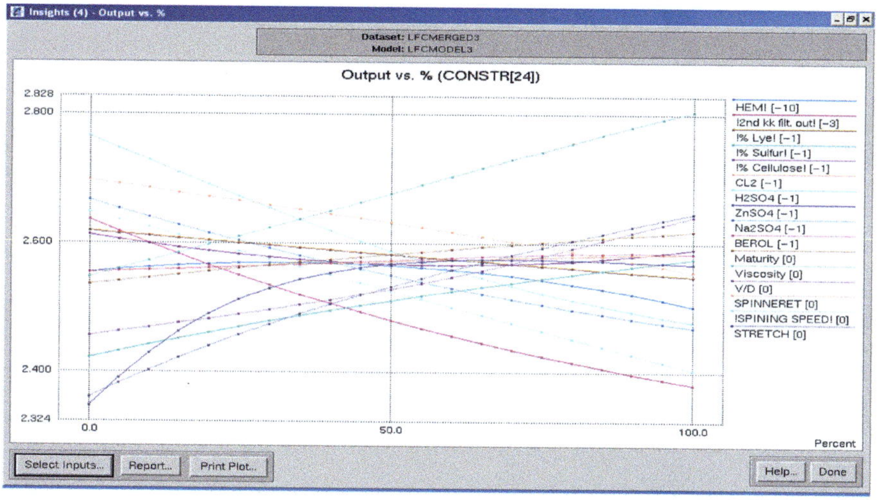

Figure 3.4 Output vs. Percent Plots of the input–output relationships.

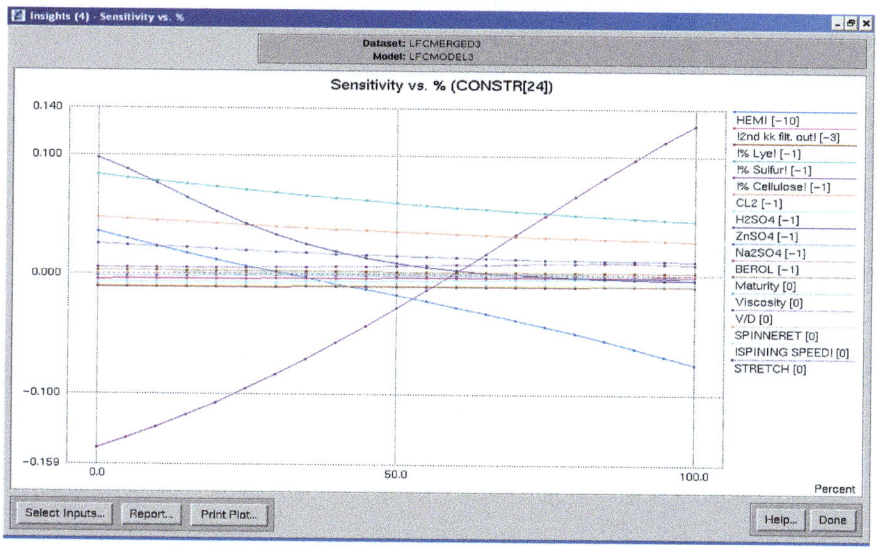

Figure 3.5 Sensitivity vs. Percent analysis plots.

The Output vs. % plot shows the details of how the output varies across the range of each input variable, Figure 3.4.

Sensitivity Report, the variables in the report are ranked according to their Average Absolute values. The sensitivity values show the average impact of the inputs on the output. Average sensitivity is the average change of the output variable as the input variable increases from its minimum to its maximum value (normalized). A positive average value indicates that, on average, the output value increases as the input variable increases. A negative average value indicates that, on average, the output value decreases as the input value increases. Average Absolute is the average of the magnitude (absolute value) of the change in the output variables the input variable increases from the minimum to its maximum value. Average Absolute is always positive. Average Absolute gives a general indication of the strength of the effect of an input on an output. Combined with Average, it can be used to tell whether the input–output relationship is linear, monotonic, or without a causal connection, Table 3.2.

Table 3.2 Sensitivity report.

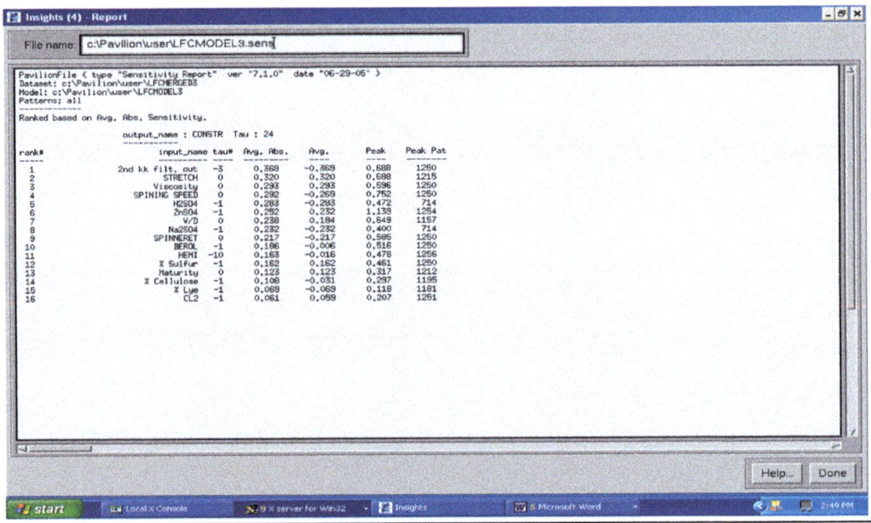

PavilionFile (type "Sensitivity Report" ver "7.1.0" date "06-29-05")
Dataset: c:\Pavilion\user\LFCMERGED3
Model: c:\Pavilion\user\LFCMODEL3
Patterns: all

Ranked based on Avg. Abs. Sensitivity.

output_name : CONSTR Tau : 24

rank#	input_name	tau#	Avg. Abs.	Avg.	Peak	Peak Pat
1	2nd kk filt. out	-3	0.369	-0.369	0.688	1250
2	STRETCH	0	0.320	0.320	0.688	1215
3	Viscosity	0	0.293	0.293	0.596	1250
4	SPINING SPEED	0	0.292	-0.269	0.752	1250
5	H2SO4	-1	0.283	-0.283	0.472	714
6	ZnSO4	-1	0.252	0.232	1.139	1254
7	V/D	0	0.238	0.184	0.649	1157
8	Na2SO4	-1	0.232	-0.232	0.400	714
9	SPINNERET	0	0.217	-0.217	0.585	1250
10	BEROL	-1	0.186	-0.006	0.516	1250
11	HEMI	-10	0.163	-0.016	0.478	1256
12	% Sulfur	-1	0.162	0.162	0.461	1250
13	Maturity	0	0.123	0.123	0.317	1212
14	% Cellulose	-1	0.108	-0.031	0.297	1195
15	% Lye	-1	0.069	-0.069	0.118	1181
16	CL2	-1	0.061	0.059	0.207	1251

It is very easy to see from the Table 3.2, the first five the most significant process factors affecting Conditioned Fiber Strength-CONSTR are:

1. 2nd KKF filterability
2. Fiber stretching
3. Viscose viscosity
4. Spinning speed
5. Spin bath H_2SO_4 concentration.

Three process factors are with negative sign: 2nd KKF filterability; spinning speed; and spin bath H_2SO_4 concentration, which means lower is better regarding fiber strength. Two other factors are with positive sign, fiber stretching and viscose viscosity, which means higher is better.

This was so-called screening step without experiment based on process historical data in a period of five years.

3.4 What-Ifs Optimization

The selected process factors have been included in optimization process using "What-Ifs," Table 3.3.

Table 3.3 Set points and What-Ifs optimization tool.

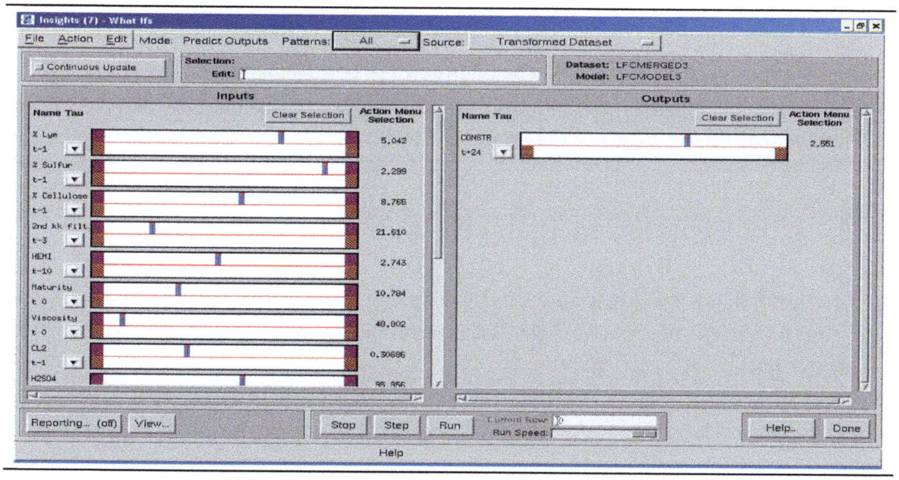

As mentioned in Introduction an analysis of historical data typically has only a 10–20% chance of success, even though hundreds/thousands of data points may be available.

Typical defects in historical data are:

- The independent variables/factors are often highly confounded, making determination of their effects difficult and ambiguous.
- Independent variables will often vary over narrow ranges, since the process has been operating with the intention of making uniform product, not for the purpose of demonstrating the effects of the factors.
- The lack of the randomization leads to large bias errors.
- The data often contain many erroneous bad data that must be corrected or deleted before further analysis.
- The proportion of data where one or more variables have missing data is often large. One missing factor disqualifies an entire row in regression analysis.

Historical databases/Data Historians are more widely available with the proliferation of on-line process computers and database management programs. In many cases, the abundance of numbers greatly exceeds the useful information contained in the data. Frequently, more than half of the effort in the analysis of historical data is required to clean up the files to a state that will yield preliminary estimates of descriptive statistics. Furthermore, the successful analysis of historical data requires the skills of statisticians experienced in the analysis of "messy data."

These remarks should not be accepted as a reason against all analyses of historical data. Such analyses are necessary in the fields of economics and social science, where direct experimentation is difficult if not impossible. But in the fields of engineering and technology, where direct experimentation is possible, the odds of success are greatly enhanced by using Design of Experiments (DOE) strategy.

Keeping in mind, the most important defect in using historical data is:

Independent variables-factors will often vary over narrow ranges, since the process has been operating with the intention of making uniform product, not for the purpose of demonstrating the effects of the factors.

3.5 DOE Experiment Using Neural Networks Model

It was decided to try to overcome this by generating data from neural network model in wider factor's range specified by DOE optimization tool. This kind of extrapolation is always with risk.

The Central Composite Rotatable Design-DOE [1] (Box-Wilson Design) was used for first three selected process factors to perform software designed experiment on a neural network model. It means 20 runs (six replicates) conducted on neural network model and predicted values for fiber strength were used to perform process optimization, Tables 3.4 and 3.5.

Table 3.4 The central composite rotatable design.

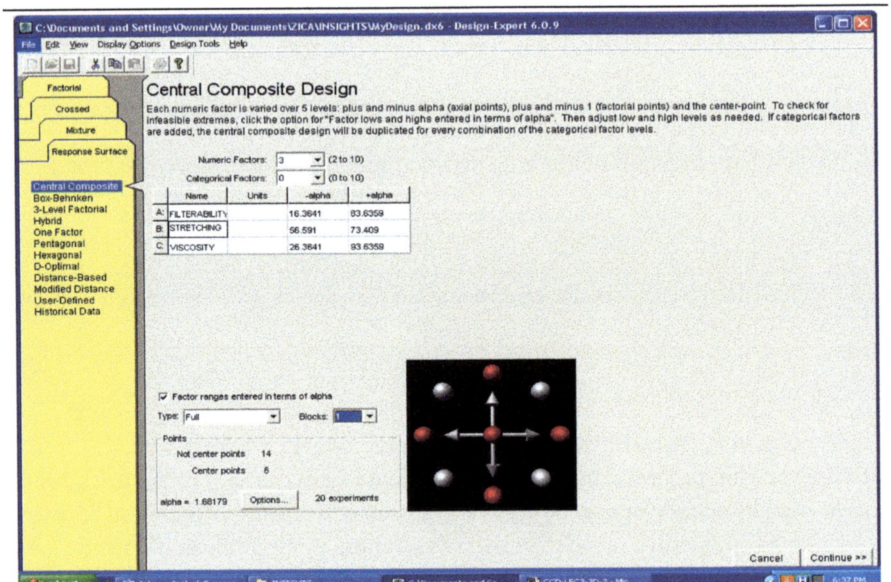

1 Software: Design Expert, "Stat-Ease, Inc", Minneapolis, MN

Table 3.5 The central composite rotatable design matrix.

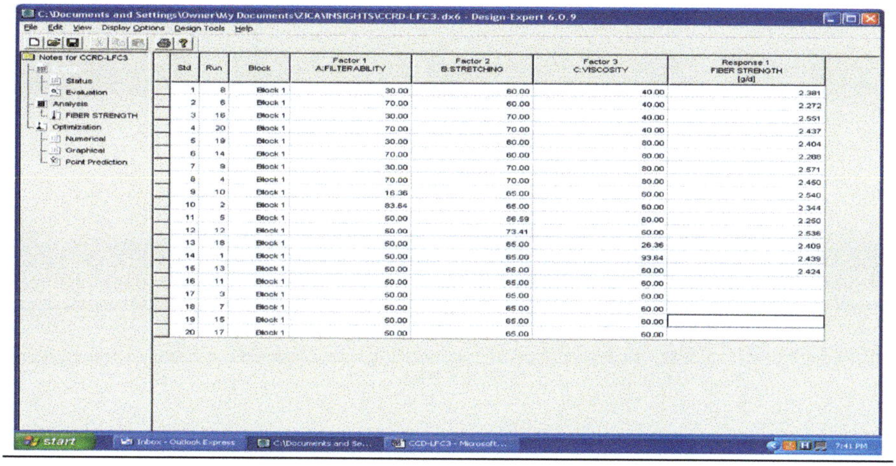

Std	Run	Block	Factor 1 A:FILTERABILITY	Factor 2 B:STRETCHING	Factor 3 C:VISCOSITY	Response 1 FIBER STRENGTH [g/d]
1	8	Block 1	30.00	60.00	40.00	2.381
2	6	Block 1	70.00	60.00	40.00	2.272
3	16	Block 1	30.00	70.00	40.00	2.551
4	20	Block 1	70.00	70.00	40.00	2.437
5	19	Block 1	30.00	60.00	80.00	2.404
6	14	Block 1	70.00	60.00	80.00	2.288
7	9	Block 1	30.00	70.00	80.00	2.571
8	4	Block 1	70.00	70.00	80.00	2.450
9	10	Block 1	16.36	65.00	60.00	2.540
10	2	Block 1	83.64	65.00	60.00	2.344
11	5	Block 1	50.00	56.59	60.00	2.250
12	12	Block 1	50.00	73.41	60.00	2.536
13	18	Block 1	50.00	65.00	26.36	2.409
14	1	Block 1	50.00	65.00	93.64	2.439
15	13	Block 1	50.00	65.00	60.00	2.424
16	11	Block 1	50.00	65.00	60.00	
17	3	Block 1	50.00	65.00	60.00	
18	7	Block 1	50.00	65.00	60.00	
19	15	Block 1	50.00	65.00	60.00	
20	17	Block 1	50.00	65.00	60.00	

Description of the factors and levels is presented in Table 3.6 below:

Table 3.6 The factor's levels.

No.	Factor name	Lower level (−)	Higher level (+)
1.	2nd KKF filterability	30	70
2.	Fiber stretching	60	70
3.	Viscose viscosity	40	80

Now, using the same model, we will use "Insights" Set points and What-Ifs tool to simulate the behavior of the process under various operating conditions specified in Table 3.5. This tool allows you to perform "What-If" simulations, or, in other words, to do "software designed experiment" rather than actually performing designed experiments in plant. This graphical interface, Table 3.3, allows you to change input factor's values and predict the output (fiber strength) for 20 new inputs.

With the aid of software, the data from fiber strength response were fitted to quadratic polynomial. Therefore, although the equation only approximates the true relationship, it will be more than adequate for empirical prediction in wider experimental space than neural network-predictive model.

It contains second-order term, which captures interactions. The relatively small but statistically significant coefficient on AC (Filterability x Viscosity) indicates significant antagonism between these factors.

Based on this model, software can produce extremely useful maps. Figure 3.6. shows 3D surfaces, with projected contour map, for the fiber strength.

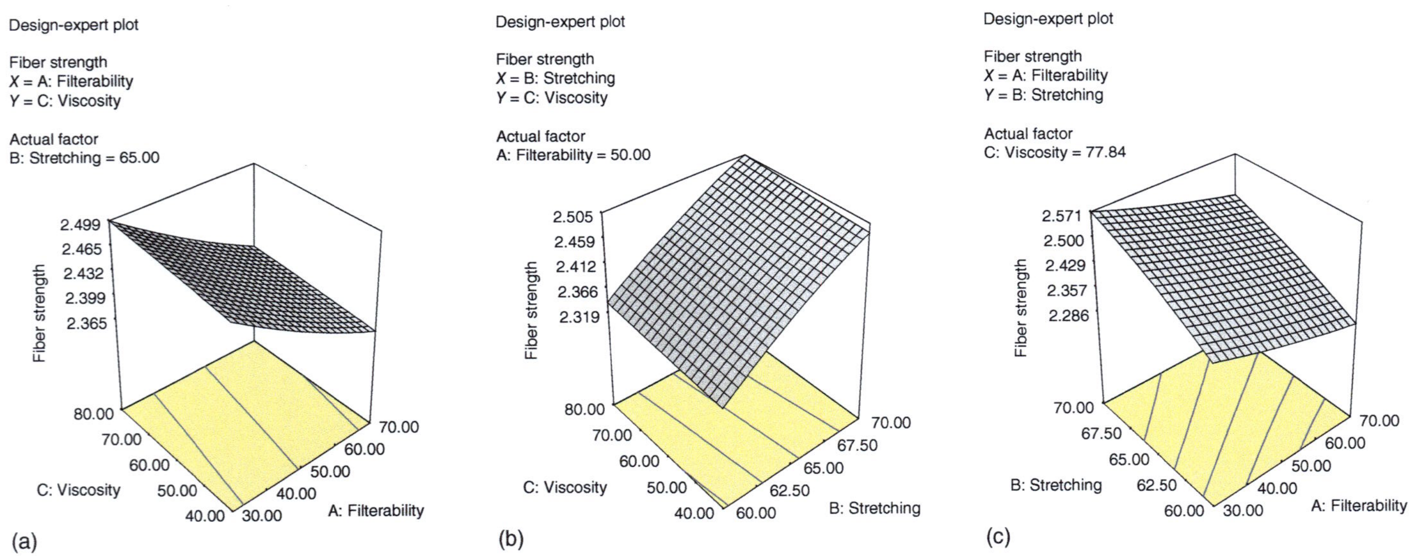

Figure 3.6 Response surfaces.

4

Evolutionary Operation – EVOP

The purpose of this book is to revisit a simple but powerful technique, which provides the experimental tools for full-scale plant optimization. It is called evolutionary operation or more briefly, EVOP [1]. It has been most widely and successfully applied in the chemical industry, but would be very useful in other types of manufacture. Evolutionary Operation is applied on the full-scale plant on the day-to-day basis. It requires no special staff and can be handled by the usual plant personnel after very brief training. Its basic philosophy is that it is inefficient to run an industrial process in such a way that only a product is produced, and that a process should be operated so as to produce not only a product but also information on how to improve the product.

Evolutionary operation is a management tool in which a continuous investigative routine becomes the basic mode of operation for the plant and replaces normal static operation.

4.1 Small-Scale and Plant-Scale Investigation

As mentioned before, EVOP is a plant-scale technique used as a basic operating method. Why should we use the full-scale plant for tuning our process and not a laboratory or pilot plant?

- Because scale-up effects may make "tuning" on other than the full-scale plant pointless,
- Because, whereas the full-scale runs will be made in any case, pilot and lab runs require additional effort and technical personnel.

4.2 Scale-up

Small-scale experimentation can provide useful information at design stage of the process and can provide first guesses at full-scale operating conditions. However, flow characteristics, heat transfer, and mixing will all change as a scale of operation is changed. It is not true that completely different characteristics are to be expected

Full Scale Plant Optimization in Chemical Engineering: A Practical Guide, First Edition. Živorad R. Lazić.

on a large scale. If this were so, design on basis of small-scale units would be useless. In transferring the process from the small scale to the plant scale, the main features of its behavior are usually preserved, but differences in detail can occur, which are important economically. In practice, we have a displacement and the distortion due to scale-up not in a two-dimensional space but in a multi-dimensional space.

Now, let's return to our second reason for using a full-scale process for tuning purposes: even if there were no scale-up difficulties, it would be usually wasteful to develop an operating process for small-scale experimentation alone. The cost of such experimentation is high, and to have sufficient laboratory, pilot facilities, and technical manpower to carry out process development in this way on all processes simultaneously would be wasteful.

In practice, such small-scale experimentation is a high-priced luxury. When we commit these facilities and their specialists to a particular project, we automatically deny them to other projects. By contrast, a full-scale plant itself and the personnel needed to run this plant are already committed in a routine production of product. If these same production runs can be used to provide needed information, it is obviously inefficient to commit other resources. The full-scale apparatus is available, the men to operate it are there and are trained in its use, the runs going to be made anyway. Why not use them? Finally, most likely improvements in plant operation will often result from modifications, which cannot be simulated at all on a small scale.

The third reason for using the full-scale process itself for tuning purposes or better to say for optimization: even where there no scale-up difficulties, there are variabilities in raw materials, energy supply fluctuation, tooling wear, gradual equipment misalignment, shifts in ambient temperature and humidity, corrosion, decreasing heat transfer, aging equipment and instrumentation, aging catalyst, etc. *This means once optimum was achieved in time, it is going to be moving target and it is very hard or impossible to use tools like DOE and data mining.*

4.3 Static and Evolutionary Operation

Routine production is normally conducted by running a plant at rigidly defined operating conditions called *standard operating procedure*. Standard operating procedure means the best conditions of operation known at the time. Manufacturing procedure, where a plant operator aims always to reproduce exactly the same set of conditions, will be called *method of static operation*.

In the EVOP method, a carefully planned cycle of minor variants became dynamic operation. The routine of plant operation then consists of running each of the variants in turn and continually repeating the cycle. The cycle of variants follows a simple pattern, the persistent repetition of which allows to collect physical properties of the product. In this way, we use routine manufacture to generate not only the product but also the information on how to improve process.

Controlled variation has been introduced into production; an effect of minor changes will be presented to the process operator. This allows the operator to see what changes ought to be made to improve production. The stream of information related to products from different production conditions is summarized in the information board. The information is set up so that the process operator can see at any time what the effects of the process changes are and decide to move the variant center to a new point, what type of change are undesirable in terms of product quality, how much it costs to run the scheme, etc. *Evolutionary Operation, however, is virtually a permanent method of running the plant.* For this reason, *only changes in the levels of factors can be permitted whose effects are virtually undetectable in individual runs,* and only techniques simple enough to run continuously by operators themselves under actual conditions of manufacture can be employed.

Suppose a new plant has just been built. In Figure 4.1, we imagine the latent improvements, which could arise from suitable adjustments to the new process measured on some scale such as profitability and arranged in descending order of magnitude. A few (five) very large effects to the left of the diagram will be quickly detected after startup. Their causes will then be tracked down and adjustment and improvement of the process will be completed.

The reason that these effects can be rapidly detected, hence, exploited, is that that "signal" they produce is large compared with the underlying level of variation of the process, sometimes called the "noise."

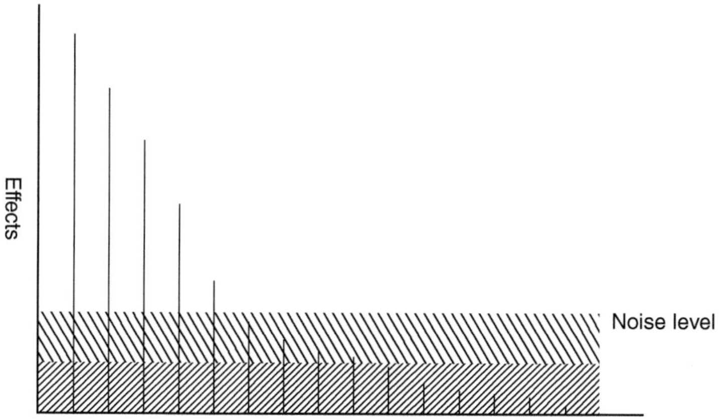

Figure 4.1 Effects and the "noise level".

Noise arises from a variety of sources such as variability of raw materials, inability to maintain precisely input factors at set levels, and from instrument and measurement error. The final variation in the measured response is a composite of all these. The magnitude of variation is measured by a quantity called the "standard deviation."

$$\sigma^2 = \frac{\sum_i (y_i - \mu)^2}{N}; \; S_y^2 = \frac{\sum_{i=1}^{n} (y_i - \bar{y})^2}{n-1} \tag{4.1}$$

At any moment of the plant's life, the size of the variation, measured by the standard deviation, has some definite magnitude that depends on the degree of process control achieved by that time. Some specified multiple of this standard deviation can be called the noise level. If changing the level of input factor produces an effect in response, which greatly exceeds noise levels, it is easy to detect. However, if the effect is much smaller than the noise level, it is unlikely to be detected and cannot be used.

To discover effects, buried in noise, we must improve signal to noise ratio. We must either decrease the effective noise level or increase the signal level. In EVOP, we do both. The signal is increased by introducing greater changes in the factors under study. The effective noise level is reduced by repetition of the changes and averaging of the response results. Figure 4.1. shows how a number of effects, previously hidden, will be shown up if we cut the noise to one-half of its previous value.

Example 4.1 It is intention in this chapter to provide a general description of the whole EVOP method and then to describe the various elements of the method in greater detail in later chapter [1]. Here we briefly describe a simple EVOP scheme for two factors, postponing more complete details of the design and analysis of the results until later.

Factors Under Study

At the particular stage of development of a certain batch process, two-process factors were being studied. These were the percentage concentration of a certain feed material and the temperature at which the reaction was conducted.

A Pattern of Variants or Design

The scheme of variants is shown in Figure 4.2. The current work process is labeled 0 and the four variants are labeled 1, 2, 3, and 4. One batch of product was made at each set of conditions, which were run successively in the order 0, 1, 2, 3, 4; 0, 1, 2, 3, 4; and so on.

Responses of Interest

Three responses were recorded.

1. The cost of manufacturing unit and weight of product. The objective was to bring this cost to the smallest value possible (cost optimization).
2. The percentage of a certain impurity. This should not exceed 0.5.
3. A measure of fluidity. This should lie between the limits of 0.55 and 80.

The Information Board

The information board in Figure 4.2. will contain all information generated from EVOP experimentation. The "phase" number at the top of the board indicates that two previous phases of EVOP have already been completed in this process. The cycle number in the top corner indicates that four cycles of this third phase of operation have been completed.

There follows a plan of the cycle of variants being run. The table below summarizes the current situation. First are the criteria for optimization that must be satisfied. These are followed by the running averages at the various production conditions set out so as to follow the plan of the variants cycle. This arrangement makes it easy to follow results.

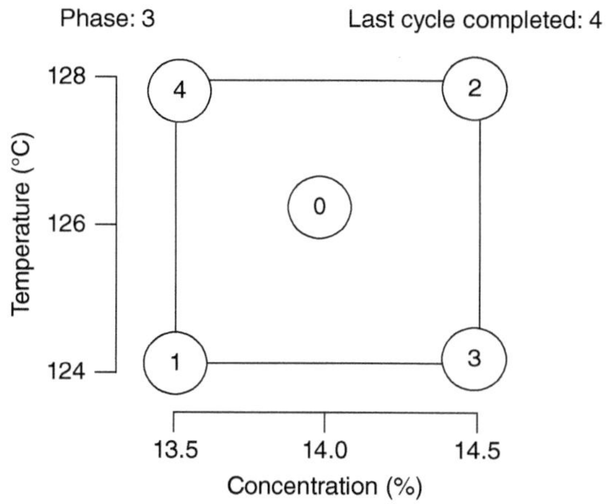

	Cost	Impurity	Fluidity
Requirement	Minimum	Less than 0.5	55 ÷ 80
Running averages	32.6 33.9 32.8 32.3 33.4	0.29 0.35 0.27 0.17 0.19	73.2 76.2 71.3 60.2 67.6
2 S.E. limits	±0.7	±0.03	±1.1

Figure 4.2 Information board after four cycles.

Error Limits

A measure of the significance of the individual running averages is supplied by two standard error limits (2S.E. limits). As we will explain in more detail later, at any given stage of the experimentation, these limits indicate a range, which almost certainly includes the true value. For example, the average of the fluidity measurements made at a 14.5% concentration and a temperature of 128 °C is 76.2. This value is subject to experimental error, but we know from the error limits that the true fluidity at these conditions is almost certainly included in the range of 76.2 ± 1.1. As each cycle is completed, the calculated error limits become narrower and narrower, reflecting the fact that the running averages become more and more reliable. Size of these error limits, at any given stage, informs process operators to what extent the apparent differences in performance, which they notice, may be more random fluctuations and to what extent they probably represent real effects.

4.4 Analysis of Information Board

After studying information on information board, operators can make one of two basic decisions.

- To allow the scheme to proceed unchanged and wait for additional information from the next cycle
- To modify operation in some way, so beginning another phase of the EVOP.

Under option two, a number of possibilities are available. Some are as follows:

- Adopt one of the variants as the new center point and run cycles with this new center point.
- Explore an indicated favorable direction of advance and run cycles about the best condition found.
- Substitute new factors for one or more of the old factors. There will normally be a large number of factors waiting to be tested with EVOP.
- Modify the pattern of variants in any way which seems desirable from the information available at the time. In particular, when after a few cycles no significant effects are found, it may be decided to vary the present factors again but over somewhat wider ranges-intervals.

That decision was, in fact, made in the example that provided the data in Figure 4.2. From the running averages, we can see that a reduction of the percentage concentration apparently reduces cost, decreases percentage impurity, and reduces fluidity. If temperature is decreased, the effect on cost is small but apparently favorable, and a marked reduction in impurity occurs. However, decrease in temperature appears to be also accompanied by major reduction in fluidity. Now, fluidity must be maintained above 55. It seems this limit is rapidly reached if the temperature drops a lot, but that a fairly large reduction in concentration could probably be made before this limit was approached. Although a large reduction in impurity, which could be achieved by a temperature decrease, would be welcome, it is not essential to meet the specification at this time. For this reason, it was decided, in fact, to reduce concentration alone. In the next phase, three sets of conditions (concentration, temperature) – (13%, 126°), (13.5%, 126°), and (14%, 126°) – were compared by running several cycles at these points. The first point gave rise to an average cost of 32.1 (a useful reduction), with an impurity level of 0.25 and a fluidity of 60.7 (still well above 55). This point (13%, 126°) then was selected as a starting point for further investigation. It was selected as the new base for further cycles of the evolutionary process.

We can see from this example that the actions to be taken in various circumstances are not precisely specified. Human judgment is a very important part of the EVOP investigation as it should be in any investigation. A properly run EVOP scheme ensures that the process operators are continuously fed with clear information about that process so they can take appropriate action.

4.5 Three-Factor Scheme

Previous example showed a typical EVOP scheme that examined two factors and provided a rough idea how the results would be examined. We now briefly describe a typical three-factor scheme. Figure 4.3 shows a cycle of variants used to examine the effects of changing the three factors [1]:

1. Rate of temperature rise
2. Air blowing
3. Carbon treatment.

The duplicated center point (runs 0 and ø) was the standard process. The variants 1, 2, 3, 4, 5, 6, 7, 8 were the eight variants obtained by combining the high and the low levels of each of the factors in all possible ways. Figure 4.3 shows the average "texture" observed after five cycles of the scheme. The objective was to reduce this measure of texture to as small a value as possible (performance optimization).

So far, we have considered simple schemes in which two or three factors were simultaneously studied. Although more factors can be examined in an EVOP scheme, it is often unwise to do so. The reason is that in practice two or three factors seem to be about the most that can be handled in a routine manner on the full-scale process. It is better to settle for a modest scheme that will actually be used than to press for a theoretically better one because excuses will be found to discontinue.

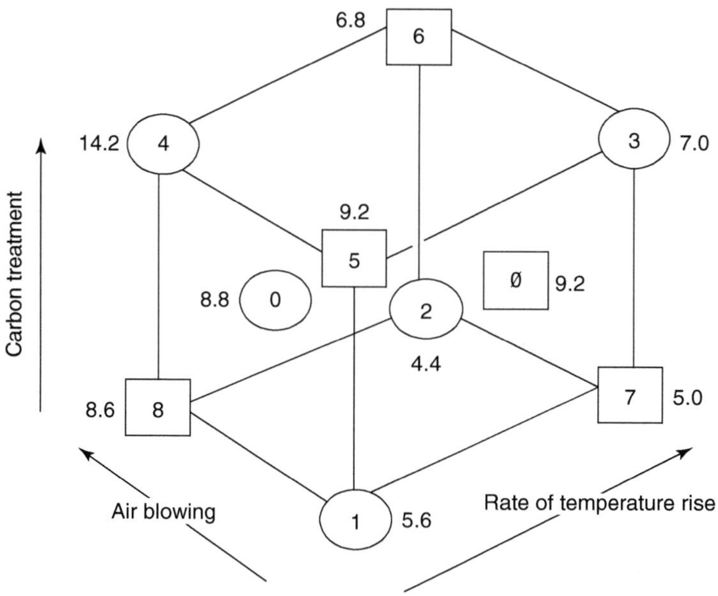

Figure 4.3 Three-factor EVOP scheme.

4.6 Current Best-Known Conditions

As an EVOP program proceeds, there will exist at any given stage, what may be called "*current best-known conditions.*" At the beginning of the program, these conditions will be those given in the standard operating procedure, but as improvements are incorporated the current best-known will change. Potentially, each new phase can establish a new set of current best-known conditions. When a new phase is begun, it is often helpful to be able to compare results directly with an appropriate reference. This can be done by returning to current best-known conditions once in every cycle. In some situations, this will mean that an additional point is added to the design, in others it will happen when the current best-known conditions correspond to one of the standard design points. It means current best-known conditions, which are outside the design, correspond to one of the standard design points or at the center of design.

The main advantage of inclusion of reference conditions is the reassurance this gives to the process operators. Anyone familiar with actual production conditions knows that the unexpected and unexplained frequently happens, even when no deliberate changes of any kind are made to the process. A quality characteristic may go almost out of specification for a period and then mysteriously return to its normal value. After the initiation of EVOP, there will be a great temptation to describe all such happenings to this program. The inclusion of reference conditions will help to clarify the situation.

When current best-known conditions are included in each cycle as a reference whether as an additional point or as one of the factorial design points, it is informative to compare the average performance achieved while running the EVOP design with the performance, which would have been obtained if all runs had been made at the reference conditions.

No matter where the reference point is located, such a comparison is always provided by the contrast:

> *"Average response over all runs in EVOP cycle - Average response at reference conditions"*

This contrast is called the *change in mean-CM* effect. Suppose n cycles have been completed in 2^2 factorial design with the current best-known condition in the center of the design, so the change in mean will be calculated by the next formula:

$$\text{Change in mean} = \frac{1}{5}(\overline{y}_1 + \overline{y}_2 + \overline{y}_3 + \overline{y}_4 + \overline{y}_0) - \overline{y}_0 = \frac{1}{5}(\overline{y}_1 + \overline{y}_2 + \overline{y}_3 + \overline{y}_4 - 4\overline{y}_0) \tag{4.2}$$

The change in mean provides a measure of the direct cost incurred by obtaining information in any particular phase. It is a measure of whether the conditions being run during the current phase yield an average result better or worse than the reference conditions. On the assumption that had we not been running EVOP, we would have been operating statically at the current best-known conditions, it measures the temporary direct cost, if any, of running the scheme and so obtaining information during the present phase.

When EVOP begins, assume that the value of factor A is not optimum, Figure 4.4a. A effect is found to be significant since the slope is steep, but change in mean effect is not significant. As later phases proceed closer to the peak, Figure 4.4b applies. Enough cycles are run to show A is significant, but the change in mean effect may or may not be significant, depending on the spread of the experiment with respect to the actual response, the curvature of the response curve, and size of experimental error-noise. But the change in mean effect is not expected to be small.

When the response maximum is finally included within the experimental space, it is possible that the factor will still have a significant effect as seen in Figure 4.4c. The change in mean effect should be significant. The usual picture is Figure 4.4d where the slope of the factor (the effect) is near zero and the change in mean effect is significant. Note the change in mean effect is negative for a convex surface. Figure 4.4e is also possible when the true maximum is a point on a relatively large plateau or when the size of the experimental pattern is too small in relation to the curvature.

Stated simply, the interaction effect measures nonlinearity or nonplanarity. This is not always true, however, since the design, which straddles a peak symmetrically, has zero interaction. In general, the existence of an interaction means that A has a different effect on the response at the low level of B than it does at the higher level of B.

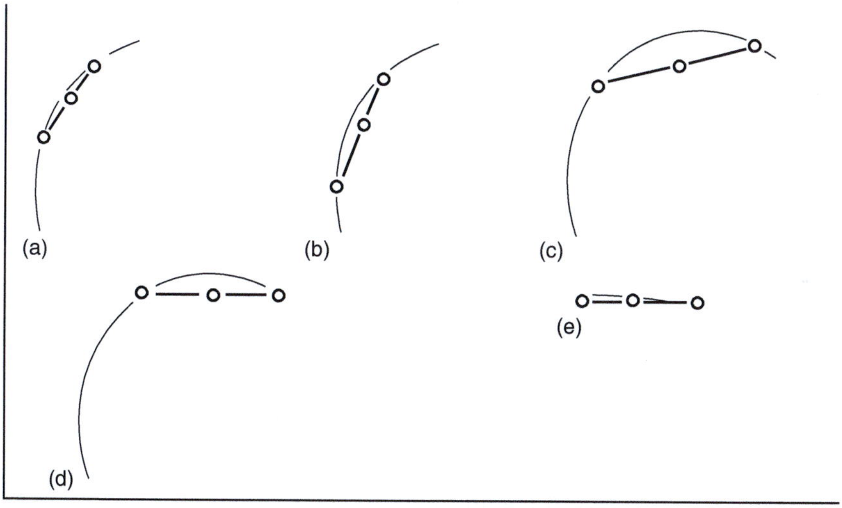

Figure 4.4 Interpretation of the change in mean effect.

It means additional quantity, which can be useful, is the "*phase mean.*" Once again, *the phase mean is the mean response over all conditions being run in the present phase.* It is estimated by the average of the results at all sets of conditions. Estimates of phase mean for the various responses being investigated will indicate the average yield, quality, etc. of the material manufactured during a particular EVOP phase. Thus for 2^2 design without additional reference conditions, the estimate of the phase mean is:

$\frac{1}{4}(\bar{y}_1 + \bar{y}_2 + \bar{y}_3 + \bar{y}_4)$. For design with additional reference conditions, it is: $\frac{1}{5}(\bar{y}_1 + \bar{y}_2 + \bar{y}_3 + \bar{y}_4 + \bar{y}_0)$. The *change in mean* is in all cases equal to *phase mean* minus the *mean* at the reference conditions.

4.7 Change in Mean for a 2^2 Factorial Design with Center Point

In general, for a 2^2 factorial design with center point, the change in mean is a measure of the surface curvature. Apart from the effect of experimental error, it will be positive near a minimum and negative near a maximum. A change in mean effect, which exceeds its 2S.E. limits and is of about *the same size numerically as the main effects, may indicate that a maximum or minimum is being closely approached.* For 2^2 factorial design, the true change in mean effect is actually a measure of the sum of the coefficients measuring quadratic curvature. We could assume that these coefficients were of the same sign (i.e. that we were approaching a "proper" minimum or maximum). A diagonally oriented saddle represented by equal coefficients of opposite sign could occur and yield a zero change in mean effect, but such a phenomenon would rarely be met in practice.

In the common EVOP situations, often more may be gained by switching the study to other factors at this point than by detailed response studies of factors, which are close to their optimal values. Again, using fluidity data of Section 4.3, the fluidity observed at the center conditions was 71.3. After four cycles, the change in mean was in accordance with formula (4.2):

$$\frac{1}{5}[60.2 + 67.6 + 73.2 + 76.2 - 4(71.3)] = -1.6$$

There is a little to be learned from this particular example except that the surface is slightly convex. The large relative magnitudes of the main effects means a maximum for fluidity is probably still fairly remote.

It would be normal statistical practice to randomize order of the runs within each cycle. The effect of randomization is to make our procedures less dependent on assumptions. In particular, randomization ensures that, if systematic pattern of trends occurs, arising from factors whose existence may not even be suspected, these will not be mistaken for effects of the EVOP factors. It also would have the effect of validating our analysis (assuming that errors within cycles are independent) in situations where the response errors are serially correlated. If runs could be randomized without too much difficulty, then this should be done. Our experience has been that a random sequence is difficult to organize under actual production for 2^2 design with added center points. Usually, the systematic sequence is as indicated by the numbering in Figure 4.2.

4.8 Standard Errors for the Effects

We have shown how data coming from 2^2 factorial design with center point may be summarized in terms of main effects, in interaction, and a change in mean effect. Using the formulas of Section 2.4. and results from this section, we can easily obtain standard errors for these quantities. These quantities are subject to sampling error.

In Example 4.1., the variance of the main effect of concentration is:

$$V(A) = V\left(\frac{1}{2}\bar{y}_4 + \frac{1}{2}\bar{y}_2 - \frac{1}{2}\bar{y}_3 - \frac{1}{2}\bar{y}_1 = \left(\frac{1}{4} + \frac{1}{4} + \frac{1}{4} + \frac{1}{4}\right)\frac{\sigma^2}{n} = \frac{\sigma^2}{n}\right.$$

And, if we have an estimate s of σ, then

$$\text{S.E.}(A) = \frac{s}{\sqrt{n}}$$

This is also the standard error for the main effect B and the interaction *AB*. The variance of the change in mean effect is:

$$V\left(\frac{1}{5}\bar{y}_1 + \frac{1}{5}\bar{y}_2 + \frac{1}{5}\bar{y}_3 + \frac{1}{5}\bar{y}_4 - \frac{4}{5}\bar{y}_0\right) = \left(\frac{1}{25} + \frac{1}{25} + \frac{1}{25} + \frac{1}{25} + \frac{16}{25}\right)\frac{\sigma^2}{n} = \frac{4}{5}\frac{\sigma^2}{n}$$

Thus,

$$\text{S.E. (change in mean)} = \frac{2}{\sqrt{5}}\frac{s}{\sqrt{n}} = 0.89\frac{s}{\sqrt{n}}$$

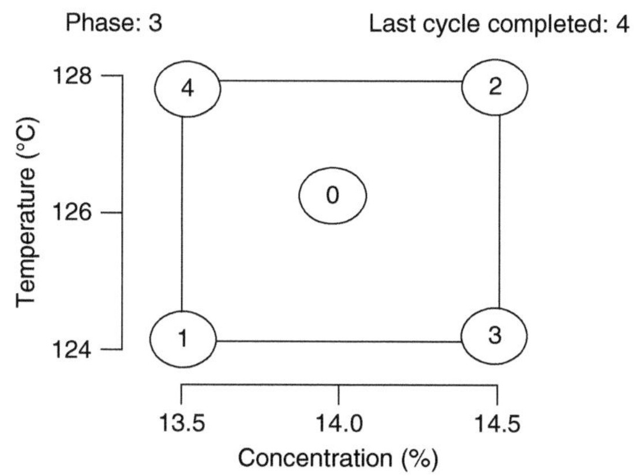

	Cost	Impurity	Fluidity
Requirement	Minimum	Less than 0.5	55 ÷ 80
Running averages	32.6 33.9 32.8 32.3 33.4	0.29 0.35 0.27 0.17 0.19	73.2 76.2 71.3 60.2 67.6
2 S.E. limits	±0.7	±0.03	±1.1
Phase mean	33.0	0.25	69.7
Concentration	1.2 ± 0.7	0.04 ± 0.03	5.2 ± 1.1
Temperature	0.4 ± 0.7	0.14 ± 0.03	10.8 ± 1.1
Connect. × Temp.	0.1 ± 0.7	0.02 ± 0.03	−2.2 ± 1.1
Change in mean	0.2 ± 0.5	−0.02 ± 0.03	−1.6 ± 1.0
Standard deviations For individual observations	0.72	0.029	1.06

Figure 4.5 Information board at the end of four cycles.

To complete calculations, we need to obtain an estimate s of the standard deviation σ.

This is most easily done by a special procedure, *using the sample range or formula for standard deviation.*

4.9 The Effects and Their Standard Errors for a 2^2 Design with Center Point

After n cycles, we could compute the following effects and standard errors (Table 4.1):

Table 4.1 The effects and standard errors.

	Effect	Standard error
Main effect of concentration	$A = 1/2(\bar{y}_2 + \bar{y}_3 - \bar{y}_4 - \bar{y}_1)$	$\frac{s}{\sqrt{n}}$
Main effect of temperature	$B = 1/2(\bar{y}_2 + \bar{y}_4 - \bar{y}_3 - \bar{y}_1)$	$\frac{s}{\sqrt{n}}$
Interaction	$AB = 1/2(\bar{y}_2 + \bar{y}_1 - \bar{y}_4 - \bar{y}_3)$	$\frac{s}{\sqrt{n}}$
Change in mean	$CM = \frac{1}{5}(\bar{y}_1 + \bar{y}_2 + \bar{y}_3 + \bar{y}_4 - 4\bar{y}_0)$	$0.89\frac{s}{\sqrt{n}}$

4.10 Analysis of Information Board for Three Responses Using Factorial Effects

In Chapter 2, Table 2.1, Figure 2.1, interpretation of the results for cost, impurity, and fluidity was based on an inspection of the individual running averages and their 2S.E. limits. We have now seen how to employ these averages to calculate the main effects, the interaction and change in mean effect, and how to obtain 2S.E. limits for these quantities. Now, let's explain how this additional analysis can be of practical value. Displayed in Figure 4.5 is the information board of Figure 2.1, but with additions showing the estimated effects of concentration and temperature, the concentration–temperature interaction and change in mean effect, as well as their 2S.E. limits for each of the three responses: cost, impurity, and fluidity. The phase mean estimates are shown in Figure 4.5 without 2S.E. limits. If limits are required for these quantities, they should be based on *an estimate of the standard deviation between cycles and not on a within-cycles estimate.*

For the responses cost and impurity, the main effects for A and B and interaction AB are small in magnitude, compared with their standard errors. Also for both these responses, changes in mean effects are small, i.e. the direct costs of information gathering or the overall curvature of the surfaces are undetectable. For these responses, changes that occur may be explained locally in terms of simple linear trends. To indicate trends, it is sometimes helpful to draw contours, as in Figure 4.6. The contours have been drawn by rough eye interpolation among the design points. It should be emphasized that no attempt needs to be made here to fit a response function formally. Where main effects dominate, as for cost and impurity, it is implied that the response function is locally planar and its contours are, therefore, equally spaced straight lines.

An appreciable interaction does occur for the response fluidity. As we already have noticed, this interaction is not of the extreme kind, where simple effects of the factors are of opposite sign, but merely indicates some lack of additivity in the effects. It is to be noted also that a negative change in mean effect is obtained of -1.6 ± 1.0. The suggestion is, therefore, that we are here dealing with a somewhat curved surface possibly approaching a maximum. Rough interpolation among the points yields approximate 75, 70, and 65 contours, shown in the figure. Since fluidity must be above 55, special interest attaches to 55 contours for fluidity. This is outside the ranges of the design, but to aid our thoughts on what ought to be done next (but certainly not for drawing definite conclusions), possible positions of 60 and 55 contours are shown in the diagram.

Inspection of the diagram for fluidity shows (55 contours for fluidity), we will do well by reducing concentration, maintaining temperature at its median level of 126 °C. As it was explained earlier, three sets of conditions (13%,126 °C), (13.5%, 126 °C), and (14%, 126 °C) were run in succession, and condition (13%, 126 °C) was used as the starting point for further investigation.

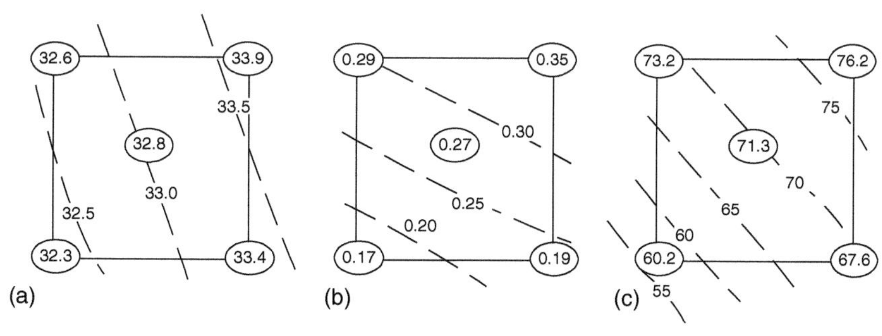

Figure 4.6 Approximate contours for (a) cost, (b) impurity, and (c) fluidity.

4.11 2^3 Factorial Design Effects, Interpretation, and Information Board

Using by-product yield from example in Section 2.5, now we consider how the various statistics may be brought together in the interpretation of results from an EVOP scheme using 2^3 design.

Table 4.2 shows the results as they would be set out on the information board. By-product averages after three cycles of operation are shown at the corners of the cube. The estimated main effects and interactions with their 2S.E. limits are given below the cube.

Table 4.2 Information board for by-product yields after three cycles.

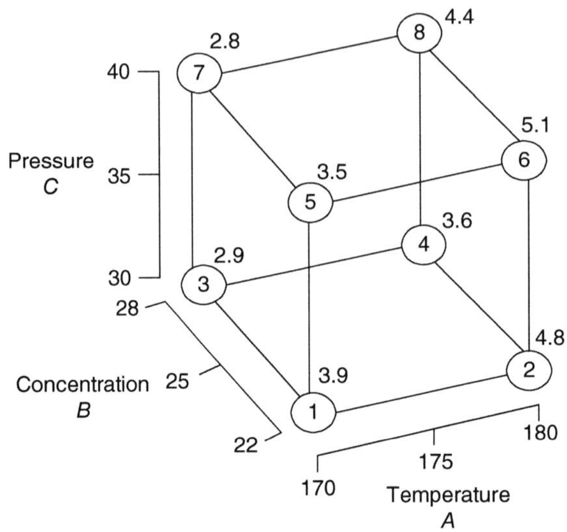

2 S.E. Limits for individual averages: ±0.44

Temperature	*A*	1.20 ± 0.31
Concentration	*B*	−0.90 ± 0.31
Pressure	*C*	0.15 ± 0.31
Temperature × concentration	*AB*	−0.05 ± 0.31
Temperature × pressure	*AC*	0.40 ± 0.31
Concentration × pressure	*BC*	0.20 ± 0.31
Temperature × concentration × pressure	*ABC*	0.05 ± 0.31

Considering this display, we apply precisely those principles discussed already for 2^2 design. Attention is focused on those effects that are reasonably large in magnitude compared with their standard errors. These are main effect of *temperature* $A = 1.20$, main effect of *concentration* $B = -0.90$, and *temperature x pressure* interaction $AC = 0.40$.

Before interpreting the main effect of a given factor, we must find if there are any significant interactions that involve this factor. If there are, then the interaction factors are interpreted together, rather than separately.

Higher temperature means increased by-product, irrespective of the level of pressure used. At the higher temperature, increased pressure probably further increases by-product yield.

In practice, usually there would be other responses to consider which might constrain possible action. If there were no such constraints, then to reduce by-product yield, we should explore conditions of lower temperature and high concentration, keeping the median level of pressure close to its previous value.

4.11.1 An Estimate of Standard Deviation

To compute standard errors of the effects, we need an estimate s of standard deviation σ of the individual observations. In our previous analysis of by-product data, there had been a concern only with the averages obtained after three cycles. Data for individual runs are shown in Table 4.3 [1].

As soon as two or more cycles have been completed, we have information on how closely results at each set of process conditions are reproduced in repeated trials. This allows us to obtain an estimate s of the standard deviation σ of a single observation from the observed reproducibility of the results. After the third cycle, there are three repetitions at each set of process conditions. We could calculate an estimate of variance from each of these groups of three observations on the assumption that the true variances are equal at the various sets of conditions, *these estimates could be averaged to produce a pooled estimate of the variance* σ^2[3].

Table 4.3 By-product data for three repetitions/cycles of a 2^3 design.

Process conditions	Cycle 1	Cycle 2	Cycle 3	Average after three cycles
1	3.9	4.2	3.6	3.90
2	4.3	5.1	5.1	4.83
3	3.4	2.5	2.7	2.87
4	3.8	3.3	3.7	3.60
5	3.7	3.3	3.6	3.53
6	5.2	5.5	4.6	5.10
7	2.9	2.8	2.7	2.80
8	4.7	4.4	4.0	4.37

Source: Box and Draper [1].

The analysis would be entirely appropriate if we could be sure that there were no systematic differences from cycle to cycle. However, changes in raw material, shifts in ambient temperature, and other causes could produce such differences. To avoid overestimation of the errors, we should eliminate such effects by removing overall differences between the column means of the table as well as differences between row means. The orthodox way of doing this is called a *two-way analysis of variance* (ANOVA) [3]. This requires more complex calculation and is not well suited to the conditions under which EVOP is usually applied. It is better to use simplified method that employs the range especially for those who do not use computer.

If there are k sets of process conditions ($k = 8$ in this example) and if n cycles have been completed, then the differences of the averages of the first n–1 cycle and the new data in the nth cycle will provide an estimate S_n in n-th cycle [1].

$$S_n = \left(\frac{n-1}{n}\right)^{1/2} w_k R_n \tag{4.3}$$

where R_n is the range, the difference of the averages of the first n–1 cycles and the new average in n-th cycle Table 4.4;

$$\text{Factor} f_{k,n} = \left(\frac{n-1}{n}\right)^{1/2} w_k$$

In our example we have k =8, for n =2, R =1.7, $f_{8,2}$ =0.25, so that: s_2 =0.25 × 1.7 = 0.42 (Table 4.5).

Table 4.4 Values of $f_{k,n}$.

Number of cycles	Number of runs in the block, k								
	2	3	4	5	6	7	8	9	10
2	0.63	0.42	0.34	0.30	0.28	0.26	0.25	0.24	0.23
3	0.72	0.48	0.40	0.35	0.32	0.30	0.29	0.27	0.26
4	0.77	0.51	0.42	0.37	0.34	0.32	0.30	0.29	0.28
5	0.79	0.53	0.43	0.38	0.35	0.33	0.31	0.30	0.29
6	0.81	0.54	0.44	0.39	0.36	0.34	0.32	0.31	0.30
7	0.82	0.55	0.45	0.40	0.37	0.34	0.33	0.31	0.30
8	0.83	0.55	0.45	0.40	0.37	0.35	0.33	0.31	0.30
9	0.84	0.56	0.46	0.44	0.37	0.35	0.33	0.32	0.31
10	0.84	0.56	0.46	0.41	0.37	0.35	0.33	0.32	0.31
15	0.86	0.57	0.47	0.42	0.38	0.36	0.34	0.33	0.31
20	0.86	0.58	0.47	0.42	0.38	0.36	0.34	0.33	0.32

Table 4.5 Calculation for S_2 and S_3.

Process conditions	Cycle 1	Cycle 2	Differences Cycle 1–Cycle 2	Average (Cycle 1 + Cycle 2)/2	Cycle 3	Differences
1	3.9	4.2	−0.3	4.05	3.6	0.45
2	4.3	5.1	−0.8*	4.70	5.1	−0.40*
3	3.4	2.5	0.9*	2.95	2.7	0.25
4	3.8	3.3	0.5	3.55	3.7	−0.15
5	3.7	3.3	0.4	3.50	3.6	−0.10
6	5.2	5.5	−0.3	5.35	4.6	0.75*
7	2.9	2.8	0.1	2.85	2.7	0.15
8	4.7	4.4	0.3	4.55	4.0	0.55

Note: * The largest range.

After two cycles, the range for the eight differences in this case is R = 0.9–(−0.8) = 1.7, $f_{8,2}$ = 0.25, so estimated standard deviation of the differences is: S_2 = 0.25 × 1.7 = 0.42. As before for n = 3, R = 0.75–(−0.40) = 1.15 and $f_{8,3}$ = 0.29, estimated standard deviation is S_3 = 0.29 × 1.15 = 0.33.

This method has a number of advantages:

- It requires only addition, subtraction, and multiplication.
- For k in the table range, the efficiency of the method is high.
- Only the quantities "old averages" and "new results," which at any given stage are immediately available, are used in the calculation.
- Cycle differences are eliminated automatically.
- Rounding errors are controlled.

A procedure similar to the one described is, of course, used in the two-factor case. See Appendix A.

4.12 Dividing the 2^3 Factorial Design Into Two Blocks

2^3 design enables studying the main effects and interactions among three factors with the same precision as would be obtained from the same number of runs applied to only two factors. The same argument suggests that it would be even more advantageous to study four or more factors simultaneously. The experience has been that the study of three factors simultaneously usually represents the practical limit for EVOP program, because EVOP must be run by the process operators themselves as a continuing routine indefinitely and does not disrupt production. With more runs in one cycle, it would be expected that more opportunity for variation would occur in a cycle of $2^3 = 8$ runs than in a cycle $2^2 = 4$ runs. Fortunately, we can gain all the advantages of the smaller cycle size with the larger 2^3 design by running it in two halves, as in two blocks of four runs each. This can be done so that all the effects and interactions of interest are calculated from comparisons made within a block. We can imagine situation where four of eight runs of 2^3 factorial design could be done in one day, and four runs the next day. The question is, how to arrange eight runs so that possible systematic day-to-day differences will not affect the main effects and two-factor interactions. Three-factor interaction ABC will usually be small and of less interest, so we use it to confound with the day-to-day difference. In other words, the interaction ABC effect also measures day-to-day effect. These two effects are said to be confounded, or confused with each other. By sacrificing information about three-factor interaction, we keep all other effects free from day-to-day differences. In general, by introducing the blocking, we can study three factors in an EVOP scheme, without introducing additional variation beyond that to be expected for a two-factor scheme.

4.13 2^3 Design with Two Center Points Run in Two Blocks

In special case where the reference conditions are at the center of the factorial design, the change in mean (average overall runs in EVOP cycles minus average at reference condition) will provide not only the estimate of the cost of information gathering, but additionally a measure of overall curvature of the response surface.

This kind of design has been shown in Figure 4.7, where eight runs of 2^3 design are split into two blocks of four runs, as previously described, and a center point is associated with each block.

The center conditions in the first block have been labeled 0 (zero) and in the second ø (zero slash).

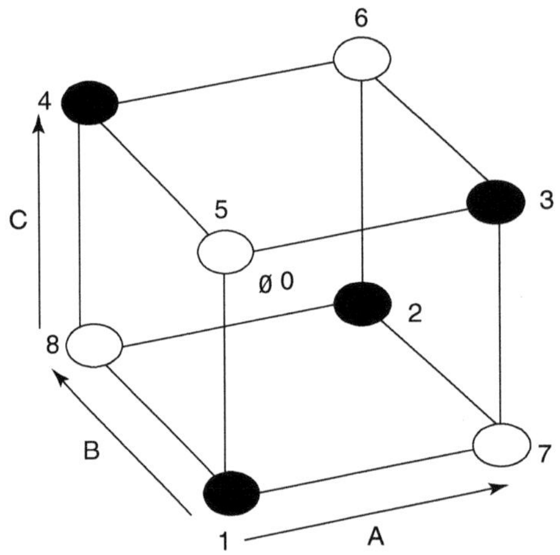

Figure 4.7 The blocks of the 2^3 design.

From the design of Figure 4.7, two measures of change in mean may be calculated, one from each block of the design. For block I, in terms of single observations,

$$\text{Change in Mean (CM)} = \frac{1}{5}(\bar{y}_1 + \bar{y}_2 + \bar{y}_3 + \bar{y}_4 - 4\bar{y}_0)$$

For block II,

$$\text{Change in Mean (CM)} = \frac{1}{5}(\bar{y}_5 + \bar{y}_6 + \bar{y}_7 + \bar{y}_8 - 4\bar{y}_{\otimes})$$

The average:

$$\frac{1}{10}(\bar{y}_1 + \bar{y}_2 + \bar{y}_3 + \bar{y}_4 - 4\bar{y}_0 + \bar{y}_5 + \bar{y}_6 + \bar{y}_7 + \bar{y}_8 - 4\bar{y}_{\otimes})$$

provides a measure of overall change in mean, free from block effect. It has precisely the same interpretation as in two-factor EVOP and is equal to the *phase mean minus the mean at the reference conditions.*

4.13.1 Two Standard Error Limits for the Overall Change in Mean

After n cycles of operation, the variance of overall change in mean would be [1]: $0.4\,\frac{\sigma^2}{n}$, and standard deviation of overall change in mean is: $0.632\,\frac{s}{\sqrt{n}}$, if the estimate of σ is s, 2 S.E. limits will be: Overall change in mean $\pm 2(0.632\frac{s}{\sqrt{n}})$.

Process in favor of EVOP:

- Batch process
- Large-scale production in longer period of time

- Process is not running on optimum
- Process factors can be adjusted easily
- Process is reaching steady conditions after upset (factors adjustment) quickly
- Factors are quantitative
- Responses measurements are quantitative
- Response measurements available in short period of time.

5

Different Techniques of EVOP

As explained in the previous chapter, Box G.E.P [1] proposed special operation as a technique for improving industrial productivity. Its basic philosophy is that it is almost always inefficient to run an industrial process to produce product alone; a process should be run to generate product plus information on how to improve the product.

Procedure consists of carefully planned cycle of minor variants on the standard work process, while routine procedure consists of running each of the variants continually repeating the cycle. Usually, the effects of these deliberate changes in the variables are masked by the large errors inherent in large-scale production units. However, since production will continue anyway, a cycle of variants, which does not affect production significantly, may be run almost indefinitely. Because of the constant repetition, the effect of small changes can be detected.

Since Box G.E.P. first proposed this method for routine use in production of chemicals, five methods (including Box Evolutionary Operation [BEVOP]) with examples and simulations will be discussed in detail:

- Box Evolutionary Operation – BEVOP
- Rotating Square Evolutionary Operation – ROVOP
- Random Evolutionary Operation – REVOP
- Simplex Evolutionary Operation – SEVOP
- Quick Start EVOP – QSEVOP.

5.1 Box EVOP – BEVOP

In EVOP, a single performance of a complete set of operating conditions is called a cycle and repeated running through a cycle of operating conditions is called a phase. A new phase of EVOP begins when new conditions are explored involving different levels of the same factors, different center point, or different factors. In practice, several responses (yield, cost, impurity, etc.) would usually be displayed

Full Scale Plant Optimization in Chemical Engineering: A Practical Guide, First Edition. Živorad R. Lazić.

on an information board. In the following example only a single response, yield (y), is followed in the calculation.

More commonly, the optimum is a compromise between several rising and falling surfaces. It would be advantageous to define a "desirability" function as a single response, expressing the several responses as a single number, Appendix E [3, 42–44]. This function would be expected to have the ideal single-peaked contour with a definable optimum. Without such function, the optimum must be defined as the most desirable combination of dependent properties, including cost and quality.

In the following examples, it is supposed that only a single batch is run at each of the sets of operating conditions within a cycle. However, if for some reason several batches (runs) are made before changing to a different set of operating conditions, the average value of the observations found for each set of conditions is treated as a single observation. An EVOP program may be run on a continuous process by allowing a suitable period of time to elapse for the process to settle down, after each change in operating conditions and then taking one or more observations before introducing a new change. The values so obtained are treated exactly like observations from a batch process.

A calculation procedure, which used the range to provide the estimate of the standard deviation and automatically removes block differences, is set out in detail here only for two- and three-factor schemes, using two-level factorial design with center points. The procedure can be adopted to any number of factors and to any kind of experimental design.

In practice, it is often the case that the standard deviation is very similar from one phase to the next one. When starting a new phase (with new factors introduced), standard deviation is available from the previous phases. At the very beginning of an EVOP program, an estimate of standard deviation would often be available from the past plant records. However, it is frequently found that one consequence of introducing EVOP is to produce more careful operation of the process and a smaller value of σ. Use of a prior estimate is often likely to lead to conservative action.

In the examples described here, *the previous estimate of σ is used in the first two cycles of each new phase*. This estimate might not be completely appropriate for the new phase. But it will usually be based on a fair number of degrees of freedom, so that sampling variation may be ignored and "limits of error", ± 2 (standard error), will provide a reasonable guide for the interpretation of the results. *After two or three cycles, a sufficiently reliable estimate of σ will be appropriate for the new phase*, and it will again make little difference if the sampling variation in this new estimate is ignored. Once again limits of error, ± 2 (standard error) are used.

5.2 Calculation Procedure for Two-Factor EVOP

Example 5.1 Suppose that the present phase of an EVOP program includes two factors, time and temperature [10]. The set of $k = 5$ operating conditions, which together form a single cycle of the EVOP operation program, are shown in Figure 5.1

Figure 5.1 2^2 EVOP with center point.

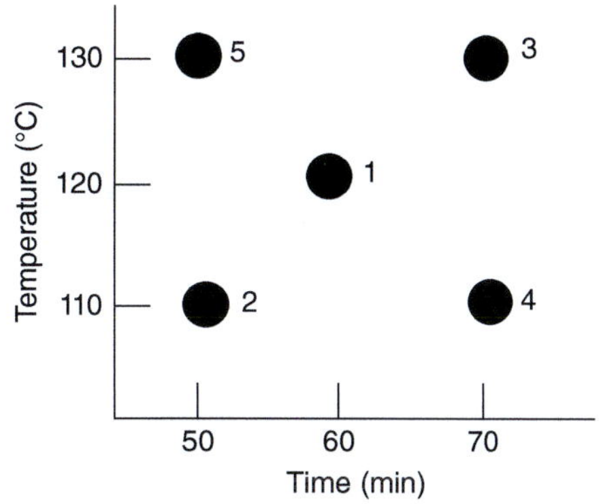

Suppose that observations shown in Table 5.1 are obtained in the first four cycles.

Table 5.1 Individual batch yields.

Conditions	1	2	3	4	5
Cycle I	63.7	62.8	63.2	67.2	60.5
Cycle II	62.1	65.8	65.5	67.6	61.3
Cycle III	59.6	62.1	62.0	65.3	64.1
Cycle IV	63.5	62.8	67.9	62.6	61.7

The calculations in work sheets in Figures 5.2–5.5 have been completed using Microsoft Excel. All detailed steps in this calculation are presented in Appendix B.

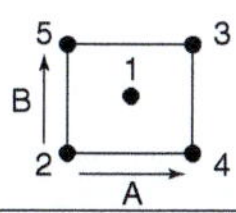

2^2 Factorial with added reference condition

Cycle n : 1
Response : YIELD

Project ______
Phase I
Date ______

Calculation of Averages						Calculation of Standard Deviation	
Operation condition	(1)	(2)	(3)	(4)	(5)	Prior estimate of σ	= 1.80
(i) Previous cycle sum						Previous sum **s**	=
(ii) Previous cycle average						Previous average **s**	=
(iii) New observations	63.70	62.80	63.20	67.20	60.50	New **s** = range × **fδ,n**	=
(iv) Differences (ii) less (iii)						Range	=
(v) New sums	63.70	62.80	63.20	67.20	60.50	New sum **s**	=
(vi) New averages : $\bar{y}_i$	63.70	62.80	63.20	67.20	60.50	New average **s** = new sum **s** / (n - 1)	=

Calculation of Effects			Calculation of 2 S.E. Limits	
Phase mean	$= 1/5\,(\bar{y}_0 + \bar{y}_1 + \bar{y}_2 + \bar{y}_3 + \bar{y}_4)$	= 63.48		
(A) effect	$= ½\,(\bar{y}_3 + \bar{y}_4 - \bar{y}_2 - \bar{y}_5)$	= 3.55	For new average **s** : $\pm (2/\sqrt{n})s$	= 3.60
(B) effect	$= ½\,(\bar{y}_3 + \bar{y}_5 - \bar{y}_2 - \bar{y}_4)$	= -3.15	For new effects : $\pm (2/\sqrt{n})s$	= 3.60
(AB) effect	$= ½\,(\bar{y}_2 + \bar{y}_3 - \bar{y}_4 - \bar{y}_5)$	= -0.85	For change in mean : $\pm (1.79/\sqrt{n})s$	= 3.22
Change in mean effect	= phase mean - $\bar{y}_0$	= -0.22		

Table of Multiplying Factors

n	1	2	3	4	5	6	7	8	9	10
fδ,n		0.30	0.35	0.37	0.38	0.39	0.40	0.40	0.40	0.41
1 / n	1.00	0.50	0.33	0.25	0.20	0.17	0.14	0.12	0.11	0.10
1 / (n - 1)		1.00	0.50	0.33	0.25	0.20	0.17	0.14	0.12	0.11
$2/\sqrt{n}$	2.00	1.41	1.15	1.00	0.89	0.82	0.76	0.71	0.67	0.63
$1.79/\sqrt{n}$	1.79	1.26	1.03	0.89	0.80	0.73	0.68	0.63	0.60	0.57

Two variable EVOP program calculation worksheet

Figure 5.2 Work sheet Cycle No. 1.

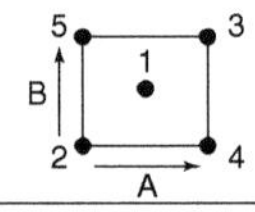

2^2 Factorial with added reference condition

Cycle n : 2
Response:YIELD

Project ______
Phase I
Date ______

Calculation of Averages						Calculation of Standard Deviation	
Operation condition	(1)	(2)	(3)	(4)	(5)	Prior estimate of σ	= 1.80
(i) Previous cycle sum	63.70	62.80	63.20	67.20	60.50	Previous sum **s**	=
(ii) Previous cycle average	63.70	62.80	63.20	67.20	60.50	Previous average **s**	=
(iii) New observations	62.10	65.80	65.50	67.60	61.30	New **s** = range × ***f δ,n***	= 1.38
(iv) Differences (ii) less (iii)	1.60	-3.00	-2.30	-0.40	-0.80	Range	= 4.60
(v) New sums	125.80	128.60	128.70	134.80	121.80	New sum **s**	= 1.38
(vi) New averages : ȳi	62.90	64.30	64.35	67.40	60.90	New average **s** = new sum **s** / (n - 1)	= 1.38

Calculation of Effects			Calculation of 2 S.E. Limits	
Phase mean	= 1/5 (ȳo + ȳ1 + ȳ2 + ȳ3 + ȳ4)	= 63.97		
(A) ———— effect	= ½ (ȳ3 + ȳ4 - ȳ2 - ȳ5)	= 3.28	For new average **s** : ± (2/√n)s	= 2.54
(B) ———— effect	= ½ (ȳ3 + ȳ5 - ȳ2 - ȳ4)	= -3.23	For new effects : ± (2/√n)s	= 2.54
(AB) ———— effect	= ½ (ȳ2 + ȳ3 - ȳ4 - ȳ5)	= 0.17	For change in mean : ± (1.79/√n)s	= 2.27
Change in mean effect	= phase mean - ȳo	= 1.07		

Table of Multiplying Factors

n	1	2	3	4	5	6	7	8	9	10
f δ,n		0.30	0.35	0.37	0.38	0.39	0.40	0.40	0.40	0.41
1 / n	1.00	0.50	0.33	0.25	0.20	0.17	0.14	0.12	0.11	0.10
1 / (n - 1)		1.00	0.50	0.33	0.25	0.20	0.17	0.14	0.12	0.11
2 / √n	2.00	1.41	1.15	1.00	0.89	0.82	0.76	0.71	0.67	0.63
1.79 / √n	1.79	1.26	1.03	0.89	0.80	0.73	0.68	0.63	0.60	0.57

Two variable EVOP program calculation worksheet

Figure 5.3 Work sheet Cycle No. 2.

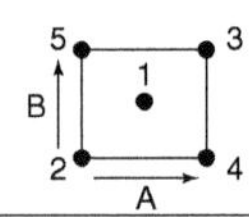

2^2 Factorial with added reference condition

Cycle n : 3
Response:YIELD

Project ______
Phase I
Date ______

Calculation of Averages						Calculation of Standard Deviation	
Operation condition	(1)	(2)	(3)	(4)	(5)	Prior estimate of σ	= 1.80
(i) Previous cycle sum	125.80	128.60	128.70	134.80	121.80	Previous sum **s**	= 1.38
(ii) Previous cycle average	62.90	64.30	64.4	67.40	60.90	Previous average **s**	= 1.38
(iii) New observations	59.60	62.10	62.00	65.30	64.10	New **s** = range × ***f δ,n***	= 2.28
(iv) Differences (ii) less (iii)	3.30	2.20	2.35	2.10	-3.20	Range	= 6.50
(v) New sums	185.40	190.70	190.70	200.10	185.90	New sum **s**	= 3.66
(vi) New averages : ȳi	61.8	63.6	63.6	66.7	62.0	New average **s** = new sum **s** / (n - 1)	= 1.83

Calculation of Effects			Calculation of 2 S.E. Limits	
Phase mean	= 1/5 (ȳo + ȳ1 + ȳ2 + ȳ3 + ȳ4)	= 63.52		
(A) ———— effect	= ½ (ȳ3 + ȳ4 - ȳ2 - ȳ5)	= 2.37	For new average **s** : ± (2/√n)s	= 2.10
(B) ———— effect	= ½ (ȳ3 + ȳ5 - ȳ2 - ȳ4)	= -2.37	For new effects : ± (2/√n)s	= 2.10
(AB) ———— effect	= ½ (ȳ2 + ȳ3 - ȳ4 - ȳ5)	= -0.77	For change in mean : ± (1.79/√n)s	= 1.88
Change in mean effect	= phase mean - ȳo	= 1.72		

Table of Multiplying Factors

n	1	2	3	4	5	6	7	8	9	10
f δ,n		0.30	0.35	0.37	0.38	0.39	0.40	0.40	0.40	0.41
1 / n	1.00	0.50	0.33	0.25	0.20	0.17	0.14	0.12	0.11	0.10
1 / (n - 1)		1.00	0.50	0.33	0.25	0.20	0.17	0.14	0.12	0.11
2 / √n	2.00	1.41	1.15	1.00	0.89	0.82	0.76	0.71	0.67	0.63
1.79 / √n	1.79	1.26	1.03	0.89	0.80	0.73	0.68	0.63	0.60	0.57

Two variable EVOP program calculation worksheet

Figure 5.4 Work sheet Cycle No. 3.

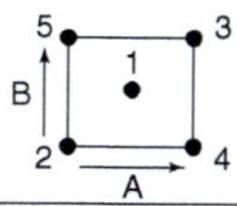

2^2 Factorial with added reference condition

Cycle n :
4
Response:YIELD

Project ______
Phase I
Date ______

Calculation of Averages						Calculation of Standard Deviation	
Operation condition	(1)	(2)	(3)	(4)	(5)	Prior estimate of σ	= 1.80
(i) Previous cycle sum	185.40	190.70	190.70	200.10	185.90	Previous sum **s**	= 3.66
(ii) Previous cycle average	61.80	63.57	63.57	66.70	61.97	Previous average **s**	= 1.83
(iii) New observations	63.50	62.80	67.90	62.60	61.70	New **s** = range × $f\delta,n$	= 3.12
(iv) Differences (ii) less (iii)	-1.70	0.77	-4.33	4.10	0.27	Range	= 8.43
(v) New sums	248.90	253.50	258.60	262.70	247.60	New sum **s**	= 6.78
(vi) New averages : $\bar{y}_i$	62.23	63.38	64.65	65.68	61.90	New average **s** = $\frac{\text{new sum } s}{n-1}$	= 2.26

Calculation of Effects			Calculation of 2 S.E. Limits	
Phase mean	$= 1/5\,(\bar{y}_0 + \bar{y}_1 + \bar{y}_2 + \bar{y}_3 + \bar{y}_4)$	= 63.57		
(A) effect	$= ½\,(\bar{y}_3 + \bar{y}_4 - \bar{y}_2 - \bar{y}_5)$	= 2.53	For new average **s** : $\pm (2/\sqrt{n})s$	= 2.26
(B) effect	$= ½\,(\bar{y}_3 + \bar{y}_5 - \bar{y}_2 - \bar{y}_4)$	= -1.25	For new effects : $\pm (2/\sqrt{n})s$	= 2.26
(AB) effect	$= ½\,(\bar{y}_2 + \bar{y}_3 - \bar{y}_4 - \bar{y}_5)$	= 0.23	For change in mean : $\pm (1.79/\sqrt{n})s$	= 2.01
Change in mean effect	= phase mean - $\bar{y}_0$	= 1.34		

Table of Multiplying Factors

n	1	2	3	4	5	6	7	8	9	10
$f\delta,n$		0.30	0.35	0.37	0.38	0.39	0.40	0.40	0.40	0.41
1 / n	1.00	0.50	0.33	0.25	0.20	0.17	0.14	0.12	0.11	0.10
1 / (n - 1)		1.00	0.50	0.33	0.25	0.20	0.17	0.14	0.12	0.11
$2/\sqrt{n}$	2.00	1.41	1.15	1.00	0.89	0.82	0.76	0.71	0.67	0.63
$1.79/\sqrt{n}$	1.79	1.26	1.03	0.89	0.80	0.73	0.68	0.63	0.60	0.57

Two variable EVOP program calculation worksheet

Figure 5.5 Work sheet Cycle No. 4.

Table 5.2 Information board.

INFORMATION BOARD AFTER PHASE NO. I
Last Cycle Completed : n = 4

	YIELD	RESPONSE No. 2	RESPONSE No. 3
REQUIREMENT :	MAXIMIZE RESPONSE		
PROCESS AVERAGES :	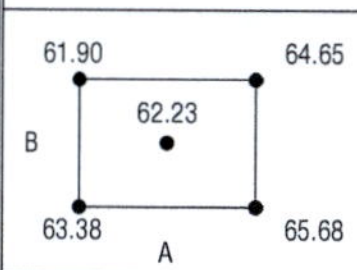		
2.S.E. LIMITS FOR AVERAGES	2.26		
EFFECTS WITH 95% ERROR LIMITS: PHASE MEAN / FACTOR A / FACTOR B / INTERACTION AB / CHANGE IN MEAN	63.57 / 2.53 ± 2.26 / -1.25 ± 2.26 / 0.23 ± 2.26 / 1.34 ± 2.01		
ESTIMATED STANDARD DEVIATION S	2.26		
PRIOR ESTIMATE OF STANDARD DEVIATION σ	1.80		

NOTE:

1. The Center of Experiment-Reference Conditions (The Current Best-Known Process Conditions)
2. The Change in Mean provides a measure of whether the conditions being run the current phase yield an average result better or worse than the reference conditions.
3. The Phase Mean is the Mean Response Over All Conditions Being Run in the present Phase.
4*. The change in Mean is a measure of the over-all Curvature of the Response Surface. It will be positive near a MINIMUM and negative near a MAXIMUM.
5. Sequencing of the Runs-Randomization, if runs can be randomized without too much difficulty, than this should be done.

*A Diagonally oriented saddle represented by equal coefficients of opposite sign could occure and yield a zero change in mean effect, but such a phenomenon would rarely be met in practice.

An appearance of the Information Board at the end of the fourth cycle is shown in Table 5.2.

Conclusions from Information Board

As we have already mentioned, usually we would have observed and considered more than one response before changing the operating conditions to new levels. It would be useless, for example, to obtain higher yields if the resulting product contained a high level of a particular impurity and was unsaleable. Considering yield alone for this example, conclusion will be to that increase in the time would be advantageous. There is an indication (suggested in earlier cycles 2 & 3) that will be beneficial of using a lower temperature, but this would need to be investigated more fully by additional cycles or in subsequent phases.

5.3 Calculation Procedure for Three-Factor EVOP

Example 5.2 The following example of the work sheets is based on data coming from the third phase of an EVOP program applied to antibiotic production [10]. In this phase, residence time, reaction temperature, and pH were being investigated. The response was yield, which was extremely variable. Figure 5.6 shows layout of the design. Data in Table 5.3 show yield results for the first three cycles.

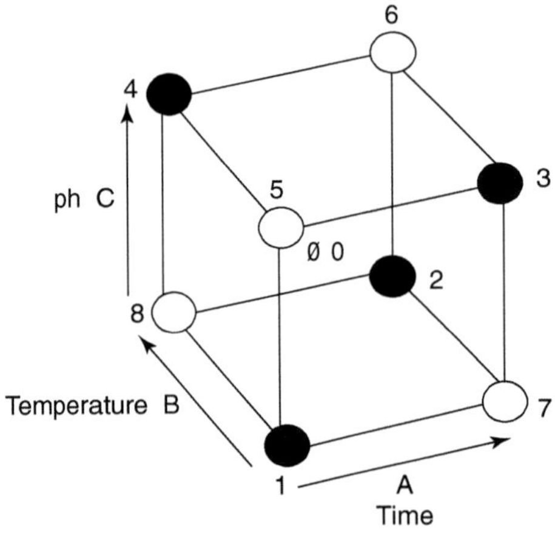

Figure 5.6 2^3 Design with center points.

Table 5.3 Individual batch yields.

Conditions	0	1	2	3	4	ø	5	6	7	8
Cycle I	78	82	63	81	88	85	79	75	78	67
Cycle II	82	75	79	96	77	69	77	80	70	84
Cycle III	65	82	68	85	79	87	96	66	82	72

A single cycle of three-factor EVOP program follows the scheme illustrated in Figure 5.6. The full cycle is broken into two blocks of five runs each, as indicated in diagram by open and filled dots. It should be noticed, since there are 10 sets of operating conditions, that we must keep tally on 10 cumulative averages. Operating conditions 0 and ø at the center of the cube are identical but are treated as a separate because they are performed in different blocks. For example, the preceding sum for condition 0 will be found on the preceding worksheet for block I and not on the worksheet for block II.

From cycle 2 onward, an estimate of standard deviation is computed from the data of each block and is combined with estimates from previous blocks. A prior estimate of σ is used until (at the end of cycle 2) two current estimates are available for averaging (asterisks indicate use of the prior estimate in the first cycle).

The calculations in work sheets in Figure 5.7–5.9 have been completed using Microsoft Excel. All detailed steps in this calculation are presented in Appendix B. The information board would appear at the end of cycle 3, after transfer of the various calculated quantities from the appropriate worksheet, as shown in Table 5.4.

According to the entries shown there, a negative temperature effect and a positive pH effect are beginning to be distinguishable from error. This should be evident in the averages displayed at the corners on the cube above the analysis. Inspection of the appropriate values on the cube confirms this. It looks reasonable to continue with more cycles.

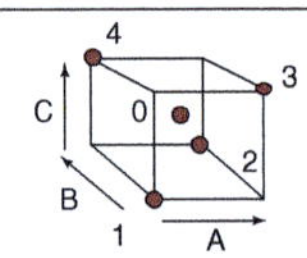

2^3 *Factorial in two blocks with reference run in each block*

BLOCK I

CYCLE 1
RESPONSE : Yield

Project ____
Phase 3
Date ____

Calculation of Averages						Calculation of Standard Deviation - Block I	
Operation condition	(0)	(1)	(2)	(3)	(4)		
(i) Previous sum for block I						Previous sum **s** (all blocks)	=
(ii) Previous average for block I							
(iii) New observations for block I	78.00	82.00	63.00	81.00	88.00	New **s** = range × ***f d,n***	=
(iv) Differences (ii) less (iii)						Range	=
(v) New sums for block I	78.00	82.00	63.00	81.00	88.00	New sum **s** (all blocks)	=
(vi) New averages for block I	78.00	82.00	63.00	81.00	88.00		

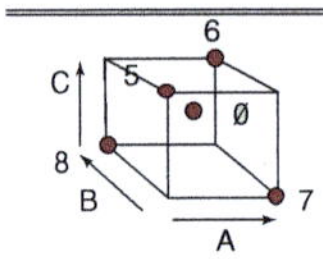

2^3 *Factorial in two blocks with reference run in each block*

BLOCK II

CYCLE 1
RESPONSE : Yield

Calculation of Averages						Calculation of Standard Deviation - Block II	
Operation condition	(Ø)	(5)	(6)	(7)	(8)	Prior estimate of **σ**	= 8.0 *
(i) Previous sum for block II						Previous sum **s** (all blocks)	=
(ii) Previous average for block II							
(iii) New observations for block II	85.00	79.00	75.00	78.00	67.00	New **s** = range × ***f d,n***	=
(iv) Differences (ii) less (iii)						Range	=
(v) New sums for block II	85.00	79.00	75.00	78.00	67.00	New sum **s** (all blocks)	=
(vi) New averages for block II	85.00	79.00	75.00	78.00	67.00	Average **s** = {1/[2(n - 1)]} × new sum s	=

Calculation of Effects Using Yates' Algorithm

	$\bar{y}$	(i)	(ii)	(iii)	Multiplier	Effect	
(1)	82.00	160.00	290.00	613.00	0.125	**76.63**	Factorial mean
(7)	78.00	130.00	323.00	-19.00	0.25	**-4.75**	A
(8)	67.00	160.00	-8.00	-27.00	0.25	**-6.75**	B
(2)	63.00	163.00	-11.00	-15.00	0.25	**-3.75**	AB
(5)	79.00	-4.00	-30.00	33.00	0.25	**8.25**	C
(3)	81.00	-4.00	3.00	-3.00	0.25	**-0.75**	AC
(4)	88.00	2.00	0.00	33.00	0.25	**8.25**	BC
(6)	75.00	-13.00	-15.00	-15.00	0.25	**-3.75**	ABC + blocks

Calculation of 2 S.E. Limits

For new average **s** :	$\pm(2/\sqrt{n})s$	= **16.00**
For new effects :	$\pm(1.41/\sqrt{n})s$	= **11.28**
For change in mean :	$\pm(1.26/\sqrt{n})s$	= **10.08**

		$\bar{y}$	**Sum**	**Multiplier**			
I	Reference (0) Conditions (Ø)	78.00 85.00	163.00	0.500	Reference mean	=	**81.50**
II	Factorial Conditions	$(1:8)\sum\bar{y}$	613.00	0.125	Factorial mean	=	**76.63**
III	All conditions	$(1:10)\sum\bar{y}$	776.00	0.100	Phase mean	=	**77.60**
IV	Change in mean	III - I			Change in mean	=	**-3.90**

Table of Multiplying Factors

n	1	2	3	4	5	6	7	8	9	10
f d,n		0.30	0.35	0.37	0.38	0.39	0.40	0.40	0.40	0.41
1 / n	1.00	0.50	0.33	0.25	0.20	0.17	0.14	0.12	0.11	0.10
1 / [2(n - 1)]		0.50	0.25	0.17	0.12	0.10	0.08	0.07	0.06	0.06
$2/\sqrt{n}$	2.00	1.41	1.15	1.00	0.89	0.82	0.76	0.71	0.67	0.63
$1.41/\sqrt{n}$	1.41	1.00	0.82	0.71	0.63	0.58	1.86	1.99	2.10	2.24
$1.26/\sqrt{n}$	1.26	0.89	0.73	0.63	0.57	0.52	0.68	0.63	0.60	0.56

Three variable EVOP program calculation worksheet.

Figure 5.7 Work sheet Cycle No. 1.

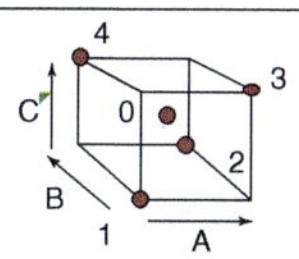

2^3 Factorial in two blocks with reference run in each block

BLOCK I

CYCLE 2
RESPONSE :

Project
Phase 3
Date

Calculation of Averages						Calculation of Standard Deviation - Block I		
Operation condition	(0)	(1)	(2)	(3)	(4)			
(i) Previous sum for block I	78.00	82.00	63.00	81.00	88.00	Previous sum **s** (all blocks)	=	
(ii) Previous average for block I	78.00	82.00	63.00	81.00	88.00			
(iii) New observations for block I	82.00	75.00	79.00	96.00	77.00	New **s** = range × ***fd,n***	=	8.10
(iv) Differences (ii) less (iii)	-4.00	7.00	-16.00	-15.00	11.00	Range	=	27.00
(v) New sums for block I	160.00	157.00	142.00	177.00	165.00	New sum **s** (all blocks)	=	8.10
(vi) New averages for block I	80.00	78.50	71.00	88.50	82.50			

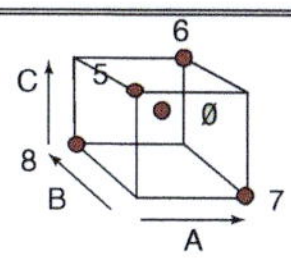

2^3 Factorial in two blocks with reference run in each block

BLOCK II

CYCLE 2
RESPONSE :

Calculation of Averages						Calculation of Standard Deviation - Block II		
Operation condition	(Ø)	(5)	(6)	(7)	(8)	Prior estimate of σ	=	8.0
(i) Previous sum for block II	85.00	79.00	75.00	78.00	67.00	Previous sum **s** (all blocks)	=	8.10
(ii) Previous average for block II	85.00	79.00	75.00	78.00	67.00			
(iii) New observations for block II	69.00	77.00	80.00	70.00	84.00	New **s** = range × ***fd,n***	=	9.90
(iv) Differences (ii) less (iii)	16.00	2.00	-5.00	8.00	-17.00	Range	=	33.00
(v) New sums for block II	154.00	156.00	155.00	148.00	151.00	New sum **s** (all blocks)	=	18.00
(vi) New averages for block II	77.00	78.00	77.50	74.00	75.50	Average **s** = {1/[2(n - 1)]} × new sum s	=	9.00

Calculation of Effects Using Yates' Algorithm

	$\bar{y}$	(i)	(ii)	(iii)	Multiplier	Effect	
(1)	78.50	152.50	299.00	625.50	0.125	**78.19**	Factorial mean
(7)	74.00	146.50	326.50	-3.50	0.25	**-0.88**	A
(8)	75.50	166.50	-9.00	-12.50	0.25	**-3.13**	B
(2)	71.00	160.00	5.50	-15.50	0.25	**-3.88**	AB
(5)	78.00	-4.50	-6.00	27.50	0.25	**6.88**	C
(3)	88.50	-4.50	-6.50	14.50	0.25	**3.63**	AC
(4)	82.50	10.50	0.00	-0.50	0.25	**-0.13**	BC
(6)	77.50	-5.00	-15.50	-15.50	0.25	**-3.88**	ABC + blocks

Calculation of 2 S.E. Limits

For new average **s** :	$\pm (2/\sqrt{n})s$	=	**12.73**
For new effects :	$\pm (1.41/\sqrt{n})s$	=	**8.97**
For change in mean :	$\pm (1.26/\sqrt{n})s$	=	**8.02**

		$\bar{y}$	**Sum**	**Multiplier**			
I	Reference (0) Conditions (Ø)	80.00 77.00	157.00	0.500	Reference mean	=	**78.50**
II	Factorial Conditions	$(1:8)\sum\bar{y}$	625.50	0.125	Factorial mean	=	**78.19**
III	All conditions	$(1:10)\sum\bar{y}$	782.50	0.100	Phase mean	=	**78.25**
IV	Change in mean	III - I			Change in mean	=	**-0.25**

Table of Multiplying Factors

n	1	2	3	4	5	6	7	8	9	10
fd,n		0.30	0.35	0.37	0.38	0.39	0.40	0.40	0.40	0.41
1 / n	1.00	0.50	0.33	0.25	0.20	0.17	0.14	0.12	0.11	0.10
1 / [2(n - 1)]		0.50	0.25	0.17	0.12	0.10	0.08	0.07	0.06	0.06
$2/\sqrt{n}$	2.00	1.41	1.15	1.00	0.89	0.82	0.76	0.71	0.67	0.63
$1.41/\sqrt{n}$	1.41	1.00	0.82	0.71	0.63	0.58	1.86	1.99	2.10	2.24
$1.26/\sqrt{n}$	1.26	0.89	0.73	0.63	0.57	0.52	0.68	0.63	0.60	0.56

Three variable EVOP program calculation worksheet.

Figure 5.8 Work sheet Cycle No. 2.

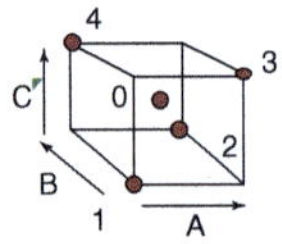

2^3 Factorial in two blocks with reference run in each block

BLOCK I

CYCLE 3

RESPONSE :

Project ______

Phase 3

Date ______

Calculation of Averages					
Operation condition	(0)	(1)	(2)	(3)	(4)
(i) Previous sum for block I	160.00	157.00	142.00	177.00	165.00
(ii) Previous average for block I	80.00	78.50	71.00	88.50	82.50
(iii) New observations for block I	65.00	82.00	68.00	85.00	79.00
(iv) Differences (ii) less (iii)	15.00	-3.50	3.00	3.50	3.50
(v) New sums for block I	225.00	239.00	210.00	262.00	244.00
(vi) New averages for block I	75.00	79.67	70.00	87.33	81.33

Calculation of Standard Deviation - Block I		
Previous sum **s** (all blocks)	=	18.00
New **s** = range × ***f d,n***	=	6.48
Range	=	18.50
New sum **s** (all blocks)	=	24.48

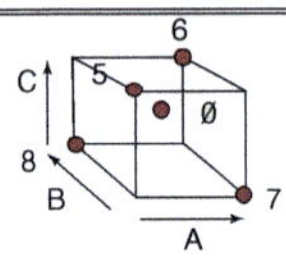

2^3 Factorial in two blocks with reference run in each block

BLOCK II

CYCLE 3

RESPONSE :

Calculation of Averages					
Operation condition	(Ø)	(5)	(6)	(7)	(8)
(i) Previous sum for block II	154.00	156.00	155.00	148.00	151.00
(ii) Previous average for block II	77.00	78.00	77.50	74.00	75.50
(iii) New observations for block II	87.00	96.00	66.00	82.00	72.00
(iv) Differences (ii) less (iii)	-10.00	-18.00	11.50	-8.00	3.50
(v) New sums for block II	241.00	252.00	221.00	230.00	223.00
(vi) New averages for block II	80.33	84.00	73.67	76.67	74.33

Calculation of Standard Deviation - Block II		
Prior estimate of σ	=	8.0
Previous sum **s** (all blocks)	=	24.48
New **s** = range × ***f d,n***	=	10.33
Range	=	29.50
New sum **s** (all blocks)	=	34.80
Average **s** = {1/[2(n - 1)]} × new sum s	=	8.70

Calculation of Effects Using Yates' Algorithm

	$\bar{y}$	(i)	(ii)	(iii)	Multiplier	Effect	
(1)	79.67	156.33	300.67	627.00	0.125	**78.38**	Factorial mean
(7)	76.67	144.33	326.33	-11.67	0.25	**-2.92**	A
(8)	74.33	171.33	-7.33	-28.33	0.25	**-7.08**	B
(2)	70.00	155.00	-4.33	-12.33	0.25	**-3.08**	AB
(5)	84.00	-3.00	-12.00	25.67	0.25	**6.42**	C
(3)	87.33	-4.33	-16.33	3.00	0.25	**0.75**	AC
(4)	81.33	3.33	-1.33	-4.33	0.25	**-1.08**	BC
(6)	73.67	-7.67	-11.00	-9.67	0.25	**-2.42**	ABC + blocks

Calculation of 2 S.E. Limits

For new average **s** :	$\pm(2/\sqrt{n})s$	= **10.05**
For new effects :	$\pm(1.41/\sqrt{n})s$	= **7.08**
For change in mean :	$\pm(1.26/\sqrt{n})s$	= **6.33**

		$\bar{y}$	Sum	Multiplier
I	Reference (0) Conditions (Ø)	75.00 80.33	155.33	0.500
II	Factorial Conditions	$(1:8)\sum\bar{y}$	627.00	0.125
III	All conditions	$(1:10)\sum\bar{y}$	782.33	0.100
IV	Change in mean	III - I		

Reference mean	=	**77.67**
Factorial mean	=	**78.38**
Phase mean	=	**78.23**
Change in mean	=	**0.57**

Table of Multiplying Factors

n	1	2	3	4	5	6	7	8	9	10
f d,n		0.30	0.35	0.37	0.38	0.39	0.40	0.40	0.40	0.41
1 / n	1.00	0.50	0.33	0.25	0.20	0.17	0.14	0.12	0.11	0.10
1 / [2(n - 1)]		0.50	0.25	0.17	0.12	0.10	0.08	0.07	0.06	0.06
$2/\sqrt{n}$	2.00	1.41	1.15	1.00	0.89	0.82	0.76	0.71	0.67	0.63
$1.41/\sqrt{n}$	1.41	1.00	0.82	0.71	0.63	0.58	1.86	1.99	2.10	2.24
$1.26/\sqrt{n}$	1.26	0.89	0.73	0.63	0.57	0.52	0.68	0.63	0.60	0.56

Three variable EVOP program calculation worksheet.

Figure 5.9 Work sheet Cycle No. 3.

Table 5.4 Information board.

2^3 INFORMATION BOARD AFTER PHASE NO. III Last Cycle Completed : n = 3			
	RESPONSE No. 1	RESPONSE No. 2	RESPONSE No. 3
REQUIREMENT :	MAXIMIZE RESPONSE		
PROCESS AVERAGES :	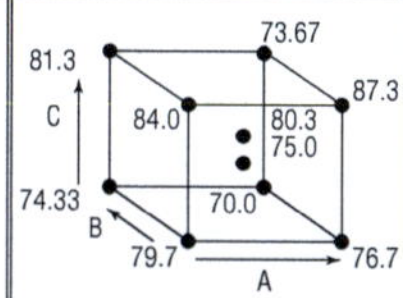		
2.S.E. LIMITS FOR AVERAGES	10.05		
EFFECTS WITH 95% ERROR LIMITS: PHASE MEAN	78.23		
FACTOR A	-2.92 ± 7.08		
FACTOR B	-7.08 ± 7.08		
FACTOR C	6.42 ± 7.08		
INTERACTION AB	-3.08 ± 7.08		
INTERACTION AC	0.75 ± 7.08		
INTERACTION BC	-1.08 ± 7.08		
INTERACTION ABC	-2.42 ± 7.08		
CHANGE IN MEAN	0.57 ± 6.33		
ESTIMATED STANDARD DEVIATION S	8.70		
PRIOR ESTIMATE OF STANDARD DEVIATION σ	8.00		

NOTE:

1. The Center of Experiment-Reference Conditions (The Current Best-Known Process Conditions)
2. The Change in Mean provides a measure of whether the conditions being run the current phase yield an average result better or worse than the reference conditions.
3. The Phase Mean is the Mean Response Over All Conditions Being Run in the present Phase.
4*. The change in Mean is a measure of the over-all Curvature of the Response Surface. It will be positive near a MINIMUM and negative near a MAXIMUM.
5. Sequencing of the Runs-Randomization, if runs can be randomized without too much difficulty, than this should be done.

*A Diagonally oriented saddle represented by equal coefficients of opposite sign could occure and yield a zero change in mean effect, but such a phenomenon would rarely be met in practice.

Example 5.3 We will illustrate EVOP using a chemical process whose yield is a function of temperature A and a reaction time B [36]. The current operating conditions are $A = 150$ [°C] and $B = 30$ [min]. EVOP procedure uses 2^2 design plus the center point shown in Figure 5.10. Notice that each of the five points in the design is numbered. The cycle is completed by running each design point in numerical order (1, 2, 3, 4, 5). Yields in the first cycle are shown in Figure 5.10.

Yields from the first cycle are entered in EVOP calculation sheet, shown in Figure 5.11. At the end of the first cycle, no estimate of standard deviation can be made. The effects and interaction for temperature and time are calculated in usual manner for 2^2 design.

A second cycle is then run and yield data are entered in another EVOP calculation sheet, shown in Figure 5.12. At the end of the second cycle, the experimental error can be estimated and the estimates of the effects can be compared to approximate 95% (two standard deviation) limits.

Notice that at the end of the second cycle, the temperature effect exceeds its error limit. Because temperature effects is positive, we should increase temperature to increase yield. Therefore, a reasonable strategy would be to begin a new EVOP phase around point $A = 155$ [°C] and $B = 30$ [min].

An important aspect of EVOP is feeding generated information back to process operators and supervisors. This is accomplished by EVOP information board. The information board for this example at the end of cycle 2 is shown in Table 5.5.

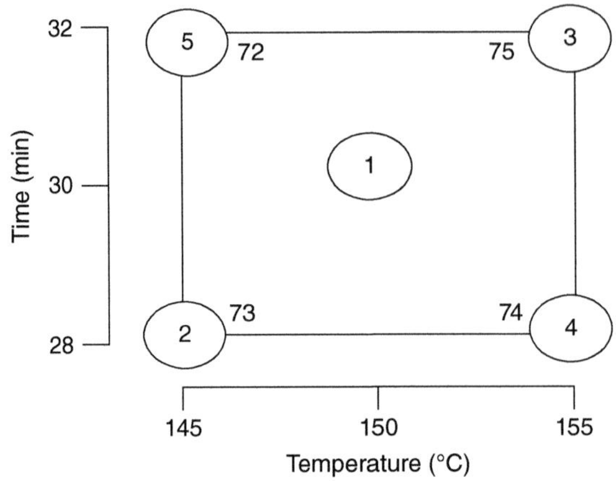

Figure 5.10 2^2 Design for EVOP.

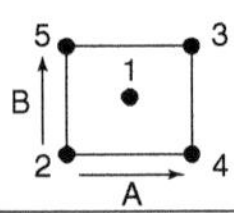

2^2 Factorial with added reference condition

Cycle n :
1
Response:YIELD

Project ________
Phase I________
Date ________

Calculation of Averages						Calculation of Standard Deviation	
Operation condition	(1)	(2)	(3)	(4)	(5)	Prior estimate of σ	=
(i) Previous cycle sum						Previous sum **s**	=
(ii) Previous cycle average						Previous average **s**	=
(iii) New observations	74.00	73.00	75.00	74.00	72.00	New **s** = range × ***fδ,n***	=
(iv) Differences (ii) less (iii)						Range	=
(v) New sums	74.00	73.00	75.00	74.00	72.00	New sum **s**	=
(vi) New averages : ӯi	74.00	73.00	75.00	74.00	72.00	New average **s** = new sum **s** / n - 1	=

Calculation of Effects			Calculation of 2 S.E. Limits	
Phase mean	= *1/5 (ӯo + ӯ1 + ӯ2 + ӯ3 + ӯ4)*	= 73.60		
(A) ———— effect	= *½ (ӯ3 + ӯ4 - ӯ2 - ӯ5)*	= 2.00	For new average **s** : ± (2/√n)s	= 0.00
(B) ———— effect	= *½ (ӯ3 + ӯ5 - ӯ2 - ӯ4)*	= 0.00	For new effects : ± (2/√n)s	= 0.00
(AB) ———— effect	= *½ (ӯ2 + ӯ3 - ӯ4 - ӯ5)*	= 1.00	For change in mean : ± (1.79/√n)s	= 0.00
Change in mean effect	= phase mean - ӯo	= -0.40		

Table of Multiplying Factors

n	1	2	3	4	5	6	7	8	9	10
fδ,n		0.30	0.35	0.37	0.38	0.39	0.40	0.40	0.40	0.41
1 / n	1.00	0.50	0.33	0.25	0.20	0.17	0.14	0.12	0.11	0.10
1 / (n - 1)		1.00	0.50	0.33	0.25	0.20	0.17	0.14	0.12	0.11
2 / √n	2.00	1.41	1.15	1.00	0.89	0.82	0.76	0.71	0.67	0.63
1.79 / √n	1.79	1.26	1.03	0.89	0.80	0.73	0.68	0.63	0.60	0.57

Two variable EVOP program calculation worksheet

Figure 5.11 Work sheet cycle No. 1.

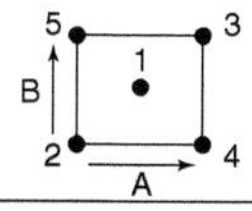

2^2 Factorial with added reference condition

Cycle n :
2
Response:YIELD

Project ______
Phase I______
Date ______

Calculation of Averages					
Operation condition	(1)	(2)	(3)	(4)	(5)
(i) Previous cycle sum	74.00	73.00	75.00	74.00	72.00
(ii) Previous cycle average	74.00	73.00	75.00	74.00	72.00
(iii) New observations	72.00	71.00	76.00	75.00	73.00
(iv) Differences (ii) less (iii)	2.00	2.00	-1.00	-1.00	-1.00
(v) New sums	146.00	144.00	151.00	149.00	145.00
(vi) New averages : $\ddot{y}i$	73.00	72.00	75.50	74.50	72.50

Calculation of Standard Deviation	
Prior estimate of σ	= 1.80
Previous sum **s**	=
Previous average **s**	=
New **s** = range × $f\delta,n$	= 0.90
Range	= 3.00
New sum **s**	= 0.90
New average **s** = $\frac{\text{new sum } s}{n-1}$	= 0.90

Calculation of Effects

Phase mean	$= 1/5\,(\ddot{y}o + \ddot{y}1 + \ddot{y}2 + \ddot{y}3 + \ddot{y}4)$	= 73.50
(A) ———— effect	$= ½\,(\ddot{y}3 + \ddot{y}4 - \ddot{y}2 - \ddot{y}5)$	= 2.75
(B) ———— effect	$= ½\,(\ddot{y}3 + \ddot{y}5 - \ddot{y}2 - \ddot{y}4)$	= 0.75
(AB) ———— effect	$= ½\,(\ddot{y}2 + \ddot{y}3 - \ddot{y}4 - \ddot{y}5)$	= 0.25
Change in mean effect	= phase mean - $\ddot{y}o$	= 0.50

Calculation of 2 S.E. Limits

For new average **s** :	$\pm (2/\sqrt{n})s$	= 1.27
For new effects :	$\pm (2/\sqrt{n})s$	= 1.27
For change in mean :	$\pm (1.79/\sqrt{n})s$	= 1.13

Table of Multiplying Factors

n	1	2	3	4	5	6	7	8	9	10
$f\delta,n$		0.30	0.35	0.37	0.38	0.39	0.40	0.40	0.40	0.41
1 / n	1.00	0.50	0.33	0.25	0.20	0.17	0.14	0.12	0.11	0.10
1 / (n - 1)		1.00	0.50	0.33	0.25	0.20	0.17	0.14	0.12	0.11
$2/\sqrt{n}$	2.00	1.41	1.15	1.00	0.89	0.82	0.76	0.71	0.67	0.63
$1.79/\sqrt{n}$	1.79	1.26	1.03	0.89	0.80	0.73	0.68	0.63	0.60	0.57

Two variable EVOP program calculation worksheet

Figure 5.12 Work sheet cycle No. 2.

Table 5.5 2^2 EVOP information board.

INFORMATION BOARD AFTER PHASE NO. I

Last Cycle Completed : **n = 2**

	YIELD	RESPONSE No. 2	RESPONSE No. 3
REQUIREMENT :	MAXIMIZE RESPONSE		
PROCESS AVERAGES :	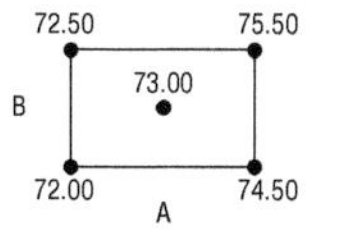		
2.S.E. LIMITS FOR AVERAGES	1.27		
EFFECTS WITH 95% ERROR LIMITS: PHASE MEAN	73.50		
FACTOR A	2.75 ± 1.27		
FACTOR B	0.75 ± 1.27		
INTERACTION AB	0.25 ± 1.27		
CHANGE IN MEAN	0.50 ± 1.13		
ESTIMATED STANDARD DEVIATION S	0.90		
PRIOR ESTIMATE OF STANDARD DEVIATION σ			

NOTE:

1. The Center of Experiment-Reference Conditions (The Current Best-Known Process Conditions)
2. The Change in Mean provides a measure of whether the conditions being run the current phase yield an average result better or worse than the reference conditions.
3. The Phase Mean is the Mean Response Over All Conditions Being Run in the present Phase.
4*. The change in Mean is a measure of the over-all Curvature of the Response Surface. It will be positive near a MINIMUM and negative near a MAXIMUM.
5. Sequencing of the Runs-Randomization, if runs can be randomized without too much difficulty, than this should be done.

*A Diagonally oriented saddle represented by equal coefficients of opposite sign could occure and yield a zero change in mean effect, but such a phenomenon would rarely be met in practice.

Example 5.4 The application of EVOP technique to optimize two inducers, namely surfactant (Tween 80) and ion ($ZnSO_4 \cdot 7H_2O$), in different concentrations in the medium for protease production has been presented here [45]. Parameters A and B are substituted by $ZnSO_4 \cdot 7H_2O$ and Tween 80, respectively. The response was Enzyme activity [mg/ml].

Experimental conditions selected for the first two cycles were as follows (Tables 5.6 and 5.7):

Table 5.6 Initial experimental conditions.

PROJECT : Example 5.4
PHASE : I
DATE :

BEVOP

Example 5.4

I. CENTER OF EXPERIMENT : Ao = 25.00; Bo = 0.10

II. FACTOR'S INTERVAL OF VARIATION : Δ a = 25.00; Δ b = 0.10

EVOP WORK SHEET

TRIALS ORDER			FACTOR A	FACTOR B	RESPONSE n = 1	n = 2	n = 3	n = 4	n = 5	n = 6	n = 7	n = 8	n = 9	n = 10
0		Block 1	25.00	0.10	26.20	29.35								
1		Block 1	0.00	0.00	16.13	18.52								
2		Block 1	50.00	0.00	26.63	24.12								
3		Block 1	0.00	0.20	37.70	41.29								
4		Block 1	50.00	0.20	30.39	31.13								

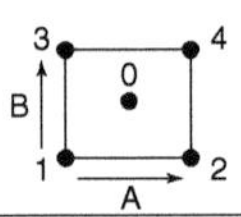

2^2 Factorial with added reference condition

Cycle n : *1*

Response:Enzyme Activity

Project Example 5.4
Phase I
Date

Calculation of Averages

Operation condition	(0)	(1)	(2)	(3)	(4)
(i) Previous cycle sum					
(ii) Previous cycle average					
(iii) New observations	26.20	16.13	26.63	37.70	30.39
(iv) Differences (ii) less (iii)					
(v) New sums	26.20	16.13	26.63	37.70	30.39
(vi) New averages : $\bar{y}_i$	26.20	16.13	26.63	37.70	30.39

Calculation of Standard Deviation

Prior estimate of σ	= 0.00
Previous sum **s**	=
Previous average **s**	=
New **s** = range × $f\delta,n$	=
Range	=
New sum **s**	=
New average **s** = new sum **s** / (n - 1)	=

Calculation of Effects

Phase mean	$= 1/5\,(\bar{y}_0 + \bar{y}_1 + \bar{y}_2 + \bar{y}_3 + \bar{y}_4)$	= 27.41
(A) effect	$= ½\,(\bar{y}_2 + \bar{y}_4 - \bar{y}_1 - \bar{y}_3)$	= 1.60
(B) effect	$= ½\,(\bar{y}_3 + \bar{y}_4 - \bar{y}_1 - \bar{y}_2)$	= 12.67
(AB) effect	$= ½\,(\bar{y}_1 + \bar{y}_4 - \bar{y}_3 - \bar{y}_2)$	= -8.91
Change in mean effect	= phase mean - $\bar{y}_0$	= 1.21

Calculation of 2 S.E. Limits

For new average **s** :	$\pm (2/\sqrt{n})s$	= 0.00
For new effects :	$\pm (2/\sqrt{n})s$	= 0.00
For change in mean :	$\pm (1.79/\sqrt{n})s$	= 0.00

Table of Multiplying Factors

n	1	2	3	4	5	6	7	8	9	10
$f\delta,n$		0.30	0.35	0.37	0.38	0.39	0.40	0.40	0.40	0.41
1 / n	1.00	0.50	0.33	0.25	0.20	0.17	0.14	0.12	0.11	0.10
1 / (n - 1)		1.00	0.50	0.33	0.25	0.20	0.17	0.14	0.12	0.11
$2/\sqrt{n}$	2.00	1.41	1.15	1.00	0.89	0.82	0.76	0.71	0.67	0.63
$1.79/\sqrt{n}$	1.79	1.26	1.03	0.89	0.80	0.73	0.68	0.63	0.60	0.57

Two variable EVOP program calculation worksheet

Figure 5.13 Work sheet cycle No. 1.

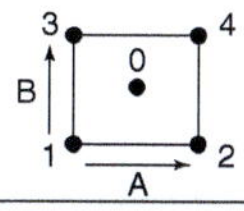

2^2 Factorial with added reference condition

Cycle n : 2

Response:Enzyme Activity

Project Example 5.4
Phase II
Date

Calculation of Averages						Calculation of Standard Deviation	
Operation condition	(0)	(1)	(2)	(3)	(4)	Prior estimate of σ	= 0.00
(i) Previous cycle sum	26.20	16.13	26.63	37.70	30.39	Previous sum **s**	=
(ii) Previous cycle average	26.20	16.13	26.63	37.70	30.39	Previous average **s**	=
(iii) New observations	29.35	18.52	24.12	41.29	31.13	New **s** = range × $f\delta,n$	= 1.83
(iv) Differences (ii) less (iii)	-3.15	-2.39	2.51	-3.59	-0.74	Range	= 6.10
(v) New sums	55.55	34.65	50.75	78.99	61.52	New sum **s**	= 1.83
(vi) New averages : $\bar{y}_i$	27.78	17.33	25.38	39.50	30.76	New average **s** = $\frac{\text{new sum } s}{n-1}$	= 1.83

Calculation of Effects			Calculation of 2 S.E. Limits	
Phase mean	$= 1/5\,(\bar{y}_0 + \bar{y}_1 + \bar{y}_2 + \bar{y}_3 + \bar{y}_4)$	= 28.15		
(A) ———— effect	$= ½\,(\bar{y}_2 + \bar{y}_4 - \bar{y}_1 - \bar{y}_3)$	= -0.34	For new average **s** : $\pm (2/\sqrt{n})s$	= 2.58
(B) ———— effect	$= ½\,(\bar{y}_3 + \bar{y}_4 - \bar{y}_1 - \bar{y}_2)$	= 13.78	For new effects : $\pm (2/\sqrt{n})s$	= 2.58
(AB) ———— effect	$= ½\,(\bar{y}_1 + \bar{y}_4 - \bar{y}_3 - \bar{y}_2)$	= -8.39	For change in mean : $\pm (1.79/\sqrt{n})s$	= 2.31
Change in mean effect	= phase mean - $\bar{y}_0$	= 0.37		

Table of Multiplying Factors

n	1	2	3	4	5	6	7	8	9	10
$f\delta,n$		0.30	0.35	0.37	0.38	0.39	0.40	0.40	0.40	0.41
1 / n	1.00	0.50	0.33	0.25	0.20	0.17	0.14	0.12	0.11	0.10
1 / (n - 1)		1.00	0.50	0.33	0.25	0.20	0.17	0.14	0.12	0.11
$2/\sqrt{n}$	2.00	1.41	1.15	1.00	0.89	0.82	0.76	0.71	0.67	0.63
$1.79/\sqrt{n}$	1.79	1.26	1.03	0.89	0.80	0.73	0.68	0.63	0.60	0.57

Two variable EVOP program calculation worksheet

Figure 5.14 Work sheet cycle No. 2.

Table 5.7 Information board.

INFORMATION BOARD AFTER PHASE NO. I

Last Cycle Completed : n = 2

		RESPONSE No. 1	RESPONSE No. 2	RESPONSE No. 3
REQUIREMENT :		MAXIMIZE RESPONSE		
PROCESS AVERAGES :		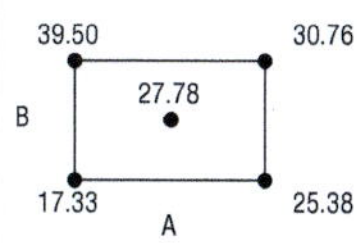		
2.S.E. LIMITS FOR AVERAGES		2.58		
EFFECTS WITH 2.S.E	PHASE MEAN FACTOR A FACTOR B INTERACTION AB CHANGE IN MEAN	28.15 -0.34 ± 2.58 13.78 ± 2.58 -8.39 ± 2.58 0.37 ± 2.31		
ESTIMATED STANDARD DEVIATION S		1.83		
PRIOR ESTIMATE OF STANDARD DEV.		0.00		

NOTE:

1. The Center of Experiment-Reference Conditions (The Current Best-Known Process Conditions)
2. The Change in Mean provides a measure of whether the conditions being run the current phase yield an average result better or worse than the reference conditions.
3. The Phase Mean is the Mean Response Over All Conditions Being Run in the present Phase.
4*. The change in Mean is a measure of the over-all Curvature of the Response Surface. It will be positive near a MINIMUM and negative near a MAXIMUM.
5. Sequencing of the Runs-Randomization, if runs can be randomized without too much difficulty, than this should be done.

*A Diagonally oriented saddle represented by equal coefficients of opposite sign could occure and yield a zero change in mean effect, but such a phenomenon would rarely be met in practice.

Conclusions

Comparison of effects with error limits indicates that the effect of Tween 80 and interaction effect are significant as they are larger than error limits. Further, as change in mean effect is small compared to error limit, an increase in concentration of Tween 80 will help to increase protease activity. Therefore, next set of experiments was designed by shifting the center point from $A_0 = 25$ and $B_0 = 0.1$ to $A_0 = 25$ and $B_0 = 0.2$ (Table 5.8, Figures 5.15 and 5.16, Table 5.9).

Table 5.8 BEVOP calculation.

PROJECT : Example 5.4
PHASE : II
DATE :

BEVOP

Example 5.4

I. CENTER OF EXPERIMENT : Ao = 25.00; Bo = 0.20

II. FACTOR'S INTERVAL OF VARIATION : Δ a = 25.00; Δ b = 0.10

EVOP WORK SHEET

TRIALS ORDER			FACTOR A	FACTOR B	RESPONSE n = 1	n = 2	n = 3	n = 4	n = 5	n = 6	n = 7	n = 8	n = 9	n = 10
0	A1	Block 1	25.00	0.20	34.89	31.86								
1	A2	Block 1	0.00	0.10	27.89	24.06								
2	A4	Block 1	50.00	0.10	31.85	32.91								
3	A5	Block 1	0.00	0.30	26.40	24.73								
4	A3	Block 1	50.00	0.30	33.30	33.93								

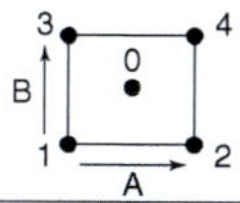

2^2 Factorial with added reference condition

Cycle n : *1*

Response:Enzyme Activity

Project Example 5.4
Phase II
Date

Calculation of Averages						Calculation of Standard Deviation	
Operation condition	(0)	(1)	(2)	(3)	(4)	Prior estimate of σ	= 0.00
(i) Previous cycle sum						Previous sum **s**	=
(ii) Previous cycle average						Previous average **s**	=
(iii) New observations	34.89	27.89	31.85	26.40	33.30	New **s** = range × **fδ,n**	=
(iv) Differences (ii) less (iii)						Range	=
(v) New sums	34.89	27.89	31.85	26.40	33.30	New sum **s**	=
(vi) New averages : $\bar{y}_i$	34.89	27.89	31.85	26.40	33.30	New average **s** = new sum **s** / (n - 1)	=

Calculation of Effects

Phase mean	$= 1/5\,(\bar{y}_0 + \bar{y}_1 + \bar{y}_2 + \bar{y}_3 + \bar{y}_4)$	= 30.87
(A) effect	$= ½\,(\bar{y}_2 + \bar{y}_4 - \bar{y}_1 - \bar{y}_3)$	= 5.43
(B) effect	$= ½\,(\bar{y}_3 + \bar{y}_4 - \bar{y}_1 - \bar{y}_2)$	= -0.02
(AB) effect	$= ½\,(\bar{y}_1 + \bar{y}_4 - \bar{y}_3 - \bar{y}_2)$	= 1.47
Change in mean effect	= phase mean - $\bar{y}_0$	= -4.02

Calculation of 2 S.E. Limits

For new average **s** :	$\pm (2/\sqrt{n})s$	= 0.00
For new effects :	$\pm (2/\sqrt{n})s$	= 0.00
For change in mean :	$\pm (1.79/\sqrt{n})s$	= 0.00

Table of Multiplying Factors

n	1	2	3	4	5	6	7	8	9	10
fδ,n		0.30	0.35	0.37	0.38	0.39	0.40	0.40	0.40	0.41
1 / n	1.00	0.50	0.33	0.25	0.20	0.17	0.14	0.12	0.11	0.10
1 / (n - 1)		1.00	0.50	0.33	0.25	0.20	0.17	0.14	0.12	0.11
$2/\sqrt{n}$	2.00	1.41	1.15	1.00	0.89	0.82	0.76	0.71	0.67	0.63
$1.79/\sqrt{n}$	1.79	1.26	1.03	0.89	0.80	0.73	0.68	0.63	0.60	0.57

Two variable EVOP program calculation worksheet

Figure 5.15 Work sheet cycle No. 1.

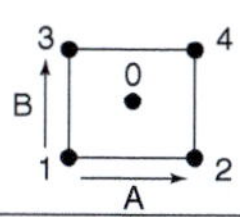

2^2 Factorial with added reference condition

Cycle n : *2*

Response:Enzyme Activity

Project Example 5.4
Phase II
Date

Calculation of Averages						Calculation of Standard Deviation	
Operation condition	(0)	(1)	(2)	(3)	(4)	Prior estimate of σ	= 0.00
(i) Previous cycle sum	34.89	27.89	31.85	26.40	33.30	Previous sum **s**	=
(ii) Previous cycle average	34.89	27.89	31.85	26.40	33.30	Previous average **s**	=
(iii) New observations	31.86	24.06	32.91	24.73	33.93	New **s** = range × **fδ,n**	= 1.47
(iv) Differences (ii) less (iii)	3.03	3.83	-1.06	1.67	-0.63	Range	= 4.89
(v) New sums	66.75	51.95	64.76	51.13	67.23	New sum **s**	= 1.47
(vi) New averages : $\bar{y}_i$	33.38	25.98	32.38	25.57	33.62	New average **s** = new sum **s** / (n - 1)	= 1.47

Calculation of Effects

Phase mean	$= 1/5\,(\bar{y}_0 + \bar{y}_1 + \bar{y}_2 + \bar{y}_3 + \bar{y}_4)$	= 30.18
(A) effect	$= ½\,(\bar{y}_2 + \bar{y}_4 - \bar{y}_1 - \bar{y}_3)$	= 7.23
(B) effect	$= ½\,(\bar{y}_3 + \bar{y}_4 - \bar{y}_1 - \bar{y}_2)$	= 0.41
(AB) effect	$= ½\,(\bar{y}_1 + \bar{y}_4 - \bar{y}_3 - \bar{y}_2)$	= 0.82
Change in mean effect	= phase mean - $\bar{y}_0$	= -3.19

Calculation of 2 S.E. Limits

For new average **s** :	$\pm (2/\sqrt{n})s$	= 2.07
For new effects :	$\pm (2/\sqrt{n})s$	= 2.07
For change in mean :	$\pm (1.79/\sqrt{n})s$	= 1.85

Table of Multiplying Factors

n	1	2	3	4	5	6	7	8	9	10
fδ,n		0.30	0.35	0.37	0.38	0.39	0.40	0.40	0.40	0.41
1 / n	1.00	0.50	0.33	0.25	0.20	0.17	0.14	0.12	0.11	0.10
1 / (n - 1)		1.00	0.50	0.33	0.25	0.20	0.17	0.14	0.12	0.11
$2/\sqrt{n}$	2.00	1.41	1.15	1.00	0.89	0.82	0.76	0.71	0.67	0.63
$1.79/\sqrt{n}$	1.79	1.26	1.03	0.89	0.80	0.73	0.68	0.63	0.60	0.57

Two variable EVOP program calculation worksheet

Figure 5.16 Work sheet cycle No. 2.

Table 5.9 Information board.

INFORMATION BOARD AFTER PHASE NO. II

Last Cycle Completed : n = 2

	RESPONSE No. 1	RESPONSE No. 2	RESPONSE No. 3
REQUIREMENT :	MAXIMIZE RESPONSE		
PROCESS AVERAGES :	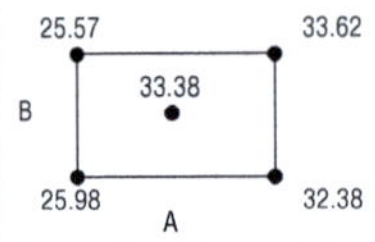		
2.S.E. LIMITS FOR AVERAGES	2.07		
EFFECTS WITH 2.S.E — PHASE MEAN	30.18		
FACTOR A	7.23 ± 2.07		
FACTOR B	0.41 ± 2.07		
INTERACTION AB	0.82 ± 2.07		
CHANGE IN MEAN	-3.19 ± 1.85		
ESTIMATED STANDARD DEVIATION S	1.47		
PRIOR ESTIMATE OF STANDARD DEV.	0.00		

NOTE:

1. The Center of Experiment-Reference Conditions (The Current Best-Known Process Conditions)
2. The Change in Mean provides a measure of whether the conditions being run the current phase yield an average result better or worse than the reference conditions.
3. The Phase Mean is the Mean Response Over All Conditions Being Run in the present Phase.
4*. The change in Mean is a measure of the over-all Curvature of the Response Surface. It will be positive near a MINIMUM and negative near a MAXIMUM.
5. Sequencing of the Runs-Randomization, if runs can be randomized without too much difficulty, than this should be done.

*A Diagonally oriented saddle represented by equal coefficients of opposite sign could occure and yield a zero change in mean effect, but such a phenomenon would rarely be met in practice.

Conclusions

The magnitude of the effect of $ZnSO_4{\cdot}7H_2O$ and change in mean effect are significant as they are larger than error limits. However, the effect of $ZnSO_4{\cdot}7H_2O$ is with positive sign, it means higher $ZnSO_4{\cdot}7H_2O$ might increase enzyme activity. Therefore, next set of experiments in phase III was designed by shifting the center point from $A_0 = 25$ and $B_0 = 0.2$ to $A_0 = 50$ and $B_0 = 0.2$ (Table 5.10, Figures 5.17 and 5.18, Table 5.11).

Table 5.10 BEVOP calculation.

PROJECT : Example 5.4
PHASE : III
DATE :

BEVOP

Example 5.4

I. CENTER OF EXPERIMENT : Ao = 50.00; Bo = 0.20

II. FACTOR'S INTERVAL OF VARIATION : Δ a = 25.00; Δ b = 0.00

EVOP WORK SHEET

TRIALS ORDER		FACTOR A	FACTOR B	RESPONSE n = 1	n = 2	n = 3	n = 4	n = 5	n = 6	n = 7	n = 8	n = 9	n = 10
0	Block 1	50.00	0.20	33.89	33.34								
1	Block 1	25.00	0.20	34.89	31.86								
2	Block 1	75.00	0.20	37.29	39.03								
3	Block 1	25.00	0.20	34.89	31.86								
4	Block 1	75.00	0.20	37.29	39.03								

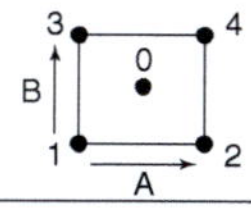

2^2 Factorial with added reference condition

Cycle n : 1

Response:Enzyme Activity

Project Example 5.4
Phase III
Date

Calculation of Averages						Calculation of Standard Deviation	
Operation condition	(0)	(1)	(2)	(3)	(4)	Prior estimate of σ	= 0.00
(i) Previous cycle sum						Previous sum **s**	=
(ii) Previous cycle average						Previous average **s**	=
(iii) New observations	33.89	34.89	37.29	34.89	37.29	New **s** = range × $f\delta,n$	=
(iv) Differences (ii) less (iii)						Range	=
(v) New sums	33.89	34.89	37.29	34.89	37.29	New sum **s**	=
(vi) New averages : $\bar{y}i$	33.89	34.89	37.29	34.89	37.29	New average **s** = new sum **s** / n - 1	=

Calculation of Effects

Phase mean	= 1/5 ($\bar{y}o + \bar{y}1 + \bar{y}2 + \bar{y}3 + \bar{y}4$)	= 35.65
(A) effect	= ½ ($\bar{y}2 + \bar{y}4 - \bar{y}1 - \bar{y}3$)	= 2.40
(B) effect	= ½ ($\bar{y}3 + \bar{y}4 - \bar{y}1 - \bar{y}2$)	= 0.00
(AB) effect	= ½ ($\bar{y}1 + \bar{y}4 - \bar{y}3 - \bar{y}2$)	= 0.00
Change in mean effect	= phase mean - $\bar{y}o$	= 1.76

Calculation of 2 S.E. Limits

For new average **s** :	± $(2/\sqrt{n})$s	= 0.00
For new effects :	± $(2/\sqrt{n})$s	= 0.00
For change in mean :	± $(1.79/\sqrt{n})$s	= 0.00

Table of Multiplying Factors

n	1	2	3	4	5	6	7	8	9	10
$f\delta,n$		0.30	0.35	0.37	0.38	0.39	0.40	0.40	0.40	0.41
1 / n	1.00	0.50	0.33	0.25	0.20	0.17	0.14	0.12	0.11	0.10
1 / (n - 1)		1.00	0.50	0.33	0.25	0.20	0.17	0.14	0.12	0.11
$2/\sqrt{n}$	2.00	1.41	1.15	1.00	0.89	0.82	0.76	0.71	0.67	0.63
$1.79/\sqrt{n}$	1.79	1.26	1.03	0.89	0.80	0.73	0.68	0.63	0.60	0.57

Two variable EVOP program calculation worksheet

Figure 5.17 Work sheet cycle No. 1.

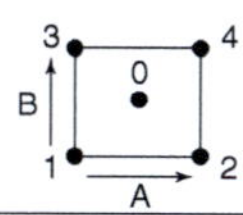

2^2 Factorial with added reference condition

Cycle n : 2

Response:Enzyme Activity

Project Example 5.4
Phase III
Date

Calculation of Averages						Calculation of Standard Deviation	
Operation condition	(0)	(1)	(2)	(3)	(4)	Prior estimate of σ	= 0.00
(i) Previous cycle sum	33.89	34.89	37.29	34.89	37.29	Previous sum **s**	=
(ii) Previous cycle average	33.89	34.89	37.29	34.89	37.29	Previous average **s**	=
(iii) New observations	33.34	31.86	39.03	31.86	39.03	New **s** = range × $f\delta,n$	= 1.43
(iv) Differences (ii) less (iii)	0.55	3.03	-1.74	3.03	-1.74	Range	= 4.77
(v) New sums	67.23	66.75	76.32	66.75	76.32	New sum **s**	= 1.43
(vi) New averages : $\bar{y}i$	33.62	33.38	38.16	33.38	38.16	New average **s** = new sum **s** / n - 1	= 1.43

Calculation of Effects

Phase mean	= 1/5 ($\bar{y}o + \bar{y}1 + \bar{y}2 + \bar{y}3 + \bar{y}4$)	= 35.34
(A) effect	= ½ ($\bar{y}2 + \bar{y}4 - \bar{y}1 - \bar{y}3$)	= 4.79
(B) effect	= ½ ($\bar{y}3 + \bar{y}4 - \bar{y}1 - \bar{y}2$)	= 0.00
(AB) effect	= ½ ($\bar{y}1 + \bar{y}4 - \bar{y}3 - \bar{y}2$)	= 0.00
Change in mean effect	= phase mean - $\bar{y}o$	= 1.72

Calculation of 2 S.E. Limits

For new average **s** :	± $(2/\sqrt{n})$s	= 2.02
For new effects :	± $(2/\sqrt{n})$s	= 2.02
For change in mean :	± $(1.79/\sqrt{n})$s	= 1.80

Table of Multiplying Factors

n	1	2	3	4	5	6	7	8	9	10
$f\delta,n$		0.30	0.35	0.37	0.38	0.39	0.40	0.40	0.40	0.41
1 / n	1.00	0.50	0.33	0.25	0.20	0.17	0.14	0.12	0.11	0.10
1 / (n - 1)		1.00	0.50	0.33	0.25	0.20	0.17	0.14	0.12	0.11
$2/\sqrt{n}$	2.00	1.41	1.15	1.00	0.89	0.82	0.76	0.71	0.67	0.63
$1.79/\sqrt{n}$	1.79	1.26	1.03	0.89	0.80	0.73	0.68	0.63	0.60	0.57

Two variable EVOP program calculation worksheet

Figure 5.18 Work sheet cycle No. 2.

Table 5.11 Information board.

INFORMATION BOARD AFTER PHASE NO. III

Last Cycle Completed : n = 2

	RESPONSE No. 1	RESPONSE No. 2	RESPONSE No. 3
REQUIREMENT :	MAXIMIZE RESPONSE		
PROCESS AVERAGES :	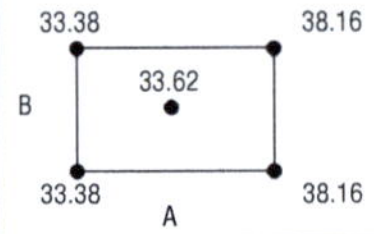		
2.S.E. LIMITS FOR AVERAGES	2.02		
EFFECTS WITH 2.S.E — PHASE MEAN	35.34		
FACTOR A	4.79 ± 2.02		
FACTOR B	0.00 ± 2.02		
INTERACTION AB	0.00 ± 2.02		
CHANGE IN MEAN	1.72 ± 1.80		
ESTIMATED STANDARD DEVIATION S	1.43		
PRIOR ESTIMATE OF STANDARD DEV.	0.00		

NOTE:

1. The Center of Experiment-Reference Conditions (The Current Best-Known Process Conditions)
2. The Change in Mean provides a measure of whether the conditions being run the current phase yield an average result better or worse than the reference conditions.
3. The Phase Mean is the Mean Response Over All Conditions Being Run in the present Phase.
4*. The change in Mean is a measure of the over-all Curvature of the Response Surface. It will be positive near a MINIMUM and negative near a MAXIMUM.
5. Sequencing of the Runs-Randomization, if runs can be randomized without too much difficulty, than this should be done.

*A Diagonally oriented saddle represented by equal coefficients of opposite sign could occure and yield a zero change in mean effect, but such a phenomenon would rarely be met in practice.

Conclusions

The effect of $ZnSO_4 \cdot 7H_2O$ on protease production is still significant as its magnitude is larger than error limit. Therefore, the next set of experiments in phase IV was designed by shifting the center point from $A_0 = 50$ and $B_0 = 0.2$ to $A_0 = 75$ and $B_0 = 0.2$ (Table 5.12, Figures 5.19 and 5.20, Table 5.13).

Table 5.12 BEVOP calculation.

PROJECT : Example 5.4
PHASE : IV
DATE :

BEVOP

Example 5.4

I. CENTER OF EXPERIMENT : Ao = 75.00; Bo = 0.20

II. FACTOR'S INTERVAL OF VARIATION : Δ a = 25.00; Δ b = 0.00

EVOP WORK SHEET

TRIALS ORDER			FACTOR A	FACTOR B	RESPONSE n = 1	n = 2	n = 3	n = 4	n = 5	n = 6	n = 7	n = 8	n = 9	n = 10
0		Block 1	75.00	0.20	37.29	39.03								
1		Block 1	50.00	0.20	33.89	33.84								
2		Block 1	100.00	0.20	41.05	42.67								
3		Block 1	50.00	0.20	33.89	33.34								
4		Block 1	100.00	0.20	41.05	42.67								

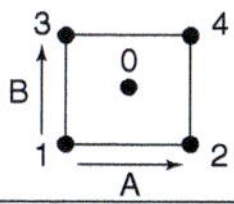

2^2 Factorial with added reference condition

Cycle n :
1
Response:Enzyme Activity

Project Example 5.4
Phase IV
Date

Calculation of Averages					
Operation condition	(0)	(1)	(2)	(3)	(4)
(i) Previous cycle sum					
(ii) Previous cycle average					
(iii) New observations	37.29	33.89	41.05	33.89	41.05
(iv) Differences (ii) less (iii)					
(v) New sums	37.29	33.89	41.05	33.89	41.05
(vi) New averages : $\bar{y}_i$	37.29	33.89	41.05	33.89	41.05

Calculation of Standard Deviation	
Prior estimate of σ	= 0.00
Previous sum **s**	=
Previous average **s**	=
New **s** = range × $f\delta,n$	=
Range	=
New sum **s**	=
New average **s** = new sum **s** / (n - 1)	=

Calculation of Effects

Phase mean	$= 1/5\,(\bar{y}_0 + \bar{y}_1 + \bar{y}_2 + \bar{y}_3 + \bar{y}_4)$	= 37.43
(A) effect	$= ½\,(\bar{y}_2 + \bar{y}_4 - \bar{y}_1 - \bar{y}_3)$	= 7.16
(B) effect	$= ½\,(\bar{y}_3 + \bar{y}_4 - \bar{y}_1 - \bar{y}_2)$	= 0.00
(AB) effect	$= ½\,(\bar{y}_1 + \bar{y}_4 - \bar{y}_3 - \bar{y}_2)$	= 0.00
Change in mean effect	= phase mean - $\bar{y}_0$	= 0.14

Calculation of 2 S.E. Limits

For new average **s** :	$\pm (2/\sqrt{n})s$	= 0.00
For new effects :	$\pm (2/\sqrt{n})s$	= 0.00
For change in mean :	$\pm (1.79/\sqrt{n})s$	= 0.00

Table of Multiplying Factors

n	1	2	3	4	5	6	7	8	9	10
$f\delta,n$		0.30	0.35	0.37	0.38	0.39	0.40	0.40	0.40	0.41
1 / n	1.00	0.50	0.33	0.25	0.20	0.17	0.14	0.12	0.11	0.10
1 / (n - 1)		1.00	0.50	0.33	0.25	0.20	0.17	0.14	0.12	0.11
$2/\sqrt{n}$	2.00	1.41	1.15	1.00	0.89	0.82	0.76	0.71	0.67	0.63
$1.79/\sqrt{n}$	1.79	1.26	1.03	0.89	0.80	0.73	0.68	0.63	0.60	0.57

Two variable EVOP program calculation worksheet

Figure 5.19 Work sheet cycle No. 1.

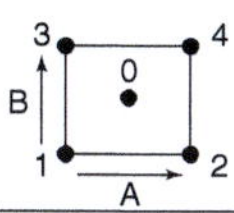

2^2 Factorial with added reference condition

Cycle n :
2
Response:Enzyme Activity

Project Example 5.4
Phase IV
Date

Calculation of Averages					
Operation condition	(0)	(1)	(2)	(3)	(4)
(i) Previous cycle sum	37.29	33.89	41.05	33.89	41.05
(ii) Previous cycle average	37.29	33.89	41.05	33.89	41.05
(iii) New observations	39.03	33.84	42.67	33.34	42.67
(iv) Differences (ii) less (iii)	-1.74	0.05	-1.62	0.55	-1.62
(v) New sums	76.32	67.73	83.72	67.23	83.72
(vi) New averages : $\bar{y}_i$	38.16	33.87	41.86	33.62	41.86

Calculation of Standard Deviation	
Prior estimate of σ	= 0.00
Previous sum **s**	=
Previous average **s**	=
New **s** = range × $f\delta,n$	= 0.69
Range	= 2.29
New sum **s**	= 0.69
New average **s** = new sum **s** / (n - 1)	= 0.69

Calculation of Effects

Phase mean	$= 1/5\,(\bar{y}_0 + \bar{y}_1 + \bar{y}_2 + \bar{y}_3 + \bar{y}_4)$	= 37.87
(A) effect	$= ½\,(\bar{y}_2 + \bar{y}_4 - \bar{y}_1 - \bar{y}_3)$	= 8.12
(B) effect	$= ½\,(\bar{y}_3 + \bar{y}_4 - \bar{y}_1 - \bar{y}_2)$	= -0.13
(AB) effect	$= ½\,(\bar{y}_1 + \bar{y}_4 - \bar{y}_3 - \bar{y}_2)$	= 0.12
Change in mean effect	= phase mean - $\bar{y}_0$	= -0.29

Calculation of 2 S.E. Limits

For new average **s** :	$\pm (2/\sqrt{n})s$	= 0.97
For new effects :	$\pm (2/\sqrt{n})s$	= 0.97
For change in mean :	$\pm (1.79/\sqrt{n})s$	= 0.87

Table of Multiplying Factors

n	1	2	3	4	5	6	7	8	9	10
$f\delta,n$		0.30	0.35	0.37	0.38	0.39	0.40	0.40	0.40	0.41
1 / n	1.00	0.50	0.33	0.25	0.20	0.17	0.14	0.12	0.11	0.10
1 / (n - 1)		1.00	0.50	0.33	0.25	0.20	0.17	0.14	0.12	0.11
$2/\sqrt{n}$	2.00	1.41	1.15	1.00	0.89	0.82	0.76	0.71	0.67	0.63
$1.79/\sqrt{n}$	1.79	1.26	1.03	0.89	0.80	0.73	0.68	0.63	0.60	0.57

Two variable EVOP program calculation worksheet

Figure 5.20 Work sheet cycle No. 2.

Table 5.13 Information board.

INFORMATION BOARD AFTER PHASE NO. IV

Last Cycle Completed : n = 2

		RESPONSE No. 1	RESPONSE No. 2	RESPONSE No. 3
REQUIREMENT :		MAXIMIZE RESPONSE		
PROCESS AVERAGES :		33.62 (B high, A low); 41.86 (B high, A high); 38.16 (center); 33.87 (B low, A low); 41.86 (B low, A high)		
2.S.E. LIMITS FOR AVERAGES		0.97		
EFFECTS WITH 2.S.E	PHASE MEAN	37.87		
	FACTOR A	8.12 ± 0.97		
	FACTOR B	-0.13 ± 0.97		
	INTERACTION AB	0.12 ± 0.97		
	CHANGE IN MEAN	-0.29 ± 0.87		
ESTIMATED STANDARD DEVIATION S		0.69		
PRIOR ESTIMATE OF STANDARD DEV.		0.00		

NOTE:

1. The Center of Experiment-Reference Conditions (The Current Best-Known Process Conditions)
2. The Change in Mean provides a measure of whether the conditions being run the current phase yield an average result better or worse than the reference conditions.
3. The Phase Mean is the Mean Response Over All Conditions Being Run in the present Phase.
4*. The change in Mean is a measure of the over-all Curvature of the Response Surface. It will be positive near a MINIMUM and negative near a MAXIMUM.
5. Sequencing of the Runs-Randomization, if runs can be randomized without too much difficulty, than this should be done.

*A Diagonally oriented saddle represented by equal coefficients of opposite sign could occure and yield a zero change in mean effect, but such a phenomenon would rarely be met in practice.

Conclusions

The effect of $ZnSO_4 \cdot 7H_2O$ on protease production is still significant as its magnitude is larger than the error limit. Therefore, the next set of experiments in phase V was designed by shifting the center point from $A_0 = 75$ and $B_0 = 0.2$ to $A_0 = 100$ and $B_0 = 0.2$ (Table 5.14, Figures 5.21 and 5.22, Table 5.15).

Table 5.14 BEVOP calculation.

PROJECT : Example 5.4
PHASE : V
DATE :

BEVOP

Example 5.4

I. CENTER OF EXPERIMENT : Ao = 100.00; Bo = 0.20

II. FACTOR'S INTERVAL OF VARIATION : Δ a = 25.00; Δ b = 0.00

EVOP WORK SHEET

TRIALS ORDER			FACTOR A	FACTOR B	RESPONSE n = 1	n = 2	n = 3	n = 4	n = 5	n = 6	n = 7	n = 8	n = 9	n = 10
0		Block 1	100.00	0.20	41.05	42.67								
1		Block 1	75.00	0.20	37.29	39.03								
2		Block 1	125.00	0.20	36.01	37.31								
3		Block 1	75.00	0.20	37.29	39.03								
4		Block 1	125.00	0.20	36.01	37.31								

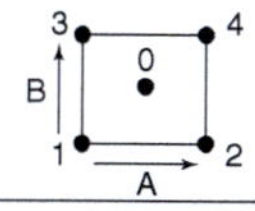

2^2 Factorial with added reference condition

Cycle n : 1
Response:Enzyme Activity

Project Example 5.4
Phase V
Date

Calculation of Averages

Operation condition	(0)	(1)	(2)	(3)	(4)
(i) Previous cycle sum					
(ii) Previous cycle average					
(iii) New observations	41.05	37.29	36.01	37.29	36.01
(iv) Differences (ii) less (iii)					
(v) New sums	41.05	37.29	36.01	37.29	36.01
(vi) New averages : ȳi	41.05	37.29	36.01	37.29	36.01

Calculation of Standard Deviation

Prior estimate of σ	= 0.00
Previous sum **s**	=
Previous average **s**	=
New **s** = range × ***f δ,n***	=
Range	=
New sum **s**	=
New average **s** = new sum **s** / (n - 1)	=

Calculation of Effects

Phase mean	= 1/5 (ȳ0 + ȳ1 + ȳ2 + ȳ3 + ȳ4)	= 37.53
(A) effect	= ½ (ȳ2 + ȳ4 - ȳ1 - ȳ3)	= -1.28
(B) effect	= ½ (ȳ3 + ȳ4 - ȳ1 - ȳ2)	= 0.00
(AB) effect	= ½ (ȳ1 + ȳ4 - ȳ3 - ȳ2)	= 0.00
Change in mean effect	= phase mean - ȳ0	= -3.52

Calculation of 2 S.E. Limits

For new average **s** :	$\pm (2/\sqrt{n})s$	= 0.00
For new effects :	$\pm (2/\sqrt{n})s$	= 0.00
For change in mean :	$\pm (1.79/\sqrt{n})s$	= 0.00

Table of Multiplying Factors

n	1	2	3	4	5	6	7	8	9	10
f δ,n		0.30	0.35	0.37	0.38	0.39	0.40	0.40	0.40	0.41
1 / n	1.00	0.50	0.33	0.25	0.20	0.17	0.14	0.12	0.11	0.10
1 / (n - 1)		1.00	0.50	0.33	0.25	0.20	0.17	0.14	0.12	0.11
$2/\sqrt{n}$	2.00	1.41	1.15	1.00	0.89	0.82	0.76	0.71	0.67	0.63
$1.79/\sqrt{n}$	1.79	1.26	1.03	0.89	0.80	0.73	0.68	0.63	0.60	0.57

Two variable EVOP program calculation worksheet

Figure 5.21 Work sheet cycle No. 1.

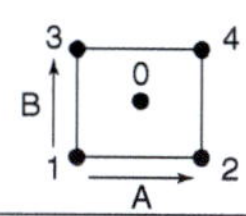

2^2 Factorial with added reference condition

Cycle n : 2
Response:Enzyme Activity

Project Example 5.4
Phase V
Date

Calculation of Averages

Operation condition	(0)	(1)	(2)	(3)	(4)
(i) Previous cycle sum	41.05	37.29	36.01	37.29	36.01
(ii) Previous cycle average	41.05	37.29	36.01	37.29	36.01
(iii) New observations	42.67	39.03	37.31	39.03	37.31
(iv) Differences (ii) less (iii)	-1.62	-1.74	-1.30	-1.74	-1.30
(v) New sums	83.72	76.32	73.32	76.32	73.32
(vi) New averages : ȳi	41.86	38.16	36.66	38.16	36.66

Calculation of Standard Deviation

Prior estimate of σ	= 0.00
Previous sum **s**	=
Previous average **s**	=
New **s** = range × ***f δ,n***	= 0.91
Range	= 3.04
New sum **s**	= 0.91
New average **s** = new sum **s** / (n - 1)	= 0.91

Calculation of Effects

Phase mean	= 1/5 (ȳ0 + ȳ1 + ȳ2 + ȳ3 + ȳ4)	= 38.30
(A) effect	= ½ (ȳ2 + ȳ4 - ȳ1 - ȳ3)	= -1.50
(B) effect	= ½ (ȳ3 + ȳ4 - ȳ1 - ȳ2)	= 0.00
(AB) effect	= ½ (ȳ1 + ȳ4 - ȳ3 - ȳ2)	= 0.00
Change in mean effect	= phase mean - ȳ0	= -3.56

Calculation of 2 S.E. Limits

For new average **s** :	$\pm (2/\sqrt{n})s$	= 1.29
For new effects :	$\pm (2/\sqrt{n})s$	= 1.29
For change in mean :	$\pm (1.79/\sqrt{n})s$	= 1.15

Table of Multiplying Factors

n	1	2	3	4	5	6	7	8	9	10
f δ,n		0.30	0.35	0.37	0.38	0.39	0.40	0.40	0.40	0.41
1 / n	1.00	0.50	0.33	0.25	0.20	0.17	0.14	0.12	0.11	0.10
1 / (n - 1)		1.00	0.50	0.33	0.25	0.20	0.17	0.14	0.12	0.11
$2/\sqrt{n}$	2.00	1.41	1.15	1.00	0.89	0.82	0.76	0.71	0.67	0.63
$1.79/\sqrt{n}$	1.79	1.26	1.03	0.89	0.80	0.73	0.68	0.63	0.60	0.57

Two variable EVOP program calculation worksheet

Figure 5.22 Work sheet cycle No. 2.

Table 5.15 Information board.

INFORMATION BOARD AFTER PHASE NO. V

Last Cycle Completed : n = 2

		RESPONSE No. 1	RESPONSE No. 2	RESPONSE No. 3
REQUIREMENT :		MAXIMIZE RESPONSE		
PROCESS AVERAGES :		38.16 36.66 41.86 B 38.16 36.66 A		
2.S.E. LIMITS FOR AVERAGES		1.29		
EFFECTS WITH 2.S.E	PHASE MEAN FACTOR A FACTOR B INTERACTION AB CHANGE IN MEAN	38.30 -1.50 ± 1.29 0.00 ± 1.29 0.00 ± 1.29 -3.56 ± 1.15		
ESTIMATED STANDARD DEVIATION S		0.91		
PRIOR ESTIMATE OF STANDARD DEV.		0.00		

NOTE:

1. The Center of Experiment-Reference Conditions (The Current Best-Known Process Conditions)
2. The Change in Mean provides a measure of whether the conditions being run the current phase yield an average result better or worse than the reference conditions.
3. The Phase Mean is the Mean Response Over All Conditions Being Run in the present Phase.
4*. The change in Mean is a measure of the over-all Curvature of the Response Surface. It will be positive near a MINIMUM and negative near a MAXIMUM.
5. Sequencing of the Runs-Randomization, if runs can be randomized without too much difficulty, than this should be done.

*A Diagonally oriented saddle represented by equal coefficients of opposite sign could occure and yield a zero change in mean effect, but such a phenomenon would rarely be met in practice.

Conclusions

The effects of all parameters are now practically insignificant. In this set, change in mean effect is negative and large compared to error limit and effect of Tween 80 and interaction effect are insignificant. Although the effect of $ZnSO_4{\cdot}7H_2O$ is of small magnitude, the experimental conditions at the center point of the Phase V have been taken as optimum ($A_0 = 100$ and $B_0 = 0.2$).

Example 5.5 The application of EVOP technique is to optimize three inducers, namely surfactant (Tween 80), metal ion ($ZnSO_4{\cdot}7H_2O$), and a vitamin (riboflavin) [46]. Different concentrations in the medium for protease production have been presented here. Parameters *A*, *B*, and *C* are to be substituted by $ZnSO_4{\cdot}7H_2O$, Tween 80, and riboflavin, respectively. The response was Enzyme activity of Rhizopus [mg/ml].

Experimental conditions selected for the first two cycles were as follows (Table 5.16, Figures 5.23 and 5.24, Table 5.17):

Table 5.16 Initial experimental conditions.

PROJECT : Example 5.5
PHASE : I
DATE :

BEVOP

I. CENTER OF EXPERIMENT :

Ao = 100.00
Bo = 0.20
Co = 10.00

II. FACTOR'S INTERVAL OF VARIATION :

Δ a = 25.00
Δ b = 0.10
Δ c = 10.00

EVOP WORK SHEET

TRIALS ORDER		Block	FACTORS			RESPONSE								
			A	B	C	n = 1	n = 2							
0		Block I	100.00	0.20	10.00	40.52	39.65							
1		Block I	75.00	0.10	0.00	28.75	29.33							
2		Block I	125.00	0.30	0.00	26.91	25.17							
3		Block I	125.00	0.10	20.00	21.68	23.58							
4		Block I	75.00	0.30	20.00	18.64	18.80							
Ø		Block II	100.00	0.20	10.00	40.52	39.65							
5		Block II	75.00	0.10	20.00	37.21	36.64							
6		Block II	125.00	0.30	20.00	36.35	41.76							
7		Block II	125.00	0.10	0.00	36.47	36.13							
8		Block II	75.00	0.30	0.00	37.91	37.86							

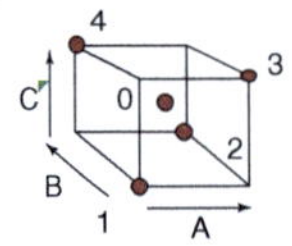

2^3 Factorial in two blocks with reference run in each block

BLOCK I

CYCLE 1
RESPONSE : Enzyme Activity

Project: Example 5.5
Phase: I
Date:

Calculation of Averages						Calculation of Standard Deviation - Block I	
Operation condition	(0)	(1)	(2)	(3)	(4)		
(i) Previous sum for block I						Previous sum **s** (all blocks)	=
(ii) Previous average for block I							
(iii) New observations for block I	40.52	28.75	26.91	21.68	18.64	New **s** = range × ***f d,n***	=
(iv) Differences (ii) less (iii)						Range	=
(v) New sums for block I	40.52	28.75	26.91	21.68	18.64	New sum **s** (all blocks)	=
(vi) New averages for block I	40.52	28.75	26.91	21.68	18.64		

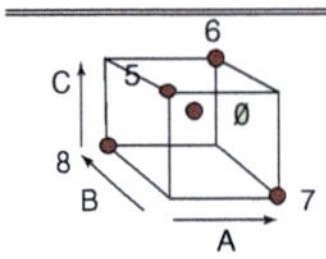

2^3 Factorial in two blocks with reference run in each block

BLOCK II

CYCLE 1
RESPONSE : Enzyme Activity

Calculation of Averages						Calculation of Standard Deviation - Block II	
Operation condition	(Ø)	(5)	(6)	(7)	(8)	Prior estimate of σ	= 0.0
(i) Previous sum for block II						Previous sum **s** (all blocks)	=
(ii) Previous average for block II							
(iii) New observations for block II	40.52	37.21	36.35	36.47	37.91	New **s** = range × ***f d,n***	=
(iv) Differences (ii) less (iii)						Range	=
(v) New sums for block II	40.52	37.21	36.35	36.47	37.91	New sum **s** (all blocks)	=
(vi) New averages for block II	40.52	37.21	36.35	36.47	37.91	Average **s** = {1/[2(n - 1)]} × new sum s	=

Calculation of Effects Using Yates' Algorithm

	$\bar{y}$	(i)	(ii)	(iii)	Multiplier	Effect	
(1)	28.75	65.22	130.04	243.92	0.125	**30.49**	Factorial mean
(7)	36.47	64.82	113.88	-1.10	0.25	**-0.27**	A
(8)	37.91	58.89	-3.28	-4.30	0.25	**-1.08**	B
(2)	26.91	54.99	2.18	14.52	0.25	**3.63**	AB
(5)	37.21	7.72	-0.40	-16.16	0.25	**-4.04**	C
(3)	21.68	-11.00	-3.90	5.46	0.25	**1.37**	AC
(4)	18.64	-15.53	-18.72	-3.50	0.25	**-0.87**	BC
(6)	36.35	17.71	33.24	51.96	0.25	**12.99**	ABC + blocks

Calculation of 2 S.E. Limits

For new average **s** :	$\pm (2/\sqrt{n})s$	= **0.00**
For new effects :	$\pm (1.41/\sqrt{n})s$	= **0.00**
For change in mean :	$\pm (1.26/\sqrt{n})s$	= **0.00**

		$\bar{y}$	Sum	Multiplier
I	Reference (0)	40.52	81.04	0.500
	Conditions (Ø)	40.52		
II	Factorial Conditions	$(1:8)\sum \bar{y}$	243.92	0.125
III	All conditions	$(1:10)\sum \bar{y}$	324.96	0.100
IV	Change in mean	III - I		

Reference mean	=	**40.52**
Factorial mean	=	**30.49**
Phase mean	=	**32.50**
Change in mean	=	**-8.02**

Table of Multiplying Factors

n	1	2	3	4	5	6	7	8	9	10
f d,n		0.30	0.35	0.37	0.38	0.39	0.40	0.40	0.40	0.41
1 / n	1.00	0.50	0.33	0.25	0.20	0.17	0.14	0.12	0.11	0.10
1 / [2(n - 1)]		0.50	0.25	0.17	0.12	0.10	0.08	0.07	0.06	0.06
$2/\sqrt{n}$	2.00	1.41	1.15	1.00	0.89	0.82	0.76	0.71	0.67	0.63
$1.41/\sqrt{n}$	1.41	1.00	0.82	0.71	0.63	0.58	1.86	1.99	2.10	2.24
$1.26/\sqrt{n}$	1.26	0.89	0.73	0.63	0.57	0.52	0.68	0.63	0.60	0.56

Three variable EVOP program calculation worksheet.

Figure 5.23 Work sheet cycle No. 1.

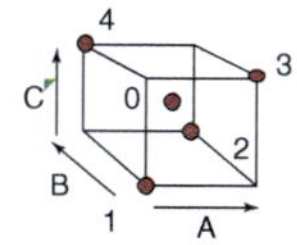

2^3 Factorial in two blocks with reference run in each block

BLOCK I

CYCLE 2
RESPONSE : Enzyme Activity

Project Example 5.5
Phase I
Date

Calculation of Averages					
Operation condition	(0)	(1)	(2)	(3)	(4)
(i) Previous sum for block I	40.52	28.75	26.91	21.68	18.64
(ii) Previous average for block I	40.52	28.75	26.91	21.68	18.64
(iii) New observations for block I	39.65	29.33	25.17	23.58	18.80
(iv) Differences (ii) less (iii)	0.87	-0.58	1.74	-1.90	-0.16
(v) New sums for block I	80.17	58.08	52.08	45.26	37.44
(vi) New averages for block I	40.09	29.04	26.04	22.63	18.72

Calculation of Standard Deviation - Block I	
Previous sum **s** (all blocks)	=
New **s** = range × ***fd,n***	= 1.09
Range	= 3.64
New sum **s** (all blocks)	= 1.09

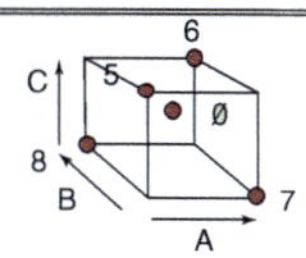

2^3 Factorial in two blocks with reference run in each block

BLOCK II

CYCLE 2
RESPONSE : Enzyme Activity

Calculation of Averages					
Operation condition	(Ø)	(5)	(6)	(7)	(8)
(i) Previous sum for block II	40.52	37.21	36.35	36.47	37.91
(ii) Previous average for block II	40.52	37.21	36.35	36.47	37.91
(iii) New observations for block II	39.65	36.64	41.76	36.13	37.86
(iv) Differences (ii) less (iii)	0.87	0.57	-5.41	0.34	0.05
(v) New sums for block II	80.17	73.85	78.11	72.60	75.77
(vi) New averages for block II	40.09	36.93	39.06	36.30	37.89

Calculation of Standard Deviation - Block II	
Prior estimate of σ	= 0.0
Previous sum **s** (all blocks)	= 1.09
New **s** = range × ***fd,n***	= 1.88
Range	= 6.28
New sum **s** (all blocks)	= 2.98
Average **s** = {1/[2(n - 1)]} × new sum s	= 1.49

Calculation of Effects Using Yates' Algorithm

	$\bar{y}$	(i)	(ii)	(iii)	Multiplier	Effect	
(1)	29.04	65.34	129.27	246.60	0.125	**30.82**	Factorial mean
(7)	36.30	63.93	117.33	1.46	0.25	**0.36**	A
(8)	37.89	59.56	-4.59	-3.20	0.25	**-0.80**	B
(2)	26.04	57.78	6.04	15.53	0.25	**3.88**	AB
(5)	36.93	7.26	-1.42	-11.94	0.25	**-2.98**	C
(3)	22.63	-11.85	-1.78	10.63	0.25	**2.66**	AC
(4)	18.72	-14.30	-19.11	-0.36	0.25	**-0.09**	BC
(6)	39.06	20.34	34.63	53.74	0.25	**13.43**	ABC + blocks

Calculation of 2 S.E. Limits

For new average **s** :	$\pm(2/\sqrt{n})s$	= **2.10**
For new effects :	$\pm(1.41/\sqrt{n})s$	= **1.48**
For change in mean :	$\pm(1.26/\sqrt{n})s$	= **1.33**

		$\bar{y}$	Sum	Multiplier
I	Reference (0) Conditions (Ø)	40.09 40.09	80.17	0.500
II	Factorial Conditions	$(1:8)\sum\bar{y}$	246.60	0.125
III	All conditions	$(1:10)\sum\bar{y}$	326.77	0.100
IV	Change in mean	III - I		

Reference mean	=	**40.09**
Factorial mean	=	**30.82**
Phase mean	=	**32.68**
Change in mean	=	**-7.41**

Table of Multiplying Factors

n	1	2	3	4	5	6	7	8	9	10
fd,n		0.30	0.35	0.37	0.38	0.39	0.40	0.40	0.40	0.41
1 / n	1.00	0.50	0.33	0.25	0.20	0.17	0.14	0.12	0.11	0.10
1 / [2(n - 1)]		0.50	0.25	0.17	0.12	0.10	0.08	0.07	0.06	0.06
$2/\sqrt{n}$	2.00	1.41	1.15	1.00	0.89	0.82	0.76	0.71	0.67	0.63
$1.41/\sqrt{n}$	1.41	1.00	0.82	0.71	0.63	0.58	1.86	1.99	2.10	2.24
$1.26/\sqrt{n}$	1.26	0.89	0.73	0.63	0.57	0.52	0.68	0.63	0.60	0.56

Three variable EVOP program calculation worksheet.

Figure 5.24 Work sheet cycle No. 2.

Table 5.17 Information board .

2^3 INFORMATION BOARD AFTER PHASE NO. I Last Cycle Completed : n = 2			
	RESPONSE No. 1	RESPONSE No. 2	RESPONSE No. 3
REQUIREMENT :	MAXIMIZE RESPONSE		
PROCESS AVERAGES :	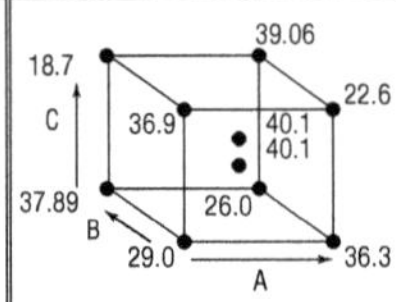		
2.S.E. LIMITS FOR AVERAGES	2.10		
EFFECTS WITH 95% ERROR LIMITS PHASE MEAN FACTOR A FACTOR B FACTOR C INTERACTION AB INTERACTION AC INTERACTION BC INTERACTION ABC CHANGE IN MEAN	32.68 0.36 ± 1.48 -0.80 ± 1.48 -2.98 ± 1.48 3.88 ± 1.48 2.66 ± 1.48 -0.09 ± 1.48 13.43 ± 1.48 -7.41 ± 1.33		
ESTIMATED STANDARD DEVIATION S	1.49		
PRIOR ESTIMATE OF STANDARD DEVIATION σ	0.00		

NOTE:

1. The Center of Experiment-Reference Conditions (The Current Best-Known Process Conditions)
2. The Change in Mean provides a measure of whether the conditions being run the current phase yield an average result better or worse than the reference conditions.
3. The Phase Mean is the Mean Response Over All Conditions Being Run in the present Phase.
4*. The change in Mean is a measure of the over-all Curvature of the Response Surface. It will be positive near a MINIMUM and negative near a MAXIMUM.
5. Sequencing of the Runs-Randomization, if runs can be randomized without too much difficulty, than this should be done.

*A Diagonally oriented saddle represented by equal coefficients of opposite sign could occure and yield a zero change in mean effect, but such a phenomenon would rarely be met in practice.

Conclusions

From analysis of the results given in Table 5.17, it may be noticed that change in mean effect is negative and large compared to error limit. This is one condition of enzyme activity maximization. Most of interaction effects are found to be positive and large compared to error limit. This does not ensure that the condition at the center of search region is the real optimum condition. Furthermore, three-factor interaction effect is very large and positive, and the effect of riboflavin is large and negative. Under the situation, two options can be tried. Either the same center point is used and the levels of metal ions and surfactants in the surrounding points increased, or the center points are shifted to a new condition with higher levels of metal ions and surfactants, keeping riboflavin at the same level. The next set of EVOP design is tried with the second option.

Therefore, next set of experiments were designed by shifting the center point from $A_0 = 100$, $B_0 = 0.20$, and $C_0 = 10$ to $A_0 = 125$, $B_0 = 0.30$, and $C_0 = 10$.

The experimental conditions and results are presented below (Table 5.18, Figures 5.25 and 5.26, Table 5.19):

Table 5.18 BEVOP calculation.

PROJECT : Example 5.5
PHASE : II
DATE :

BEVOP

I. CENTER OF EXPERIMENT :

Ao = 125.00
Bo = 0.30
Co = 10.00

II. FACTOR'S INTERVAL OF VARIATION :

Δ a = 25.00
Δ b = 0.10
Δ c = 10.00

EVOP WORK SHEET

TRIALS ORDER		Block	FACTORS			RESPONSE									
			A	B	C	n = 1	n = 2								
0	A1	Block I	125.00	0.30	10.00	53.26	50.17								
1	A2	Block I	100.00	0.20	0.00	21.66	24.83								
2	A3	Block I	150.00	0.40	0.00	33.60	33.58								
3	A4	Block I	150.00	0.20	20.00	40.49	39.20								
4	A5	Block I	100.00	0.40	20.00	24.94	26.05								
Ø	A6	Block II	125.00	0.30	10.00	53.26	50.17								
5	A7	Block II	100.00	0.20	20.00	27.55	28.46								
6	A8	Block II	150.00	0.40	20.00	35.12	33.73								
7	A9	Block II	150.00	0.20	0.00	24.99	23.25								
8	A10	Block II	100.00	0.40	0.00	37.62	35.27								

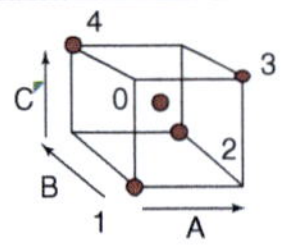

2³ Factorial in two blocks with reference run in each block

BLOCK I

CYCLE 1
RESPONSE : Enzyme Activity

Project Example 5.5
Phase II
Date

Calculation of Averages					
Operation condition	(0)	(1)	(2)	(3)	(4)
(i) Previous sum for block I					
(ii) Previous average for block I					
(iii) New observations for block I	53.26	21.66	33.60	40.49	24.94
(iv) Differences (ii) less (iii)					
(v) New sums for block I	53.26	21.66	33.60	40.49	24.94
(vi) New averages for block I	53.26	21.66	33.60	40.49	24.94

Calculation of Standard Deviation - Block I	
Previous sum **s** (all blocks)	=
New **s** = range × ***f d,n***	=
Range	=
New sum **s** (all blocks)	=

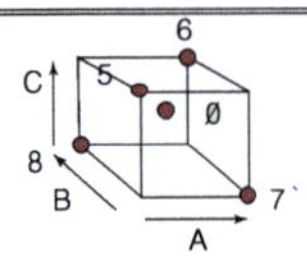

2³ Factorial in two blocks with reference run in each block

BLOCK II

CYCLE 1
RESPONSE : Enzyme Activity

Calculation of Averages					
Operation condition	(Ø)	(5)	(6)	(7)	(8)
(i) Previous sum for block II					
(ii) Previous average for block II					
(iii) New observations for block II	53.26	27.55	35.12	24.99	37.62
(iv) Differences (ii) less (iii)					
(v) New sums for block II	53.26	27.55	35.12	24.99	37.62
(vi) New averages for block II	53.26	27.55	35.12	24.99	37.62

Calculation of Standard Deviation - Block II	
Prior estimate of σ	= 0.0
Previous sum **s** (all blocks)	=
New **s** = range × ***f d,n***	=
Range	=
New sum **s** (all blocks)	=
Average **s** = {1/[2(n - 1)]} × new sum s	=

Calculation of Effects Using Yates' Algorithm

	$\bar{y}$	(i)	(ii)	(iii)	Multiplier	Effect	
(1)	21.66	46.65	117.87	245.97	0.125	**30.75**	Factorial mean
(7)	24.99	71.22	128.10	22.43	0.25	**5.61**	A
(8)	37.62	68.04	-0.69	16.59	0.25	**4.15**	B
(2)	33.60	60.06	23.12	-10.11	0.25	**-2.53**	AB
(5)	27.55	3.33	24.57	10.23	0.25	**2.56**	C
(3)	40.49	-4.02	-7.98	23.81	0.25	**5.95**	AC
(4)	24.94	12.94	-7.35	-32.55	0.25	**-8.14**	BC
(6)	35.12	10.18	-2.76	4.59	0.25	**1.15**	ABC + blocks

Calculation of 2 S.E. Limits

For new average **s** :	$\pm (2/\sqrt{n})s$	= **0.00**
For new effects :	$\pm (1.41/\sqrt{n})s$	= **0.00**
For change in mean :	$\pm (1.26/\sqrt{n})s$	= **0.00**

		$\bar{y}$	**Sum**	**Multiplier**
I	Reference (0) Conditions (Ø)	53.26 53.26	106.52	0.500
II	Factorial Conditions	$(1:8)\sum\bar{y}$	245.97	0.125
III	All conditions	$(1:10)\sum\bar{y}$	352.49	0.100
IV	Change in mean	III - I		

Reference mean	=	**53.26**
Factorial mean	=	**30.75**
Phase mean	=	**35.25**
Change in mean	=	**-18.01**

Table of Multiplying Factors

n	1	2	3	4	5	6	7	8	9	10
f d,n		0.30	0.35	0.37	0.38	0.39	0.40	0.40	0.40	0.41
1 / n	1.00	0.50	0.33	0.25	0.20	0.17	0.14	0.12	0.11	0.10
1 / [2(n - 1)]		0.50	0.25	0.17	0.12	0.10	0.08	0.07	0.06	0.06
$2/\sqrt{n}$	2.00	1.41	1.15	1.00	0.89	0.82	0.76	0.71	0.67	0.63
$1.41/\sqrt{n}$	1.41	1.00	0.82	0.71	0.63	0.58	1.86	1.99	2.10	2.24
$1.26/\sqrt{n}$	1.26	0.89	0.73	0.63	0.57	0.52	0.68	0.63	0.60	0.56

Three variable EVOP program calculation worksheet.

Figure 5.25 Work sheet cycle No. 1.

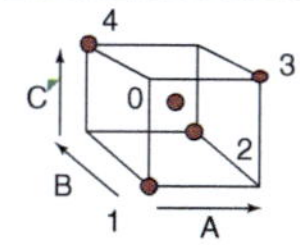

2^3 Factorial in two blocks with reference run in each block

BLOCK I

CYCLE 2
RESPONSE : Enzyme Activity

Project Example 5.5
Phase II
Date

Calculation of Averages					
Operation condition	(0)	(1)	(2)	(3)	(4)
(i) Previous sum for block I	53.26	21.66	33.60	40.49	24.94
(ii) Previous average for block I	53.26	21.66	33.60	40.49	24.94
(iii) New observations for block I	50.17	24.83	33.58	39.20	26.05
(iv) Differences (ii) less (iii)	3.09	-3.17	0.02	1.29	-1.11
(v) New sums for block I	103.43	46.49	67.18	79.69	50.99
(vi) New averages for block I	51.72	23.25	33.59	39.85	25.50

Calculation of Standard Deviation - Block I	
Previous sum **s** (all blocks)	=
New **s** = range × ***fd,n***	= 1.88
Range	= 6.26
New sum **s** (all blocks)	= 1.88

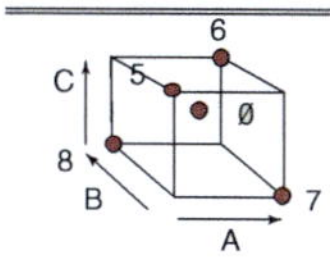

2^3 Factorial in two blocks with reference run in each block

BLOCK II

CYCL 2
RESPONSE : Enzyme Activity

Calculation of Averages					
Operation condition	(Ø)	(5)	(6)	(7)	(8)
(i) Previous sum for block II	53.26	27.55	35.12	24.99	37.62
(ii) Previous average for block II	53.26	27.55	35.12	24.99	37.62
(iii) New observations for block II	50.17	28.46	33.73	23.25	35.27
(iv) Differences (ii) less (iii)	3.09	-0.91	1.39	1.74	2.35
(v) New sums for block II	103.43	56.01	68.85	48.24	72.89
(vi) New averages for block II	51.72	28.01	34.43	24.12	36.45

Calculation of Standard Deviation - Block II	
Prior estimate of σ	= 0.0
Previous sum **s** (all blocks)	= 1.88
New **s** = range × ***fd,n***	= 1.20
Range	= 4.00
New sum **s** (all blocks)	= 3.08
Average **s** = {1/[2(n - 1)]} × new sum s	= 1.54

Calculation of Effects Using Yates' Algorithm

	$\bar{y}$	(i)	(ii)	(iii)	Multiplier	Effect	
(1)	23.25	47.37	117.40	245.17	0.125	**30.65**	Factorial mean
(7)	24.12	70.04	127.77	18.79	0.25	**4.70**	A
(8)	36.45	67.85	-1.98	14.74	0.25	**3.69**	B
(2)	33.59	59.92	20.77	-6.64	0.25	**-1.66**	AB
(5)	28.01	0.88	22.67	10.37	0.25	**2.59**	C
(3)	39.85	-2.86	-7.93	22.75	0.25	**5.69**	AC
(4)	25.50	11.84	-3.73	-30.60	0.25	**-7.65**	BC
(6)	34.43	8.93	-2.91	0.82	0.25	**0.20**	ABC + blocks

Calculation of 2 S.E. Limits

For new average **s** :	$\pm(2/\sqrt{n})s$	= **2.18**
For new effects :	$\pm(1.41/\sqrt{n})s$	= **1.53**
For change in mean :	$\pm(1.26/\sqrt{n})s$	= **1.37**

		$\bar{y}$	**Sum**	**Multiplier**
I	Reference (0) Conditions (Ø)	51.72 51.72	103.43	0.500
II	Factorial Conditions	$(1:8)\sum\bar{y}$	245.17	0.125
III	All conditions	$(1:10)\sum\bar{y}$	348.60	0.100
IV	Change in mean	III - I		

Reference mean	=	**51.72**
Factorial mean	=	**30.65**
Phase mean	=	**34.86**
Change in mean	=	**-16.86**

Table of Multiplying Factors

n	1	2	3	4	5	6	7	8	9	10
fd,n		0.30	0.35	0.37	0.38	0.39	0.40	0.40	0.40	0.41
1 / n	1.00	0.50	0.33	0.25	0.20	0.17	0.14	0.12	0.11	0.10
1 / [2(n - 1)]		0.50	0.25	0.17	0.12	0.10	0.08	0.07	0.06	0.06
$2/\sqrt{n}$	2.00	1.41	1.15	1.00	0.89	0.82	0.76	0.71	0.67	0.63
$1.41/\sqrt{n}$	1.41	1.00	0.82	0.71	0.63	0.58	1.86	1.99	2.10	2.24
$1.26/\sqrt{n}$	1.26	0.89	0.73	0.63	0.57	0.52	0.68	0.63	0.60	0.56

Three variable EVOP program calculation worksheet.

Figure 5.26 Work sheet cycle No. 2.

Table 5.19 Information board.

2^3 INFORMATION BOARD AFTER PHASE NO. II Last Cycle Completed : n = 2			
	RESPONSE No. 1	RESPONSE No. 2	RESPONSE No. 3
REQUIREMENT :	MAXIMIZE RESPONSE		
PROCESS AVERAGES :	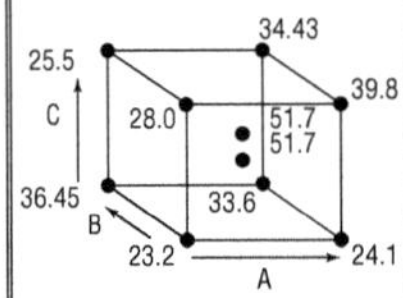		
2.S.E. LIMITS FOR AVERAGES	2.18		
EFFECTS WITH 95% ERROR LIMITS PHASE MEAN	34.86		
FACTOR A	4.70 ± 1.53		
FACTOR B	3.69 ± 1.53		
FACTOR C	2.59 ± 1.53		
INTERACTION AB	-1.66 ± 1.53		
INTERACTION AC	5.69 ± 1.53		
INTERACTION BC	-7.65 ± 1.53		
INTERACTION ABC	0.20 ± 1.37		
CHANGE IN MEAN	-16.86 ± 1.37		
ESTIMATED STANDARD DEVIATION S	1.54		
PRIOR ESTIMATE OF STANDARD DEVIATION σ	0.00		

NOTE:

1. The Center of Experiment-Reference Conditions (The Current Best-Known Process Conditions)
2. The Change in Mean provides a measure of whether the conditions being run the current phase yield an average result better or worse than the reference conditions.
3. The Phase Mean is the Mean Response Over All Conditions Being Run in the present Phase.
4*. The change in Mean is a measure of the over-all Curvature of the Response Surface. It will be positive near a MINIMUM and negative near a MAXIMUM.
5. Sequencing of the Runs-Randomization, if runs can be randomized without too much difficulty, than this should be done.

*A Diagonally oriented saddle represented by equal coefficients of opposite sign could occure and yield a zero change in mean effect, but such a phenomenon would rarely be met in practice.

Conclusions

Interesting results can be observed from Table 5.19. Here, also, change in mean effect is very large and negative. It means that, according to decision-making rule, maximum enzyme activity has been reached. However, no clear-cut conclusion can be drawn as other conditions (effects should be less than error limit) are not fulfilled. Although three-factor interaction effect is negligibly small, the other effects are either positive or negative but large. To make a clear-cut decision in this situation is difficult.

Example 5.6 In the present study, the combined effects of biotin (vitamin)-A, calcium chloride (metal ion)-B, and 1-naphthanelic acetic acid (hormone)-C, on protease synthesis were investigated using Evolutionary Operation – EVOP technique [48].

Experimental conditions selected for the first two cycles were as follows (Table 5.20, Figures 5.27 and 5.28, Table 5.21):

Table 5.20 BEVOP calculation.

PROJECT : Example 5.6
PHASE : I
DATE :

BEVOP

I. CENTER OF EXPERIMENT :

Ao = 2.50
Bo = 50.00
Co = 7.50

II. FACTOR'S INTERVAL OF VARIATION :

Δ a = 0.50
Δ b = 20.00
Δ c = 2.50

EVOP WORK SHEET

TRIALS ORDER		Block	FACTORS			RESPONSE									
			A	B	C	n = 1	n = 2								
0		Block I	2.50	50.00	7.50	360.00	367.00								
1		Block I	2.00	30.00	5.00	305.00	301.00								
2		Block I	3.00	70.00	5.00	255.00	253.00								
3		Block I	3.00	30.00	10.00	380.00	385.00								
4		Block I	2.00	70.00	10.00	215.00	209.00								
Ø		Block II	2.50	50.00	7.50	365.00	358.00								
5		Block II	2.00	30.00	10.00	307.00	305.00								
6		Block II	3.00	70.00	10.00	315.00	317.00								
7		Block II	3.00	30.00	5.00	290.00	282.00								
8		Block II	2.00	70.00	5.00	190.00	198.00								

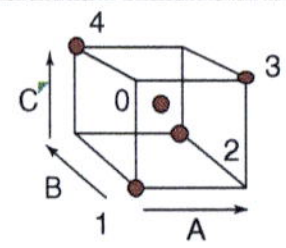

2^3 Factorial in two blocks with reference run in each block

BLOCK I

CYCLE 1
RESPONSE : Enzyme Activity

Project Example 5.6
Phase I
Date

Calculation of Averages					
Operation condition	(0)	(1)	(2)	(3)	(4)
(i) Previous sum for block I					
(ii) Previous average for block I					
(iii) New observations for block I	360.00	305.00	255.00	380.00	215.00
(iv) Differences (ii) less (iii)					
(v) New sums for block I	360.00	305.00	255.00	380.00	215.00
(vi) New averages for block I	360.00	305.00	255.00	380.00	215.00

Calculation of Standard Deviation - Block I	
Previous sum **s** (all blocks)	=
New **s** = range × ***f d,n***	=
Range	=
New sum **s** (all blocks)	=

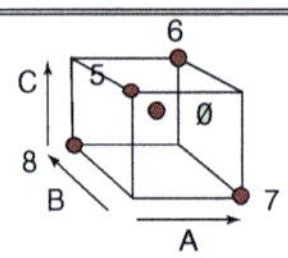

2^3 Factorial in two blocks with reference run in each block

BLOCK II

CYCLE 1
RESPONSE : Enzyme Activity

Calculation of Averages					
Operation condition	(Ø)	(5)	(6)	(7)	(8)
(i) Previous sum for block II					
(ii) Previous average for block II					
(iii) New observations for block II	365.00	307.00	315.00	290.00	190.00
(iv) Differences (ii) less (iii)					
(v) New sums for block II	365.00	307.00	315.00	290.00	190.00
(vi) New averages for block II	365.00	307.00	315.00	290.00	190.00

Calculation of Standard Deviation - Block II	
Prior estimate of **σ**	= 0.0
Previous sum **s** (all blocks)	=
New **s** = range × ***f d,n***	=
Range	=
New sum **s** (all blocks)	=
Average **s** = {1/[2(n - 1)]} × new sum s	=

Calculation of Effects Using Yates' Algorithm

	$\bar{y}$	(i)	(ii)	(iii)	Multiplier	Effect	
(1)	305.00	595.00	1040.00	2257.00	0.125	**282.13**	Factorial mean
(7)	290.00	445.00	1217.00	223.00	0.25	**55.75**	A
(8)	190.00	687.00	50.00	-307.00	0.25	**-76.75**	B
(2)	255.00	530.00	173.00	107.00	0.25	**26.75**	AB
(5)	307.00	-15.00	-150.00	177.00	0.25	**44.25**	C
(3)	380.00	65.00	-157.00	123.00	0.25	**30.75**	AC
(4)	215.00	73.00	80.00	-7.00	0.25	**-1.75**	BC
(6)	315.00	100.00	27.00	-53.00	0.25	**-13.25**	ABC + blocks

Calculation of 2 S.E. Limits

For new average **s** :	$\pm(2/\sqrt{n})s$	= **0.00**
For new effects :	$\pm(1.41/\sqrt{n})s$	= **0.00**
For change in mean :	$\pm(1.26/\sqrt{n})s$	= **0.00**

		$\bar{y}$	Sum	Multiplier
I	Reference (0)	360.00	725.00	0.500
	Conditions (Ø)	360.00		
II	Factorial Conditions	$(1:8)\sum\bar{y}$	2257.00	0.125
III	All conditions	$(1:10)\sum\bar{y}$	2982.00	0.100
IV	Change in mean	III - I		

Reference mean	= **362.50**
Factorial mean	= **282.13**
Phase mean	= **298.20**
Change in mean	= **-64.30**

Table of Multiplying Factors

n	1	2	3	4	5	6	7	8	9	10
f d,n		0.30	0.35	0.37	0.38	0.39	0.40	0.40	0.40	0.41
1 / n	1.00	0.50	0.33	0.25	0.20	0.17	0.14	0.12	0.11	0.10
1 / [2(n - 1)]		0.50	0.25	0.17	0.12	0.10	0.08	0.07	0.06	0.06
$2/\sqrt{n}$	2.00	1.41	1.15	1.00	0.89	0.82	0.76	0.71	0.67	0.63
$1.41/\sqrt{n}$	1.41	1.00	0.82	0.71	0.63	0.58	1.86	1.99	2.10	2.24
$1.26/\sqrt{n}$	1.26	0.89	0.73	0.63	0.57	0.52	0.68	0.63	0.60	0.56

Three variable EVOP program calculation worksheet.

Figure 5.27 Work sheet cycle No. 1.

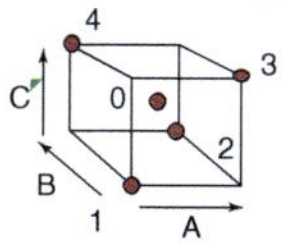

2³ Factorial in two blocks with reference run in each block

BLOCK I

CYCLE 2
RESPONSE : Enzyme Activity

Project Example 5.6
Phase I
Date

Calculation of Averages						Calculation of Standard Deviation - Block I
Operation condition	(0)	(1)	(2)	(3)	(4)	
(i) Previous sum for block I	360.00	305.00	255.00	380.00	215.00	Previous sum **s** (all blocks) =
(ii) Previous average for block I	360.00	305.00	255.00	380.00	215.00	
(iii) New observations for block I	367.00	301.00	253.00	385.00	209.00	New **s** = range × ***fd,n*** = 3.90
(iv) Differences (ii) less (iii)	-7.00	4.00	2.00	-5.00	6.00	Range = 13.00
(v) New sums for block I	727.00	606.00	508.00	765.00	424.00	New sum **s** (all blocks) = 3.90
(vi) New averages for block I	363.50	303.00	254.00	382.50	212.00	

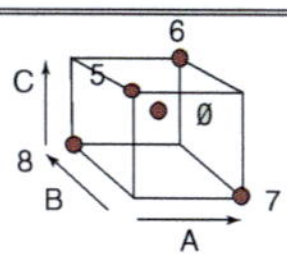

2³ Factorial in two blocks with reference run in each block

BLOCK II

CYCLE 2
RESPONSE : Enzyme Activity

Calculation of Averages						Calculation of Standard Deviation - Block II
Operation condition	(Ø)	(5)	(6)	(7)	(8)	Prior estimate of **σ** = 0.0
(i) Previous sum for block II	365.00	307.00	315.00	290.00	190.00	Previous sum **s** (all blocks) = 3.90
(ii) Previous average for block II	365.00	307.00	315.00	290.00	190.00	
(iii) New observations for block II	358.00	305.00	317.00	282.00	198.00	New **s** = range × ***fd,n*** = 4.80
(iv) Differences (ii) less (iii)	7.00	2.00	-2.00	8.00	-8.00	Range = 16.00
(v) New sums for block II	723.00	612.00	632.00	572.00	388.00	New sum **s** (all blocks) = 8.70
(vi) New averages for block II	361.50	306.00	316.00	286.00	194.00	Average **s** = {1/[2(n - 1)]} × new sum s = 4.35

Calculation of Effects Using Yates' Algorithm

	$\bar{y}$	(i)	(ii)	(iii)	Multiplier	Effect	
(1)	303.00	589.00	1037.00	2253.50	0.125	**281.69**	Factorial mean
(7)	286.00	448.00	1216.50	223.50	0.25	**55.88**	A
(8)	194.00	688.50	43.00	-301.50	0.25	**-75.38**	B
(2)	254.00	528.00	180.50	104.50	0.25	**26.13**	AB
(5)	306.00	-17.00	-141.00	179.50	0.25	**44.88**	C
(3)	382.50	60.00	-160.50	137.50	0.25	**34.38**	AC
(4)	212.00	76.50	77.00	-19.50	0.25	**-4.88**	BC
(6)	316.00	104.00	27.50	-49.50	0.25	**-12.38**	ABC + blocks

Calculation of 2 S.E. Limits

For new average **s** :	$\pm (2/\sqrt{n})s$	= **6.15**
For new effects :	$\pm (1.41/\sqrt{n})s$	= **4.34**
For change in mean :	$\pm (1.26/\sqrt{n})s$	= **3.88**

		$\bar{y}$	**Sum**	**Multiplier**		
I	Reference (0) Conditions (Ø)	363.50 361.50	725.00	0.500	Reference mean =	**362.50**
II	Factorial Conditions	$(1:8)\sum\bar{y}$	2253.50	0.125	Factorial mean =	**281.69**
III	All conditions	$(1:10)\sum\bar{y}$	2978.50	0.100	Phase mean =	**297.85**
IV	Change in mean	III - I			Change in mean =	**-64.65**

Table of Multiplying Factors

n	1	2	3	4	5	6	7	8	9	10
fd,n		0.30	0.35	0.37	0.38	0.39	0.40	0.40	0.40	0.41
1 / n	1.00	0.50	0.33	0.25	0.20	0.17	0.14	0.12	0.11	0.10
1 / [2(n - 1)]		0.50	0.25	0.17	0.12	0.10	0.08	0.07	0.06	0.06
$2/\sqrt{n}$	2.00	1.41	1.15	1.00	0.89	0.82	0.76	0.71	0.67	0.63
$1.41/\sqrt{n}$	1.41	1.00	0.82	0.71	0.63	0.58	1.86	1.99	2.10	2.24
$1.26/\sqrt{n}$	1.26	0.89	0.73	0.63	0.57	0.52	0.68	0.63	0.60	0.56

Three variable EVOP program calculation worksheet.

Figure 5.28 Work sheet cycle No. 2.

Table 5.21 Information board.

2^3 INFORMATION BOARD AFTER PHASE NO. I

Last Cycle Completed : n = 2

		RESPONSE No. 1			RESPONSE No. 2	RESPONSE No. 3
REQUIREMENT :		MAXIMIZE RESPONSE				
PROCESS AVERAGES :		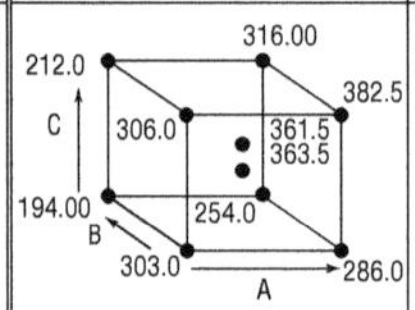				
2.S.E. LIMITS FOR AVERAGES		6.15				
EFFECTS WITH 95% ERROR LIMITS	PHASE MEAN	297.85				
	FACTOR A	55.88	±	4.34		
	FACTOR B	-75.38	±	4.34		
	FACTOR C	44.88	±	4.34		
	INTERACTION AB	26.13	±	4.34		
	INTERACTION AC	34.38	±	4.34		
	INTERACTION BC	-4.88	±	4.34		
	INTERACTION ABC	-12.38	±	4.34		
	CHANGE IN MEAN	-64.65	±	3.88		
ESTIMATED STANDARD DEVIATION S		4.35				
PRIOR ESTIMATE OF STANDARD DEVIATION σ		0.00				

NOTE:

1. The Center of Experiment-Reference Conditions (The Current Best-Known Process Conditions)
2. The Change in Mean provides a measure of whether the conditions being run the current phase yield an average result better or worse than the reference conditions.
3. The Phase Mean is the Mean Response Over All Conditions Being Run in the present Phase.
4*. The change in Mean is a measure of the over-all Curvature of the Response Surface. It will be positive near a MINIMUM and negative near a MAXIMUM.
5. Sequencing of the Runs-Randomization, if runs can be randomized without too much difficulty, than this should be done.

*A Diagonally oriented saddle represented by equal coefficients of opposite sign could occure and yield a zero change in mean effect, but such a phenomenon would rarely be met in practice.

Conclusions

Comparison of effects with error limits indicates that the effect of $CaCl_2 \cdot 2H_2O$ (B) is negative and highly significant, effect of biotin (A) is positive and significant, and effect of 1-naphthanelic acetic acid (C) is positive and significant. Interaction effects are found to be positive and negative and large compared to error limit. It may be noticed that change in mean effect is negative and large compared to error limit. This is one condition of enzyme activity maximization. Even in such situation, magnitudes of the effects being large compared to error limit do not ensure that the real optimum was reached. Therefore, next set of experiments, phase II, was designed by shifting the center point from $A_0 = 2.5$, $B_0 = 50.0$, and $C_0 = 7.5$ to $A_0 = 3.0$, $B_0 = 30.0$, and $C_0 = 10.0$ (Table 5.22, Figures 5.29 and 5.30, Table 5.23).

Table 5.22 BEVOP calculation.

PROJECT : Example 5.6
PHASE : II
DATE :

BEVOP

I. CENTER OF EXPERIMENT :

Ao = 3.00
Bo = 30.00
Co = 10.00

II. FACTOR'S INTERVAL OF VARIATION :

Δ a = 0.50
Δ b = 10.00
Δ c = 2.50

EVOP WORK SHEET

TRIALS ORDER	Block	FACTORS			RESPONSE	
		A	B	C	n = 1	n = 2
0	Block I	**3.00**	**30.00**	**10.00**	375.00	385.00
1	Block I	2.50	20.00	7.50	**305.00**	**305.00**
2	Block I	3.50	40.00	7.50	**275.00**	**285.00**
3	Block I	3.50	20.00	12.50	**310.00**	**305.00**
4	Block I	2.50	40.00	12.50	**314.00**	**317.00**
Ø	Block II	**3.00**	**30.00**	**10.00**	**384.00**	**377.00**
5	Block II	2.50	20.00	12.50	**399.00**	**401.00**
6	Block II	3.50	40.00	12.50	**393.00**	**398.00**
7	Block II	3.50	20.00	7.50	**333.00**	**328.00**
8	Block II	2.50	40.00	7.50	**297.00**	**304.00**

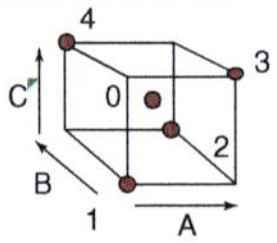

2^3 Factorial in two blocks with reference run in each block

BLOCK I
CYCLE 1
RESPONSE : Enzyme Activity

Project Example 5.6
Phase II
Date

Calculation of Averages						Calculation of Standard Deviation - Block I	
Operation condition	(0)	(1)	(2)	(3)	(4)		
(i) Previous sum for block I						Previous sum **s** (all blocks)	=
(ii) Previous average for block I							
(iii) New observations for block I	375.00	305.00	275.00	310.00	314.00	New **s** = range × ***f d,n***	=
(iv) Differences (ii) less (iii)						Range	=
(v) New sums for block I	375.00	305.00	275.00	310.00	314.00	New sum **s** (all blocks)	=
(vi) New averages for block I	375.00	305.00	275.00	310.00	314.00		

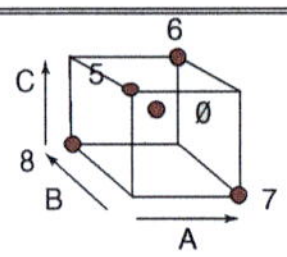

2^3 Factorial in two blocks with reference run in each block

BLOCK II
CYCLE 1
RESPONSE : Enzyme Activity

Calculation of Averages						Calculation of Standard Deviation - Block II	
Operation condition	(Ø)	(5)	(6)	(7)	(8)	Prior estimate of σ	= 0.0
(i) Previous sum for block II						Previous sum **s** (all blocks)	=
(ii) Previous average for block II							
(iii) New observations for block II	384.00	399.00	393.00	333.00	297.00	New **s** = range × ***f d,n***	=
(iv) Differences (ii) less (iii)						Range	=
(v) New sums for block II	384.00	399.00	393.00	333.00	297.00	New sum **s** (all blocks)	=
(vi) New averages for block II	384.00	399.00	393.00	333.00	297.00	Average **s** = {1/[2(n - 1)]} × new sum s	=

Calculation of Effects Using Yates' Algorithm

	$\bar{y}$	(i)	(ii)	(iii)	Multiplier	Effect	
(1)	305.00	638.00	1210.00	2626.00	0.125	**328.25**	Factorial mean
(7)	333.00	572.00	1416.00	-4.00	0.25	**-1.00**	A
(8)	297.00	709.00	6.00	-68.00	0.25	**-17.00**	B
(2)	275.00	707.00	-10.00	118.00	0.25	**29.50**	AB
(5)	399.00	28.00	-66.00	206.00	0.25	**51.50**	C
(3)	310.00	-22.00	-2.00	-16.00	0.25	**-4.00**	AC
(4)	314.00	-89.00	-50.00	64.00	0.25	**16.00**	BC
(6)	393.00	79.00	168.00	218.00	0.25	**54.50**	ABC + blocks

Calculation of 2 S.E. Limits

For new average **s** :	$\pm (2/\sqrt{n})s$	= **0.00**
For new effects :	$\pm (1.41/\sqrt{n})s$	= **0.00**
For change in mean :	$\pm (1.26/\sqrt{n})s$	= **0.00**

		$\bar{y}$	Sum	Multiplier		
I	Reference (0) Conditions (Ø)	375.00 384.00	759.00	0.500	Reference mean =	**379.50**
II	Factorial Conditions	$(1:8)\sum\bar{y}$	2626.00	0.125	Factorial mean =	**328.25**
III	All conditions	$(1:10)\sum\bar{y}$	3385.00	0.100	Phase mean =	**338.50**
IV	Change in mean	III - I			Change in mean =	**-41.00**

Table of Multiplying Factors

n	1	2	3	4	5	6	7	8	9	10
f d,n		0.30	0.35	0.37	0.38	0.39	0.40	0.40	0.40	0.41
1 / n	1.00	0.50	0.33	0.25	0.20	0.17	0.14	0.12	0.11	0.10
1 / [2(n - 1)]		0.50	0.25	0.17	0.12	0.10	0.08	0.07	0.06	0.06
$2/\sqrt{n}$	2.00	1.41	1.15	1.00	0.89	0.82	0.76	0.71	0.67	0.63
$1.41/\sqrt{n}$	1.41	1.00	0.82	0.71	0.63	0.58	1.86	1.99	2.10	2.24
$1.26/\sqrt{n}$	1.26	0.89	0.73	0.63	0.57	0.52	0.68	0.63	0.60	0.56

Three variable EVOP program calculation worksheet.

Figure 5.29 Work sheet cycle No. 1.

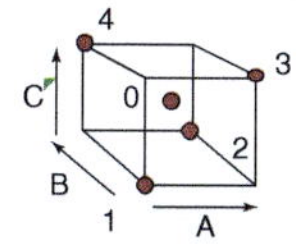

2^3 Factorial in two blocks with reference run in each block

BLOCK I

CYCLE 2
RESPONSE : Enzyme Activity

Project Example 5.6
Phase II
Date

Calculation of Averages					
Operation condition	(0)	(1)	(2)	(3)	(4)
(i) Previous sum for block I	375.00	305.00	275.00	310.00	314.00
(ii) Previous average for block I	375.00	305.00	275.00	310.00	314.00
(iii) New observations for block I	385.00	305.00	285.00	305.00	317.00
(iv) Differences (ii) less (iii)	-10.00	0.00	-10.00	5.00	-3.00
(v) New sums for block I	760.00	610.00	560.00	615.00	631.00
(vi) New averages for block I	380.00	305.00	280.00	307.50	315.50

Calculation of Standard Deviation - Block I		
Previous sum **s** (all blocks)	=	
New **s** = range × ***f d,n***	=	4.50
Range	=	15.00
New sum **s** (all blocks)	=	4.50

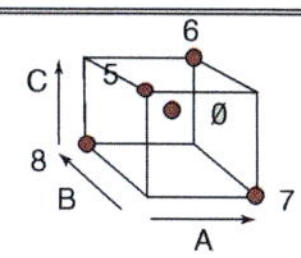

2^3 Factorial in two blocks with reference run in each block

BLOCK II

CYCLE 2
RESPONSE : Enzyme Activity

Calculation of Averages					
Operation condition	(Ø)	(5)	(6)	(7)	(8)
(i) Previous sum for block II	384.00	399.00	393.00	333.00	297.00
(ii) Previous average for block II	384.00	399.00	393.00	333.00	297.00
(iii) New observations for block II	377.00	401.00	398.00	328.00	304.00
(iv) Differences (ii) less (iii)	7.00	-2.00	-5.00	5.00	-7.00
(v) New sums for block II	761.00	800.00	791.00	661.00	601.00
(vi) New averages for block II	380.50	400.00	395.50	330.50	300.50

Calculation of Standard Deviation - Block II		
Prior estimate of σ	=	0.0
Previous sum **s** (all blocks)	=	4.50
New **s** = range × ***f d,n***	=	4.20
Range	=	14.00
New sum **s** (all blocks)	=	8.70
Average **s** = {1/[2(n - 1)]} × new sum s	=	4.35

Calculation of Effects Using Yates' Algorithm

	$\bar{y}$	(i)	(ii)	(iii)	Multiplier	Effect	
(1)	305.00	635.50	1216.00	2634.50	0.125	**329.31**	Factorial mean
(7)	330.50	580.50	1418.50	-7.50	0.25	**-1.88**	A
(8)	300.50	707.50	5.00	-51.50	0.25	**-12.88**	B
(2)	280.00	711.00	-12.50	126.50	0.25	**31.63**	AB
(5)	400.00	25.50	-55.00	202.50	0.25	**50.63**	C
(3)	307.50	-20.50	3.50	-17.50	0.25	**-4.38**	AC
(4)	315.50	-92.50	-46.00	58.50	0.25	**14.63**	BC
(6)	395.50	80.00	172.50	218.50	0.25	**54.63**	ABC + blocks

Calculation of 2 S.E. Limits

For new average **s** :	$\pm (2/\sqrt{n})s$	= **6.15**
For new effects :	$\pm (1.41/\sqrt{n})s$	= **4.34**
For change in mean :	$\pm (1.26/\sqrt{n})s$	= **3.88**

		$\bar{y}$	**Sum**	**Multiplier**
I	Reference (0) Conditions (Ø)	380.00 380.50	760.50	0.500
II	Factorial Conditions	$(1:8)\sum\bar{y}$	2634.50	0.125
III	All conditions	$(1:10)\sum\bar{y}$	3395.00	0.100
IV	Change in mean	III - I		

Reference mean	=	**380.25**
Factorial mean	=	**329.31**
Phase mean	=	**339.50**
Change in mean	=	**-40.75**

Table of Multiplying Factors

n	1	2	3	4	5	6	7	8	9	10
f d,n		0.30	0.35	0.37	0.38	0.39	0.40	0.40	0.40	0.41
1 / n	1.00	0.50	0.33	0.25	0.20	0.17	0.14	0.12	0.11	0.10
1 / [2(n - 1)]		0.50	0.25	0.17	0.12	0.10	0.08	0.07	0.06	0.06
$2/\sqrt{n}$	2.00	1.41	1.15	1.00	0.89	0.82	0.76	0.71	0.67	0.63
$1.41/\sqrt{n}$	1.41	1.00	0.82	0.71	0.63	0.58	1.86	1.99	2.10	2.24
$1.26/\sqrt{n}$	1.26	0.89	0.73	0.63	0.57	0.52	0.68	0.63	0.60	0.56

Three variable EVOP program calculation worksheet.

Figure 5.30 Work sheet cycle No. 2.

Table 5.23 Information board.

2^3 INFORMATION BOARD AFTER PHASE NO. II Last Cycle Completed : n = 2			
	RESPONSE No. 1	RESPONSE No. 2	RESPONSE No. 3
REQUIREMENT :	MAXIMIZE RESPONSE		
PROCESS AVERAGES :	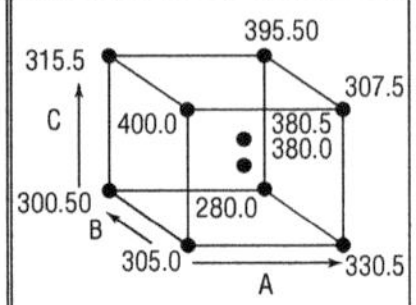		
2.S.E. LIMITS FOR AVERAGES	6.15		
EFFECTS WITH 95% ERROR LIMITS PHASE MEAN FACTOR A FACTOR B FACTOR C INTERACTION AB INTERACTION AC INTERACTION BC INTERACTION ABC CHANGE IN MEAN	339.50 -1.88 ± 4.34 -12.88 ± 4.34 50.63 ± 4.34 31.63 ± 4.34 -4.38 ± 4.34 14.63 ± 4.34 54.63 ± 4.34 -40.75 ± 3.88		
ESTIMATED STANDARD DEVIATION S	4.35		
PRIOR ESTIMATE OF STANDARD DEVIATION σ	0.00		

NOTE:

1. The Center of Experiment-Reference Conditions (The Current Best-Known Process Conditions)
2. The Change in Mean provides a measure of whether the conditions being run the current phase yield an average result better or worse than the reference conditions.
3. The Phase Mean is the Mean Response Over All Conditions Being Run in the present Phase.
4*. The change in Mean is a measure of the over-all Curvature of the Response Surface. It will be positive near a MINIMUM and negative near a MAXIMUM.
5. Sequencing of the Runs-Randomization, if runs can be randomized without too much difficulty, than this should be done.

*A Diagonally oriented saddle represented by equal coefficients of opposite sign could occure and yield a zero change in mean effect, but such a phenomenon would rarely be met in practice.

Conclusions

The effect of $CaCl_2{\cdot}2H_2O$ (B) is highly significant and negative. Effect of biotin (A) is not significant, and effect of 1-naphthanelic acetic acid (C) is positive and significant. Interaction effects are found to be positive and negative and large compared to the error limit. It may be noticed that the change in mean effect is negative and large compared to the error limit. Under the above situation, a third phase of experiments was designed by shifting the center point from $A_0 = 3.0$, $B_0 = 30.0$, and $C_0 = 10.0$ to $A_0 = 2.5$, $B_0 = 20.0$, and $C_0 = 12.50$ (Table 5.24, Figures 5.31 and 5.32, Table 5.25).

Table 5.24 BEVOP calculation.

PROJECT : Example 5.6
PHASE : III
DATE :

BEVOP

I. CENTER OF EXPERIMENT :

Ao = 2.50
Bo = 20.00
Co = 12.50

II. FACTOR'S INTERVAL OF VARIATION :

Δ a = 0.50
Δ b = 10.00
Δ c = 2.50

EVOP WORK SHEET

TRIALS ORDER		Block	FACTORS A	B	C	RESPONSE n = 1	n = 2							
0		Block I	2.50	20.00	12.50	408.00	410.00							
1		Block I	2.00	10.00	10.00	385.00	380.00							
2		Block I	3.00	30.00	10.00	390.00	395.00							
3		Block I	3.00	10.00	15.00	400.00	401.00							
4		Block I	2.00	30.00	15.00	380.00	390.00							
Ø		Block II	2.50	20.00	12.50	410.00	400.00							
5		Block II	2.00	10.00	15.00	385.00	390.00							
6		Block II	3.00	30.00	15.00	397.00	395.00							
7		Block II	3.00	10.00	10.00	395.00	395.00							
8		Block II	2.00	30.00	10.00	398.00	400.00							

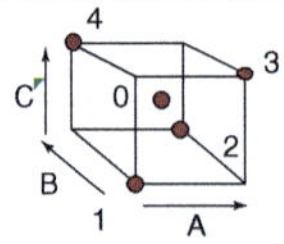

2^3 Factorial in two blocks with reference run in each block

BLOCK I
CYCLE 1
RESPONSE : Enzyme Activity

Project Example 5.6
Phase III
Date

Calculation of Averages						Calculation of Standard Deviation - Block I	
Operation condition	(0)	(1)	(2)	(3)	(4)		
(i) Previous sum for block I						Previous sum s (all blocks)	=
(ii) Previous average for block I							
(iii) New observations for block I	408.00	385.00	390.00	400.00	380.00	New s = range × fd,n	=
(iv) Differences (ii) less (iii)						Range	=
(v) New sums for block I	408.00	385.00	390.00	400.00	380.00	New sum s (all blocks)	=
(vi) New averages for block I	408.00	385.00	390.00	400.00	380.00		

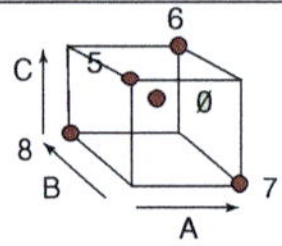

2^3 Factorial in two blocks with reference run in each block

BLOCK II
CYCLE 1
RESPONSE : Enzyme Activity

Calculation of Averages						Calculation of Standard Deviation - Block II	
Operation condition	(Ø)	(5)	(6)	(7)	(8)	Prior estimate of σ	= 0.0
(i) Previous sum for block II						Previous sum s (all blocks)	=
(ii) Previous average for block II							
(iii) New observations for block II	410.00	385.00	397.00	395.00	398.00	New s = range × fd,n	=
(iv) Differences (ii) less (iii)						Range	=
(v) New sums for block II	410.00	385.00	397.00	395.00	398.00	New sum s (all blocks)	=
(vi) New averages for block II	410.00	385.00	397.00	395.00	398.00	Average s = {1/[2(n - 1)]} × new sum s	=

Calculation of Effects Using Yates' Algorithm

	$\bar{y}$	(i)	(ii)	(iii)	Multiplier	Effect	
(1)	385.00	780.00	1568.00	3130.00	0.125	**391.25**	Factorial mean
(7)	395.00	788.00	1562.00	34.00	0.25	**8.50**	A
(8)	398.00	785.00	2.00	0.00	0.25	**0.00**	B
(2)	390.00	777.00	32.00	-16.00	0.25	**-4.00**	AB
(5)	385.00	10.00	8.00	-6.00	0.25	**-1.50**	C
(3)	400.00	-8.00	-8.00	30.00	0.25	**7.50**	AC
(4)	380.00	15.00	-18.00	-16.00	0.25	**-4.00**	BC
(6)	397.00	17.00	2.00	20.00	0.25	**5.00**	ABC + blocks

Calculation of 2 S.E. Limits

For new average s :	$\pm (2/\sqrt{n})s$	= **0.00**
For new effects :	$\pm (1.41/\sqrt{n})s$	= **0.00**
For change in mean :	$\pm (1.26/\sqrt{n})s$	= **0.00**

		$\bar{y}$	Sum	Multiplier			
I	Reference (0) Conditions (Ø)	408.00 410.50	818.00	0.500	Reference mean	=	**409.00**
II	Factorial Conditions	$(1:8)\sum\bar{y}$	3130.00	0.125	Factorial mean	=	**391.25**
III	All conditions	$(1:10)\sum\bar{y}$	3948.00	0.100	Phase mean	=	**394.80**
IV	Change in mean	III - I			Change in mean	=	**-14.20**

Table of Multiplying Factors

n	1	2	3	4	5	6	7	8	9	10
fd,n		0.30	0.35	0.37	0.38	0.39	0.40	0.40	0.40	0.41
1 / n	1.00	0.50	0.33	0.25	0.20	0.17	0.14	0.12	0.11	0.10
1 / [2(n - 1)]		0.50	0.25	0.17	0.12	0.10	0.08	0.07	0.06	0.06
$2/\sqrt{n}$	2.00	1.41	1.15	1.00	0.89	0.82	0.76	0.71	0.67	0.63
$1.41/\sqrt{n}$	1.41	1.00	0.82	0.71	0.63	0.58	1.86	1.99	2.10	2.24
$1.26/\sqrt{n}$	1.26	0.89	0.73	0.63	0.57	0.52	0.68	0.63	0.60	0.56

Three variable EVOP program calculation worksheet.

Figure 5.31 Work sheet cycle No. 1.

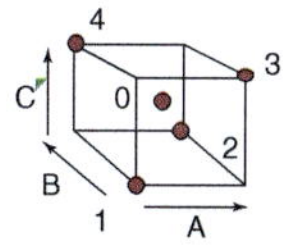

2^3 Factorial in two blocks with reference run in each block

BLOCK I

CYCLE 2
RESPONSE : Enzyme Activity

Project Example 5.6
Phase III
Date

Calculation of Averages						Calculation of Standard Deviation - Block I		
Operation condition	(0)	(1)	(2)	(3)	(4)			
(i) Previous sum for block I	408.00	385.00	390.00	400.00	380.00	Previous sum **s** (all blocks)	=	
(ii) Previous average for block I	408.00	385.00	390.00	400.00	380.00			
(iii) New observations for block I	410.00	380.00	395.00	401.00	390.00	New **s** = range × ***f d,n***	=	4.50
(iv) Differences (ii) less (iii)	-2.00	5.00	-5.00	-1.00	-10.00	Range	=	15.00
(v) New sums for block I	818.00	765.00	785.00	801.00	770.00	New sum **s** (all blocks)	=	4.50
(vi) New averages for block I	409.00	382.50	392.50	400.50	385.00			

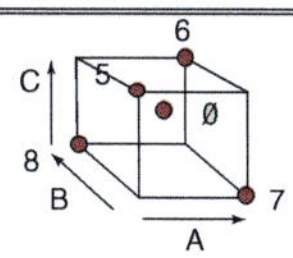

2^3 Factorial in two blocks with reference run in each block

BLOCK II

CYCLE 2
RESPONSE : Enzyme Activity

Calculation of Averages						Calculation of Standard Deviation - Block II		
Operation condition	(Ø)	(5)	(6)	(7)	(8)	Prior estimate of σ	=	0.0
(i) Previous sum for block II	410.00	385.00	397.00	395.00	398.00	Previous sum **s** (all blocks)	=	4.50
(ii) Previous average for block II	410.00	385.00	397.00	395.00	398.00			
(iii) New observations for block II	400.00	390.00	395.00	395.00	400.00	New **s** = range × ***f d,n***	=	4.50
(iv) Differences (ii) less (iii)	10.00	-5.00	2.00	0.00	-2.00	Range	=	15.00
(v) New sums for block II	810.00	775.00	792.00	790.00	798.00	New sum **s** (all blocks)	=	9.00
(vi) New averages for block II	405.00	387.50	396.00	395.00	399.00	Average **s** = {1/[2(n - 1)]} × new sum s	=	4.50

Calculation of Effects Using Yates' Algorithm

	$\bar{y}$	(i)	(ii)	(iii)	Multiplier	Effect	
(1)	382.50	777.50	1569.00	3138.00	0.125	**392.25**	Factorial mean
(7)	395.00	791.50	1569.00	30.00	0.25	**7.50**	A
(8)	399.00	788.00	6.00	7.00	0.25	**1.75**	B
(2)	392.50	781.00	24.00	-21.00	0.25	**-5.25**	AB
(5)	387.50	12.50	14.00	0.00	0.25	**0.00**	C
(3)	400.50	-6.50	-7.00	18.00	0.25	**4.50**	AC
(4)	385.00	13.00	-19.00	-21.00	0.25	**-5.25**	BC
(6)	396.00	11.00	-2.00	17.00	0.25	**4.25**	ABC + blocks

Calculation of 2 S.E. Limits

For new average **s** :	$\pm(2/\sqrt{n})s$	= **6.36**
For new effects :	$\pm(1.41/\sqrt{n})s$	= **4.49**
For change in mean :	$\pm(1.26/\sqrt{n})s$	= **4.01**

		$\bar{y}$	Sum	Multiplier			
I	Reference (0) Conditions (Ø)	409.00 405.00	814.00	0.500	Reference mean	=	**407.00**
II	Factorial Conditions	$(1:8)\sum\bar{y}$	3138.00	0.125	Factorial mean	=	**392.25**
III	All conditions	$(1:10)\sum\bar{y}$	3952.00	0.100	Phase mean	=	**395.20**
IV	Change in mean	III - I			Change in mean	=	**-11.80**

Table of Multiplying Factors

n	1	2	3	4	5	6	7	8	9	10
fd,n		0.30	0.35	0.37	0.38	0.39	0.40	0.40	0.40	0.41
1 / n	1.00	0.50	0.33	0.25	0.20	0.17	0.14	0.12	0.11	0.10
1 / [2(n - 1)]		0.50	0.25	0.17	0.12	0.10	0.08	0.07	0.06	0.06
$2/\sqrt{n}$	2.00	1.41	1.15	1.00	0.89	0.82	0.76	0.71	0.67	0.63
$1.41/\sqrt{n}$	1.41	1.00	0.82	0.71	0.63	0.58	1.86	1.99	2.10	2.24
$1.26/\sqrt{n}$	1.26	0.89	0.73	0.63	0.57	0.52	0.68	0.63	0.60	0.56

Three variable EVOP program calculation worksheet.

Figure 5.32 Work sheet cycle No. 2.

Table 5.25 Information board.

2^3 INFORMATION BOARD AFTER PHASE NO. III

Last Cycle Completed : n = 2

	RESPONSE No. 1	RESPONSE No. 2	RESPONSE No. 3
REQUIREMENT :	MAXIMIZE RESPONSE		
PROCESS AVERAGES :	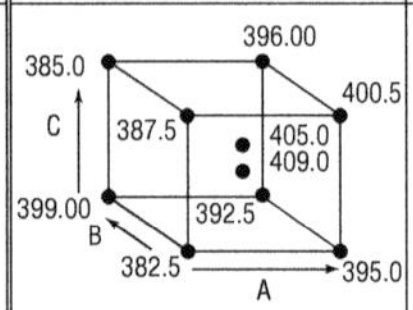		
2.S.E. LIMITS FOR AVERAGES	6.36		
EFFECTS WITH 95% ERROR LIMITS PHASE MEAN FACTOR A FACTOR B FACTOR C INTERACTION AB INTERACTION AC INTERACTION BC INTERACTION ABC CHANGE IN MEAN	 395.20 7.50 ± 4.49 1.75 ± 4.49 0.00 ± 4.49 -5.25 ± 4.49 4.50 ± 4.49 -5.25 ± 4.49 4.25 ± 4.49 -11.80 ± 4.01		
ESTIMATED STANDARD DEVIATION S	4.50		
PRIOR ESTIMATE OF STANDARD DEVIATION σ	0.00		

NOTE:

1. The Center of Experiment-Reference Conditions (The Current Best-Known Process Conditions)
2. The Change in Mean provides a measure of whether the conditions being run the current phase yield an average result better or worse than the reference conditions.
3. The Phase Mean is the Mean Response Over All Conditions Being Run in the present Phase.
4*. The change in Mean is a measure of the over-all Curvature of the Response Surface. It will be positive near a MINIMUM and negative near a MAXIMUM.
5. Sequencing of the Runs-Randomization, if runs can be randomized without too much difficulty, than this should be done.

*A Diagonally oriented saddle represented by equal coefficients of opposite sign could occure and yield a zero change in mean effect, but such a phenomenon would rarely be met in practice.

Conclusions

The effect of biotin (A) is found to be positive and large. The effects of $CaCl_2 \cdot 2H_2O$ (B) and 1-naphthanelic acetic acid (C) being lower than the error limit (are not significant) indicate that their concentrations are around optimum. That means, its concentration should have been higher. The change in mean effect is negative and large compared to error limit, and this satisfies one requirement of optimization. Interestingly, the center of the experiment gave the maximum enzyme activity. To arrive at the real optimum condition, a few more sets of experiments are required with the level of factors being varied over a smaller range.

5.4 BEVOP in Plant-Scale Experiments

Answering questions about a process is generally difficult, as three things operate against us:

- Inevitable variability: There is always noise and this tends to cloud the picture and makes it difficult to determine the value of response.
- General complexity of physical processes: Relationship between the process factors and response is not a simple one.
- Economics: There is a limit to the funds to be invested in an experimental program. These three elements impose restraints on an experimental program, depending upon the context of the experimentation.
  When we choose to do our experiments on the plant production equipment, we undertake additional responsibility, i.e. for protecting the company's investment in the equipment and raw materials we are going to use. Engineering knowledge of the process and its equipment is a must to prevent damage to the equipment and injury to the people running it. Also, we must remember that investment in equipment was not made for running experiments, but for manufacturing a product for sale.
  Unlike a pilot plant or laboratory, our primary interest here is in the product. If we can improve the product while making it, so much the better. But, on the other hand, if we have selected an experimental pattern resulting in a lot of poor-quality products, we have failed in our responsibility and our actions may affect the economics of the entire company. Before tampering with a production process, *we should better know what we are doing*. And before starting an EVOP program, plant supervision usually wants to be convinced of this. To do that we can use a pilot plant or prototype equipment, when available, to demonstrate capabilities of an EVOP program. If those two options are out of question, the easiest way will be to use *software simulations demonstrated in this book*. However, if we cannot demonstrate the value of our program and our own competence, then we have no right to ask plant supervision to back an experiment on a plant scale where the stakes are high. Once plant supervision has been convinced that its investment will be protected and that the program has potential gain, we should dig deep into everything we know about process. Early in the procedure, we must determine which process

factors are to be studied. Basic variability in our observations is inevitable, we must select those factors which in a reasonable time will cause sufficient change so that the noise of natural variability does not disguise the effect of our changes.

Previous experience in a pilot plant may indicate which factors are important. But care must be exercised here because of the factors, not looked at in the pilot plant, may exist, which are equally important. Other factors may have changed in importance since scale-up to the commercial plant.

Sometimes, major factors may be screened out from production log records. However, this procedure is likely to be more successful if deliberate changes have been made so that what we see is more than noise. If production data are not reliable, perhaps an engineer should record settings of factors over a period of time. Also, a close look at the process chemistry and mechanism should indicate candidates for test factors. Two other items have to be considered: *How much* and *How Often* should the factors be changed?

The how much should the factors be changed aspect is the one that can really hurt if we are wrong. When information about expectation is not available, it's best to be a little bit conservative. Perhaps a reasonable change is one equal to a degree of control for that factor (*factor's fixing error*). For example, if a temperature can be controlled to ±2 [°C], then we would make changes of 4 [°C] (factor's interval of variation). Perhaps, larger changes could be made. In any event, as much thought should go into this part of setting up a program as any other part.

How-often aspect depends on the process itself and the data collection system. There are two main principles to remember:

- The process should return to *equilibrium before* data are collected for any set of operating conditions. For mechanical processes, this is usually no problem because the system usually will return to equilibrium within a couple of hours. For some chemical processes, it may take days or weeks for the full effect of a change made early in a process to be felt.
- The set of conditions should be run long enough to obtain sufficient information about response for that set of conditions.

If these two principles are met, then *the changes can be made at any convenient interval: once per shift, once per day, or once per week.*

To this point, as we mentioned earlier, we have considered only one response, but obviously in any practical case, there are many responses like yield, productivity, cost, and profit. Some care must be given to a choice of the response that is going to be maximized. The point at which productivity is a maximum is probably not the point at which yield is maximum or profit highest. In such cases, we are dealing with a *constrained optimization* problem, an output function is to be maximized, but subject to the restrictions imposed by the other outputs. Converted into practice, the simultaneous consideration of all outputs means that we have to:

- Combine all output responses into a single response, which is an index of performance, and optimize with respect to this index. It would be advantageous to define a "desirability" function as single response, Appendix E [3, 42–44], expressing several responses as a single number. This function would be expected to have an ideal single-peaked contour with a definable optimum. Without such function, the optimum must be defined as the most desirable combination of dependent properties, including cost and quality.

- Make observations on all responses and follow each closely to avoid taking direction, which leads to the wrong consequences for any of them.
- Abandon the maximum-hunting type of experimentation and use another approach. For some chemical processes, a system may be so complex that a computer is needed to assist in seeking and maintaining optimization.

In connection with measuring output response, we should perhaps go more into detail. May be more measurements on one response are needed because of high analytical variability. The special tools from *Six-Sigma* methodology like *Gage R&R* will provide answer if analytical measurement contributes significantly to total process variability. If that is the case, we may even have to develop or install a new measure, which is more sensitive to changes in the product.

Since our actions are based on values of the observations and what we do with them, we must ensure their reliability and validity. This is an important phase of overall program. On the other hand, we must be certain to provide for following the product through the entire process, especially in multistep processes. Then observations may be correctly related to a specific set of operating conditions.

Also, we must not forget another concept for improving the economics of manufacturing processes like statistical process control (SPC). Because of high variability, it is necessary to operate at each experimental point over a relatively long time, even though operation may not be consecutive. Therefore, it is important to have a process in a state of statistical control.

We must have, at least to a first approximation, the same error distribution at each point and the same distribution each time we run the same point. If not, then unidentified sources of variation exist and if their effect cannot be predicted, we may be led to false conclusions about the factors we are changing. There are factors that have not been considered and we have not insisted on their control. There may even be factors over which we have no control. These could be factors such as humidity or raw material quality.

An EVOP program usually covers a considerable length of time, and often it is impossible to isolate the process completely from other experimental work. It is impractical to insist that all other technical effort on the process be stopped. Therefore, there is another potential source for noise resulting from other experimental work not under EVOP control.

They represent changes in the process from other technical effort, which may not be related to what we are doing, but can nevertheless affect our results. In that case, simple statistical analysis as described may not be adequate and a more sophisticated analysis may be needed. For most industrial processes, EVOP program will be successful, provided that rules of the game are followed. This assumes that once started, EVOP program is allowed to continue. If plant supervision, assured of our competence initially, has given permission for the program to run, why should they want to stop the operation? There are several possible reasons, none of which should be justifiable:

- The laboratory complains of too much additional analytical measurements.
- Process supervision claims it upsets their daily routine.
- Bad product is being made.

  That a bad product is being made should not be a reason for termination, unless we have made mistake. All personnel in supervision related to the process under study must be acquainted with the facts of life concerning experimental programs. They must realize that for every set of conditions which improves response, there is another set that makes it worse, unless we are close to optimum. Supervision must be assured that we will recommend a change when the results of our statistical analysis indicate that the differences in response, we have been observing between sets of conditions, are real differences.

  Agreement must be obtained prior to an EVOP start-up that we have control over the conduct of the experimentation. It means we must keep everyone informed of the progress. This can be done by information board, weekly meetings, or a discussion of results after each complete cycle. The supervisor directly responsible for the process should have up-to-date chart or summary of the status of the program. There must be very close cooperation between the supervisor and EVOP experimenters. It is the supervisor's process.

  In summary, we wish to emphasize that EVOP is not a new technique. Rather, it is a straightforward application of sound, well-tested principles of experimentation to an area where experimenters encounter more than usual restraints. It is a new and different approach to process improvement [47].

  Comparison between DOE and EVOP approaches to process improvement is listed below:

DOE	EVOP
Breakthrough	Evolutionary
Off-line	During production
Off-spec product required	No off-spec product allowed
Wide factor ranges-intervals	Narrow factor ranges-intervals
Fast data collection	Slow data collection
Iterative by choice	Iterative by design
Collaboration is useful	Collaboration is critical

5.5 BEVOP Applications

Dow Chemical Company has a plant with a number of film extruders. They have eight heat zones with friction of the screw adding its share of heat. Heat zones and screw r.p.m.'s can be set at a fantastic number of combinations, so without a definite technique for determining optimum settings, operators becomes an expert on their own. Thus, it may not be unusual for temperature chart and production record of an extruder to show that everything is running smoothly and then suddenly a log may change drastically, heat on zone 3 may decrease 10 [°], and that on zone 2 increase 12 [°]. This was puzzling until it was found that the change occurred at the precise time when a new operator came on duty.

It would be wrong to say if plastics molder or an extruder adopts EVOP that its troubles would vanish, though in Dow Chemical plant, it did help.

Subsequently EVOP techniques were applied to one machine, important zones were identified, and optimum settings were determined. In a short time, this machine increased its output by 20 [%] above other machines. As the study continued, yield continued to improve until undesirable characteristics appeared. At the cutoff point, output had increased by 37 [%].

One step at a time, classical procedure has achieved many good results, but what happens when one takes a certain step and then gets defeated heavily. That actually happened in one instance when use of recycle material was stepped up and off-quality of the product increased significantly. With EVOP, several factors were studied and the amount of used recycle material was increased, while nothing adverse happened. Ultimately, the ratio of recycle material could be doubled, while another factor, product quality, actually improved. In addition to having a better product, estimated cost savings amounted to some $4,000/month [12, 13].

A problem in a styrene plant was to find optimal temperature, ethylbenzene feed rate, and steam rate.

Three-factor EVOP study was set up and the amount of styrene produced was established for each of several combinations of the three factors. With this information, the superintendent could maximize production as far as these three factors are concerned.

However, in this instance, each combination of the three factors varied not only the amount of styrene produced, but also amounts of by-products: benzene and toluene. Then in the next step, net cost per pound of styrene and value of by-products have to be included in savings.

In an EVOP study of another process, one full phase of an operation was eliminated, and cycle time of another step was reduced; this reduced polymerization time by 52 [%]. In such case where lack of capacity is a factor, such a gain can be equated in terms of added capacity. If $1 million is invested in equipment, which can produce 10,000 lb/month of a certain material, increasing the output to 20,000 lb/month with EVOP would be equivalent to doubling capital investment without EVOP.

Monsanto Company's, through EVOP savings of $250,000/year for two products, have been realized because of reduced raw material usage and reduced time cycles in a batch operation [14]. Monsanto has also applied EVOP in the operation of an ammonia plant [15].

American Cyanamid Company has used EVOP extensively and successfully. EVOP has contributed substantially to the understanding of their hydrogen cyanide process [16]. In the production of hydrogen cyanide, by converting ammonia, methane, and oxygen over a platinum gauze catalyst, those factors of critical importance from a control standpoint were discovered. By directing additional attention to the control of these factors, the hydrogen cyanide yield was increased by 4%.

The factors entering most frequently into Imperial Chemical Industries EVOP programs have been temperature, time, and concentration [17, 18]. The most common responses have been yield, purity, or a composite cost function. The author describes the EVOP of a batch dyestuff process through ten phases of two to three factors at a time. In the first ten years of operation, a yield increase greater than 10% was obtained. The author also discusses the operation of a continuous catalytic reactor. Varying the ratio of two reactants and temperature for this reactor resulted in an efficiency increase of 1%, which, because of the volume of throughput and the value of the product, represented considerable savings.

The application of EVOP has resulted in a substantial savings for Tennessee Eastman Company. A batch process is described [19] in which chemicals react to produce a desired product as well as an acid, which is neutralized after a given length of time. The weight of acid-producing chemical and neutralization time were varied, resulting in a yield improvement after the fourth phase of operation from 562 to 588 lb. This represents a savings of $25,000/year.

Monsanto Company's savings of $250,000/year with two products have been realized using EVOP. DeBusk [20] discusses Tennessee Eastman Company's problems with EVOP applications. He also mentions the application of three-factor EVOP on an established process, which (after three phases) resulted in $27,500 savings/year. He points out that EVOP is used in approximately fifteen processes, with savings from $15,000 to $50,000/year/process.

Wilson [21] discusses the operation of a unit designed with a thousand-fold scale-up factor. From laboratory studies it was found that a step in the process, involving a condensation reaction, affected total cost more than any other step. For EVOP study reported, temperature and feed rate were the factors considered and cost, yield and production rate were the responses. After four cycles, the product cost was reduced by five cents per pound and a 26 % increase in productivity was achieved.

The "Chemstrand" Corporation has also achieved success in EVOP applications [22]. Riordan [23] has discussed both administrative problems and technical rewards. He presented several examples of how EVOP led to equipment modification and how EVOP yielded results, appearing completely erroneous, thus stimulated an investigation, which uncovered situations not previously suspected. Apart from the several $100,000/ year in overall savings and increased plant capacity, he points out the benefit of increased process understanding through EVOP.

Several other industries have had success with EVOP applications. These applications include automotive industry, where EVOP has been applied to resistance welding of automotive sheet metal [24] and other manufacturing operations [25]. It has been reported that Phillips Petroleum Company has successfully applied EVOP program [26]. Standard Oil of Ohio's lubricating oil plants successfully applied EVOP to one of their catalytic cracking units [27].

Food industry also has its examples of successful EVOP applications. It has been suggested that large companies such as Swift and Canada Packers Limited use EVOP. Swift has had successful application in meat-smoking and -curing units, vegetable oil extraction mills, and in certain other processes [28].

A. E. Staley Manufacturing Company has applied EVOP to corn-grinding operation [29]. The use of EVOP for increasing the productivity of fermentation process has been discussed [30]. An EVOP example in "Novozymes" Danish company in 2005 is presented, where the method was used in a fermentation process of an industrial enzyme. In the particular example, process yield was improved by 45% [34]. Many applications of EVOP in Mayamy Chemical's plant operations include saccharin production, a biocide for sea lampreys, isatoic anhydride, anthranilic acid, and benzotriazole processes [31]. EVOP might be proved useful for the following if DOE is not an option:

- Metal die casting.
- Metal heat treating.
- Food manufacturing.
- Plastics, extrusion, injection molding, and thermoforming.
- Non-woven fabric production.

5.6 BEVOP Advantages and Disadvantages

The advantages of these designs are:

1. They are simple to understand, perform, and analyze.
2. They are short in operation so that each cycle may be completed in a reasonable time.
3. They form convenient blocks for eliminating the effects of uncontrolled factors, such as time trends.
4. They may be used for qualitative or quantitative factors.
5. The central points permit comparison with the standard process, at least initially, so that a value and a cost of an EVOP may be estimated.
6. By comparing the relative magnitudes of the simple main effects with the "change in mean" and the interaction effects, it is possible to obtain some idea of the complexity of the response surface.
7. They can be used as the basis for more elaborate designs, like the composite second-order rotatable designs [3].

Disadvantages of the designs are:

1. Determination of the effect of a factor may depend on the size of variant used.
2. The procedure offers no indication of the number of cycles required in each phase.
3. No guidance on movement of factor values from one phase to the next.
4. For more than three factors the number of points per cycle can be very large; for k factors 2^k points are needed. The design tends to become complicated and interactions may be calculated but cannot be resolved.

The approximate method of estimating the number of cycles in BEVOP is shown in Appendix D.

5.7 BEVOP Simulation

5.7.1 2^2 BEVOP Simulation

5.7.1.1 Simulation No. 1

At a particular stage of development of a certain batch process, two process factors were being studied. These were Temperature (A) and Pressure (B) at which the reaction was conducted. The current best process conditions are: $A_0 = 10$ [°C], $B_0 = 10$ [bar] and corresponding product yield is around $y = 36$ [%]. Theoretically, yield range is: $0.0 < y < 100$ [%]. Start simulation with factor's intervals: $\Delta a = 5.0$; $\Delta b = 5.0$. The current best process condition, factor's interval, and process noise measured by standard deviation can be set on different value to test EVOP efficiency. Experimental limits for temperature are $0.0 < A < 100$ [°C] and for the pressure $0.0 < A < 100$ [bar]. The appropriate mathematical function was used to calculate product response numbers plus noise (Standard Deviation – STDEV multiplied by Random Normal Deviates) to mask response and simulate real plant yield data. The objective of this study is to maximize product yield using EVOP.

Phase I: The settings (Center Point, Factor's Interval, and Process Noise) and yield for different process conditions have been calculated using Microsoft Excel simulation in Table 5.26. The corresponding calculations for ten cycles, phase I, have been shown in work sheets in Figure 5.33–5.35. The information board that would appear at the end of cycle 10, after transfer of the various calculated quantities from the appropriate worksheet, is shown in Table 5.27.

This simulation started with small factor's change-interval and the large noise (STDEV).

Table 5.26 2^2 BEVOP simulation.

PROJECT : S-1
PHASE : I
DATE :

BEVOP SIMULATION

S-1

Gen Random

I. CENTER OF EXPERIMENT :	Ao =	10.00
	Bo =	10.00
II. FACTOR'S INTERVAL OF VARIATION :	Δ a =	0.50
	Δ b =	0.50
III. PROCESS (PROPERTY-RESPONSE) NOISE :	STDV =	10.00

EVOP WORK SHEET

TRIALS ORDER			FACTOR		RESPONSE									
			A	B	n = 1	n = 2	n = 3	n = 4	n = 5	n = 6	n = 7	n = 8	n = 9	n = 10
0		Block 1	**10.00**	**10.00**	**37.38**	**23.68**	**26.83**	**43.17**	**25.07**	**38.56**	**21.44**	**41.62**	**27.37**	**31.95**
1		Block 1	9.50	9.50	**44.83**	22.45	45.75	29.49	29.29	36.46	47.16	27.83	34.40	**37.02**
2		Block 1	10.50	9.50	**21.97**	22.36	37.44	43.63	33.81	36.64	48.81	45.69	44.45	**30.02**
3		Block 1	9.50	10.50	**45.18**	58.71	38.57	36.63	39.34	40.77	50.24	59.89	61.10	**26.21**
4		Block 1	10.50	10.50	**41.06**	45.70	51.48	14.80	21.94	28.90	42.08	54.40	33.41	**47.01**

Table 5.26 (Continued)

PROJECT : S-1
PHASE : I
DATE :

BEVOP SIMULATION

S-1

I. CENTER OF EXPERIMENT : Ao = 10.00; Bo = 10.00

II. FACTOR'S INTERVAL OF VARIATION : Δ a = 0.50; Δ b = 0.50

III. PROCESS (PROPERTY-RESPONSE) NOISE : STDV = 10.00

Gen Random

0.1377	1.044325	-1.402482	0.9195	0.347071
[illegible]	1.19366	-1.363082	2.2722	0.811203
[illegible]	1.136214	0.144742	0.2534	1.388507
0.7173	-0.48987	0.764356	0.0636	-2.27938
-1.093	-0.51007	-0.217858	0.3347	-1.56485
0.2559	0.207217	0.064572	0.4784	-0.86879
1.456	1.277338	1.281501	1.4251	0.449381
0.5617	-0.65566	0.969642	2.39	1.681246
-0.863	0.001032	0.845923	2.5114	-0.41761
-0.405	0.263376	-0.596747	-0.978	0.942339

EVOP WORK SHEET

TRIALS ORDER			FACTOR		RESPONSE									
			A	B	n = 1	n = 2	n = 3	n = 4	n = 5	n = 6	n = 7	n = 8	n = 9	n = 10
0		Block 1	10.00	10.00	37.38	23.68	26.83	43.17	25.07	38.56	21.44	41.62	27.37	31.95
1		Block 1	9.50	9.50	44.83	22.45	45.75	29.49	29.29	36.46	47.16	27.83	34.40	37.02
2		Block 1	10.50	9.50	21.97	22.36	37.44	43.63	33.81	36.64	48.81	45.69	44.45	30.02
3		Block 1	9.50	10.50	45.18	58.71	38.57	36.63	39.34	40.77	50.24	59.89	61.10	26.21
4		Block 1	10.50	10.50	41.06	45.70	51.48	14.80	21.94	28.90	42.08	54.40	33.41	47.01

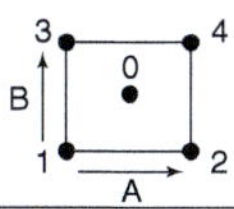

2^2 Factorial with added reference condition

Cycle n :
1
Response:YIELD

Project S-1
Phase I
Date

Calculation of Averages					
Operation condition	(0)	(1)	(2)	(3)	(4)
(i) Previous cycle sum					
(ii) Previous cycle average					
(iii) New observations	37.38	44.83	21.97	45.18	41.06
(iv) Differences (ii) less (iii)					
(v) New sums	37.38	44.83	21.97	45.18	41.06
(vi) New averages : $\bar{y}i$	37.38	44.83	21.97	45.18	41.06

Calculation of Standard Deviation	
Prior estimate of σ	= 10.00
Previous sum **s**	=
Previous average **s**	=
New **s** = range × $f\delta,n$	=
Range	=
New sum **s**	=
New average **s** = new sum **s** / (n - 1)	=

Calculation of Effects		
Phase mean	$= 1/5\ (\bar{y}o + \bar{y}1 + \bar{y}2 + \bar{y}3 + \bar{y}4)$	= 38.08
(A) ———— effect	$= ½\ (\bar{y}2 + \bar{y}4 - \bar{y}1 - \bar{y}3)$	= -13.50
(B) ———— effect	$= ½\ (\bar{y}3 + \bar{y}4 - \bar{y}1 - \bar{y}2)$	= 9.72
(AB) ———— effect	$= ½\ (\bar{y}1 + \bar{y}4 - \bar{y}3 - \bar{y}2)$	= 9.37
Change in mean effect	= phase mean - $\bar{y}o$	= 0.71

Calculation of 2 S.E. Limits		
For new average **s** :	$\pm (2/\sqrt{n})s$	= 20.00
For new effects :	$\pm (2/\sqrt{n})s$	= 20.00
For change in mean :	$\pm (1.79/\sqrt{n})s$	= 17.90

Table of Multiplying Factors

n	1	2	3	4	5	6	7	8	9	10
$f\delta,n$		0.30	0.35	0.37	0.38	0.39	0.40	0.40	0.40	0.41
1 / n	1.00	0.50	0.33	0.25	0.20	0.17	0.14	0.12	0.11	0.10
1 / (n - 1)		1.00	0.50	0.33	0.25	0.20	0.17	0.14	0.12	0.11
$2/\sqrt{n}$	2.00	1.41	1.15	1.00	0.89	0.82	0.76	0.71	0.67	0.63
$1.79/\sqrt{n}$	1.79	1.26	1.03	0.89	0.80	0.73	0.68	0.63	0.60	0.57

Two variable EVOP program calculation worksheet

Figure 5.33 Work sheet cycle No. 1.

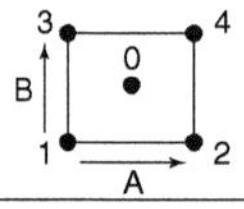

2^2 Factorial with added reference condition

Cycle n :
3
Response:YIELD

Project S-1
Phase I
Date

Calculation of Averages						Calculation of Standard Deviation	
Operation condition	(0)	(1)	(2)	(3)	(4)	Prior estimate of σ	= 10.00
(i) Previous cycle sum	61.05	67.29	44.32	103.90	86.76	Previous sum **s**	= 10.77
(ii) Previous cycle average	30.53	33.64	22.16	51.95	43.38	Previous average **s**	= 10.77
(iii) New observations	26.83	45.75	37.44	38.57	51.48	New **s** = range × $f\delta,n$	= 10.03
(iv) Differences (ii) less (iii)	3.69	-12.11	-15.28	13.37	-8.09	Range	= 28.65
(v) New sums	57.36	79.40	59.60	90.52	94.86	New sum **s**	= 20.80
(vi) New averages : $\bar{y}i$	28.68	39.70	29.80	45.26	47.43	New average **s** = new sum **s** / n - 1	= 10.40

Calculation of Effects			Calculation of 2 S.E. Limits	
Phase mean	= 1/5 ($\bar{y}o$ + $\bar{y}1$ + $\bar{y}2$ + $\bar{y}3$ + $\bar{y}4$)	= 38.17		
(A) ———— effect	= ½ ($\bar{y}2$ + $\bar{y}4$ - $\bar{y}1$ - $\bar{y}3$)	= -3.87	For new average **s** : ± ($2/\sqrt{n}$)s	= 11.96
(B) ———— effect	= ½ ($\bar{y}3$ + $\bar{y}4$ - $\bar{y}1$ - $\bar{y}2$)	= 11.60	For new effects : ± ($2/\sqrt{n}$)s	= 11.96
(AB) ———— effect	= ½ ($\bar{y}1$ + $\bar{y}4$ - $\bar{y}3$ - $\bar{y}2$)	= 6.03	For change in mean : ± ($1.79/\sqrt{n}$)s	= 10.71
Change in mean effect	= phase mean - $\bar{y}o$	= 9.49		

Table of Multiplying Factors

n	1	2	3	4	5	6	7	8	9	10
$f\delta,n$		0.30	0.35	0.37	0.38	0.39	0.40	0.40	0.40	0.41
1 / n	1.00	0.50	0.33	0.25	0.20	0.17	0.14	0.12	0.11	0.10
1 / (n - 1)		1.00	0.50	0.33	0.25	0.20	0.17	0.14	0.12	0.11
$2/\sqrt{n}$	2.00	1.41	1.15	1.00	0.89	0.82	0.76	0.71	0.67	0.63
$1.79/\sqrt{n}$	1.79	1.26	1.03	0.89	0.80	0.73	0.68	0.63	0.60	0.57

Two variable EVOP program calculation worksheet

Figure 5.34 Work sheet cycle No. 3.

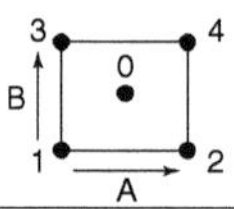

2^2 Factorial with added reference condition

Cycle n :
10
Response:YIELD

Project S-1
Phase I
Date

Calculation of Averages						Calculation of Standard Deviation	
Operation condition	(0)	(1)	(2)	(3)	(4)	Prior estimate of σ	= 10.00
(i) Previous cycle sum	62.17	68.66	88.48	113.72	78.06	Previous sum **s**	= 78.17
(ii) Previous cycle average	31.09	34.33	44.24	56.86	39.03	Previous average **s**	= 9.77
(iii) New observations	31.95	37.02	30.02	26.21	47.01	New **s** = range × $f\delta,n$	= 15.84
(iv) Differences (ii) less (iii)	−0.86	−2.69	14.22	30.65	−7.98	Range	= 38.63
(v) New sums	63.03	71.35	74.26	83.07	86.05	New sum **s**	= 94.01
(vi) New averages : $\bar{y}i$	31.52	35.68	37.13	41.54	43.02	New average **s** = new sum **s** / n − 1	= 10.45

Calculation of Effects			Calculation of 2 S.E. Limits	
Phase mean	= 1/5 ($\bar{y}o$ + $\bar{y}1$ + $\bar{y}2$ + $\bar{y}3$ + $\bar{y}4$)	= 37.78		
(A) ———— effect	= ½ ($\bar{y}2$ + $\bar{y}4$ − $\bar{y}1$ − $\bar{y}3$)	= 1.47	For new average **s** : ± ($2/\sqrt{n}$)s	= 6.58
(B) ———— effect	= ½ ($\bar{y}3$ + $\bar{y}4$ − $\bar{y}1$ − $\bar{y}2$)	= 5.88	For new effects : ± ($2/\sqrt{n}$)s	= 6.58
(AB) ———— effect	= ½ ($\bar{y}1$ + $\bar{y}4$ − $\bar{y}3$ − $\bar{y}2$)	= 0.02	For change in mean : ± ($1.79/\sqrt{n}$)s	= 5.95
Change in mean effect	= phase mean − $\bar{y}o$	= 6.26		

Table of Multiplying Factors

n	1	2	3	4	5	6	7	8	9	10
$f\delta,n$		0.30	0.35	0.37	0.38	0.39	0.40	0.40	0.40	0.41
1 / n	1.00	0.50	0.33	0.25	0.20	0.17	0.14	0.12	0.11	0.10
1 / (n − 1)		1.00	0.50	0.33	0.25	0.20	0.17	0.14	0.12	0.11
$2/\sqrt{n}$	2.00	1.41	1.15	1.00	0.89	0.82	0.76	0.71	0.67	0.63
$1.79/\sqrt{n}$	1.79	1.26	1.03	0.89	0.80	0.73	0.68	0.63	0.60	0.57

Two variable EVOP program calculation worksheet

Figure 5.35 Work sheet cycle No. 10.

Table 5.27 2^2 BEVOP information board.

INFORMATION BOARD AFTER PHASE NO. I **Last Cycle Completed : n = 10**			
	RESPONSE No. 1	**RESPONSE No. 2**	**RESPONSE No. 3**
REQUIREMENT :	**MAXIMIZE RESPONSE**		
PROCESS AVERAGES :	41.54 43.02 31.52 B 35.68 37.13 A		
2.S.E. LIMITS FOR AVERAGES	6.58		
EFFECTS WITH 2.S.E **PHASE MEAN** **FACTOR A** **FACTOR B** **INTERACTION AB** **CHANGE IN MEAN**	37.78 1.47 ± 6.58 5.88 ± 6.58 0.02 ± 6.58 6.26 ± 5.95		
ESTIMATED STANDARD DEVIATION S **PRIOR ESTIMATE OF STANDARD DEV.**	10.45 10.00		

NOTE:

1. The Center of Experiment-Reference Conditions (The Current Best-Known Process Conditions)
2. The Change in Mean provides a measure of whether the conditions being run the current phase yield an average result better or worse than the reference conditions.
3. The Phase Mean is the Mean Response Over All Conditions Being Run in the present Phase.
4*. The change in Mean is a measure of the over-all Curvature of the Response Surface. It will be positive near a MINIMUM and negative near a MAXIMUM.
5. Sequencing of the Runs-Randomization, if runs can be randomized without too much difficulty, than this should be done.

*A Diagonally oriented saddle represented by equal coefficients of opposite sign could occure and yield a zero change in mean effect, but such a phenomenon would rarely be met in practice.

Conclusions

- There is no significant factor's effect due to high noise after ten cycles.
- The *change in mean* effect is significant and is a measure of overall curvature of the Response Surface. It will be positive near a MINIMUM and negative near a MAXIMUM.
- The estimated standard deviation is very close to the prior estimate of the standard deviation.

Recommendation for Phase II

- It is recommended to reduce noise from STDEV=10 to STDEV=3.0.
- To enlarge factor's variation from 0.5 to 5.0.

Phase II: The settings (Center Point , Factor's Interval, and Process Noise) and yield for different process conditions have been completed using Microsoft Excel simulation in Table 5.28. The corresponding calculations for ten cycles, phase II, have been shown in work sheets in Figure 5.36–5.38. The information board that would appear at the end of cycle 10, after transfer of the various calculated quantities from the appropriate worksheet, is shown in Table 5.29.

This simulation started with larger factor's change (interval) and the lower noise (STDEV).

Table 5.28 2^2 BEVOP simulation.

PROJECT : S-1
PHASE : II
DATE :

BEVOP SIMULATION

S-1

Gen Random

I. CENTER OF EXPERIMENT : Ao = 10.00; Bo = 10.00

II. FACTOR'S INTERVAL OF VARIATION : Δa = 5.00; Δb = 5.00

III. PROCESS (PROPERTY-RESPONSE) NOISE : STDV = 3.00

EVOP WORK SHEET

TRIALS ORDER			FACTOR		RESPONSE									
			A	B	n = 1	n = 2	n = 3	n = 4	n = 5	n = 6	n = 7	n = 8	n = 9	n = 10
0		Block 1	10.00	10.00	36.41	32.30	33.25	38.15	32.72	36.77	31.63	37.69	33.41	34.78
1		Block 1	5.00	5.00	22.13	15.42	22.41	17.53	17.47	19.62	22.83	17.03	19.00	19.79
2		Block 1	15.00	5.00	30.79	30.91	35.43	37.29	34.35	35.19	38.84	37.91	37.54	33.21
3		Block 1	5.00	15.00	37.76	41.82	35.78	35.19	36.00	36.44	39.28	42.17	42.53	32.07
4		Block 1	15.00	15.00	52.04	53.43	55.17	44.16	46.31	48.39	52.35	56.04	49.75	53.83

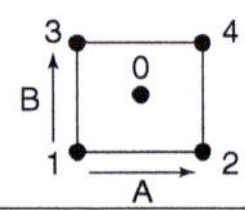

2^2 *Factorial with added reference condition*

Cycle n : *1*

Response:YIELD

Project S-1
Phase II
Date

Calculation of Averages						Calculation of Standard Deviation	
Operation condition	(0)	(1)	(2)	(3)	(4)	Prior estimate of σ	= 3.00
(i) Previous cycle sum						Previous sum **s**	=
(ii) Previous cycle average						Previous average **s**	=
(iii) New observations	36.41	22.13	30.79	37.76	52.04	New **s** = range × ***f δ,n***	=
(iv) Differences (ii) less (iii)						Range	=
(v) New sums	36.41	22.13	30.79	37.76	52.04	New sum **s**	=
(vi) New averages : *ȳi*	36.41	22.13	30.79	37.76	52.04	New average **s** = new sum **s** / (n − 1)	=

Calculation of Effects

Phase mean = *1/5 (ȳo + ȳ1 + ȳ2 + ȳ3 + ȳ4)* = 35.83

(A) effect = *½ (ȳ2 + ȳ4 − ȳ1 − ȳ3)* = 11.47

(B) effect = *½ (ȳ3 + ȳ4 − ȳ1 − ȳ2)* = 18.44

(AB) effect = *½ (ȳ1 + ȳ4 − ȳ3 − ȳ2)* = 2.81

Change in mean effect = phase mean − *ȳo* = −0.59

Calculation of 2 S.E. Limits

For new average **s** : $\pm (2/\sqrt{n})s$ = 6.00

For new effects : $\pm (2/\sqrt{n})s$ = 6.00

For change in mean : $\pm (1.79/\sqrt{n})s$ = 5.37

Table of Multiplying Factors

n	1	2	3	4	5	6	7	8	9	10
f δ,n		0.30	0.35	0.37	0.38	0.39	0.40	0.40	0.40	0.41
1 / n	1.00	0.50	0.33	0.25	0.20	0.17	0.14	0.12	0.11	0.10
1 / (n − 1)		1.00	0.50	0.33	0.25	0.20	0.17	0.14	0.12	0.11
$2/\sqrt{n}$	2.00	1.41	1.15	1.00	0.89	0.82	0.76	0.71	0.67	0.63
$1.79/\sqrt{n}$	1.79	1.26	1.03	0.89	0.80	0.73	0.68	0.63	0.60	0.57

Two variable EVOP program calculation worksheet

Figure 5.36 Work sheet cycle No. 1.

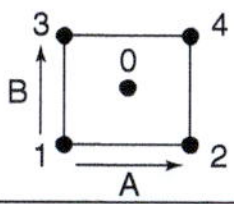

2^2 Factorial with added reference condition

Cycle n : *3*
Response:YIELD

Project S-1
Phase II
Date

Calculation of Averages

Operation condition	(0)	(1)	(2)	(3)	(4)
(i) Previous cycle sum	68.72	37.55	61.70	79.58	105.47
(ii) Previous cycle average	34.36	18.78	30.85	39.79	52.74
(iii) New observations	33.25	22.41	35.43	35.78	55.17
(iv) Differences (ii) less (iii)	1.11	−3.63	−4.58	4.01	−2.43
(v) New sums	67.61	41.18	66.29	75.56	107.90
(vi) New averages : $\bar{y}_i$	33.80	20.59	33.14	37.78	53.95

Calculation of Standard Deviation

Prior estimate of σ	= 3.00
Previous sum **s**	= 3.23
Previous average **s**	= 3.23
New **s** = range × $f\delta,n$	= 3.01
Range	= 8.59
New sum **s**	= 6.24
New average **s** = $\frac{\text{new sum } s}{n-1}$	= 3.12

Calculation of Effects

Phase mean	$= 1/5\,(\bar{y}_0 + \bar{y}_1 + \bar{y}_2 + \bar{y}_3 + \bar{y}_4)$	= 35.85
(A) effect	$= ½\,(\bar{y}_2 + \bar{y}_4 - \bar{y}_1 - \bar{y}_3)$	= 14.36
(B) effect	$= ½\,(\bar{y}_3 + \bar{y}_4 - \bar{y}_1 - \bar{y}_2)$	= 19.00
(AB) effect	$= ½\,(\bar{y}_1 + \bar{y}_4 - \bar{y}_3 - \bar{y}_2)$	= 1.81
Change in mean effect	= phase mean − $\bar{y}_0$	= 2.05

Calculation of 2 S.E. Limits

For new average **s** :	$\pm(2/\sqrt{n})s$	= 3.59
For new effects :	$\pm(2/\sqrt{n})s$	= 3.59
For change in mean :	$\pm(1.79/\sqrt{n})s$	= 3.21

Table of Multiplying Factors

n	1	2	3	4	5	6	7	8	9	10
$f\delta,n$		0.30	0.35	0.37	0.38	0.39	0.40	0.40	0.40	0.41
1 / n	1.00	0.50	0.33	0.25	0.20	0.17	0.14	0.12	0.11	0.10
1 / (n − 1)		1.00	0.50	0.33	0.25	0.20	0.17	0.14	0.12	0.11
$2/\sqrt{n}$	2.00	1.41	1.15	1.00	0.89	0.82	0.76	0.71	0.67	0.63
$1.79/\sqrt{n}$	1.79	1.26	1.03	0.89	0.80	0.73	0.68	0.63	0.60	0.57

Two variable EVOP program calculation worksheet

Figure 5.37 Work sheet cycle No. 3.

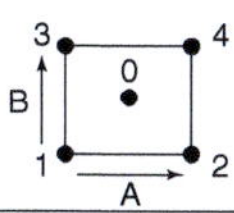

2^2 Factorial with added reference condition

Cycle n : *10*
Response:YIELD

Project S-1
Phase II
Date

Calculation of Averages

Operation condition	(0)	(1)	(2)	(3)	(4)
(i) Previous cycle sum	69.05	37.96	74.95	82.52	102.87
(ii) Previous cycle average	34.53	18.98	37.48	41.26	51.43
(iii) New observations	34.78	19.79	33.21	32.07	53.83
(iv) Differences (ii) less (iii)	−0.26	−0.81	4.27	9.20	−2.39
(v) New sums	69.31	38.77	70.68	73.33	105.26
(vi) New averages : $\bar{y}_i$	34.65	19.39	35.34	36.66	52.63

Calculation of Standard Deviation

Prior estimate of σ	= 3.00
Previous sum **s**	= 23.45
Previous average **s**	= 2.93
New **s** = range × $f\delta,n$	= 4.75
Range	= 11.59
New sum **s**	= 28.20
New average **s** = $\frac{\text{new sum } s}{n-1}$	= 3.13

Calculation of Effects

Phase mean	$= 1/5\,(\bar{y}_0 + \bar{y}_1 + \bar{y}_2 + \bar{y}_3 + \bar{y}_4)$	= 35.74
(A) effect	$= ½\,(\bar{y}_2 + \bar{y}_4 - \bar{y}_1 - \bar{y}_3)$	= 15.96
(B) effect	$= ½\,(\bar{y}_3 + \bar{y}_4 - \bar{y}_1 - \bar{y}_2)$	= 17.28
(AB) effect	$= ½\,(\bar{y}_1 + \bar{y}_4 - \bar{y}_3 - \bar{y}_2)$	= 0.00
Change in mean effect	= phase mean − $\bar{y}_0$	= 1.08

Calculation of 2 S.E. Limits

For new average **s** :	$\pm(2/\sqrt{n})s$	= 1.97
For new effects :	$\pm(2/\sqrt{n})s$	= 1.97
For change in mean :	$\pm(1.79/\sqrt{n})s$	= 1.79

Table of Multiplying Factors

n	1	2	3	4	5	6	7	8	9	10
$f\delta,n$		0.30	0.35	0.37	0.38	0.39	0.40	0.40	0.40	0.41
1 / n	1.00	0.50	0.33	0.25	0.20	0.17	0.14	0.12	0.11	0.10
1 / (n − 1)		1.00	0.50	0.33	0.25	0.20	0.17	0.14	0.12	0.11
$2/\sqrt{n}$	2.00	1.41	1.15	1.00	0.89	0.82	0.76	0.71	0.67	0.63
$1.79/\sqrt{n}$	1.79	1.26	1.03	0.89	0.80	0.73	0.68	0.63	0.60	0.57

Two variable EVOP program calculation worksheet

Figure 5.38 Work sheet cycle No. 10.

Table 5.29 2^2 BEVOP information board.

INFORMATION BOARD AFTER PHASE NO. II Last Cycle Completed : n = 10				
		RESPONSE No. 1	RESPONSE No. 2	RESPONSE No. 3
REQUIREMENT :		MAXIMIZE RESPONSE		
PROCESS AVERAGES :		36.66 52.63 34.65 B 19.39 35.34 A		
2.S.E. LIMITS FOR AVERAGES		1.97		
EFFECTS WITH 2.S.E	PHASE MEAN FACTOR A FACTOR B INTERACTION AB CHANGE IN MEAN	35.74 15.96 ± 1.97 17.28 ± 1.97 0.00 ± 1.97 1.08 ± 1.79		
ESTIMATED STANDARD DEVIATION S		3.13		
PRIOR ESTIMATE OF STANDARD DEV.		3.00		

NOTE:

1. The Center of Experiment-Reference Conditions (The Current Best-Known Process Conditions)
2. The Change in Mean provides a measure of whether the conditions being run the current phase yield an average result better or worse than the reference conditions.
3. The Phase Mean is the Mean Response Over All Conditions Being Run in the present Phase.
4*. The change in Mean is a measure of the over-all Curvature of the Response Surface. It will be positive near a MINIMUM and negative near a MAXIMUM.
5. Sequencing of the Runs-Randomization, if runs can be randomized without too much difficulty, than this should be done.

*A Diagonally oriented saddle represented by equal coefficients of opposite sign could occure and yield a zero change in mean effect, but such a phenomenon would rarely be met in practice.

Conclusions

- Two factors are highly significant with positive sign.
- The estimated standard deviation is very close to the prior estimate of the standard deviation.

Recommendation for Phase III

- It is recommended to move the center of the experiment ($A_0 = 10$; $B_0 = 10$) with increase in both factors ($A_0 = 30$; $B_0 = 30$).
- To keep the same factor's variation 5.0.
- To keep the same noise level STDEV = 3.0 (Table 5.30, Figures 5.39–5.41, Table 5.31).

Table 5.30 2^2 BEVOP simulation.

PROJECT : S-1
PHASE : III
DATE :

SIMULATION

S-1

I. CENTER OF EXPERIMENT : A_o = 30.00, B_o = 30.00 [Gen Random]

II. FACTOR'S INTERVAL OF VARIATION : Δa = 5.00, Δb = 5.00

III. PROCESS (PROPERTY-RESPONSE) NOISE : STDV = 3.00

EVOP WORK SHEET

TRIALS ORDER			FACTOR		RESPONSE									
			A	B	n = 1	n = 2	n = 3	n = 4	n = 5	n = 6	n = 7	n = 8	n = 9	n = 10
0		Block 1	30.00	30.00	89.62	79.73	82.81	88.33	79.13	84.07	86.71	86.45	87.30	79.14
1		Block 1	25.00	25.00	78.43	76.00	80.15	75.52	73.44	78.74	77.15	76.22	76.01	76.75
2		Block 1	35.00	25.00	79.63	84.81	81.33	77.10	84.01	84.08	85.78	78.03	81.79	80.17
3		Block 1	25.00	35.00	76.66	82.68	84.25	76.75	86.97	81.10	87.36	85.98	83.73	84.86
4		Block 1	35.00	35.00	86.11	97.97	91.58	90.92	91.79	91.67	90.89	88.88	87.90	94.01

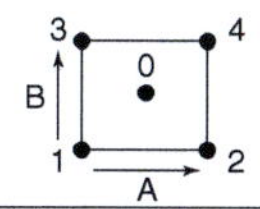

2^2 Factorial with added reference condition

Cycle n : *1*
Response:YIELD

Project S-1
Phase III
Date

Calculation of Averages

Operation condition	(0)	(1)	(2)	(3)	(4)
(i) Previous cycle sum					
(ii) Previous cycle average					
(iii) New observations	89.62	78.43	79.63	76.66	86.11
(iv) Differences (ii) less (iii)					
(v) New sums	89.62	78.43	79.63	76.66	86.11
(vi) New averages : $\bar{y}i$	89.62	78.43	79.63	76.66	86.11

Calculation of Standard Deviation

Prior estimate of σ	= 3.00
Previous sum **s**	=
Previous average **s**	=
New **s** = range × $f\delta,n$	=
Range	=
New sum **s**	=
New average **s** = $\frac{\text{new sum } s}{n-1}$	=

Calculation of Effects

Phase mean	$= 1/5\,(\bar{y}o + \bar{y}1 + \bar{y}2 + \bar{y}3 + \bar{y}4)$	= 82.09
(A) effect	$= \frac{1}{2}(\bar{y}2 + \bar{y}4 - \bar{y}1 - \bar{y}3)$	= 5.33
(B) effect	$= \frac{1}{2}(\bar{y}3 + \bar{y}4 - \bar{y}1 - \bar{y}2)$	= 2.36
(AB) effect	$= \frac{1}{2}(\bar{y}1 + \bar{y}4 - \bar{y}3 - \bar{y}2)$	= 4.12
Change in mean effect	= phase mean - $\bar{y}o$	= -7.53

Calculation of 2 S.E. Limits

For new average **s** :	$\pm (2/\sqrt{n})s$	= 6.00
For new effects :	$\pm (2/\sqrt{n})s$	= 6.00
For change in mean :	$\pm (1.79/\sqrt{n})s$	= 5.37

Table of Multiplying Factors

n	1	2	3	4	5	6	7	8	9	10
$f\delta,n$		0.30	0.35	0.37	0.38	0.39	0.40	0.40	0.40	0.41
1 / n	1.00	0.50	0.33	0.25	0.20	0.17	0.14	0.12	0.11	0.10
1 / (n - 1)		1.00	0.50	0.33	0.25	0.20	0.17	0.14	0.12	0.11
$2/\sqrt{n}$	2.00	1.41	1.15	1.00	0.89	0.82	0.76	0.71	0.67	0.63
$1.79/\sqrt{n}$	1.79	1.26	1.03	0.89	0.80	0.73	0.68	0.63	0.60	0.57

Two variable EVOP program calculation worksheet

Figure 5.39 Work sheet cycle No. 1.

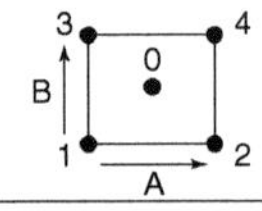

2^2 Factorial with added reference condition

Cycle n :
3
Response:YIELD

Project S-1
Phase III
Date

Calculation of Averages					
Operation condition	(0)	(1)	(2)	(3)	(4)
(i) Previous cycle sum	169.35	154.43	164.45	159.34	184.08
(ii) Previous cycle average	84.68	77.22	82.22	79.67	92.04
(iii) New observations	82.81	80.15	81.33	84.25	91.58
(iv) Differences (ii) less (iii)	1.87	-2.94	0.89	-4.58	0.46
(v) New sums	167.48	157.37	163.56	163.92	183.62
(vi) New averages : $\bar{y}_i$	83.74	78.68	81.78	81.96	91.81

Calculation of Standard Deviation	
Prior estimate of σ	= 3.00
Previous sum **s**	= 6.52
Previous average **s**	= 6.52
New **s** = range × $f\delta,n$	= 2.26
Range	= 6.45
New sum **s**	= 8.78
New average **s** = $\frac{\text{new sum } s}{n-1}$	= 4.39

Calculation of Effects		
Phase mean	$= 1/5\,(\bar{y}_0 + \bar{y}_1 + \bar{y}_2 + \bar{y}_3 + \bar{y}_4)$	= 83.60
(A) effect	$= ½\,(\bar{y}_2 + \bar{y}_4 - \bar{y}_1 - \bar{y}_3)$	= 6.47
(B) effect	$= ½\,(\bar{y}_3 + \bar{y}_4 - \bar{y}_1 - \bar{y}_2)$	= 6.65
(AB) effect	$= ½\,(\bar{y}_1 + \bar{y}_4 - \bar{y}_3 - \bar{y}_2)$	= 3.38
Change in mean effect	$=$ phase mean $- \bar{y}_0$	= -0.15

Calculation of 2 S.E. Limits		
For new average **s** :	$\pm (2/\sqrt{n})s$	= 5.05
For new effects :	$\pm (2/\sqrt{n})s$	= 5.05
For change in mean :	$\pm (1.79/\sqrt{n})s$	= 4.52

Table of Multiplying Factors

n	1	2	3	4	5	6	7	8	9	10
$f\delta,n$		0.30	0.35	0.37	0.38	0.39	0.40	0.40	0.40	0.41
1 / n	1.00	0.50	0.33	0.25	0.20	0.17	0.14	0.12	0.11	0.10
1 / (n - 1)		1.00	0.50	0.33	0.25	0.20	0.17	0.14	0.12	0.11
$2/\sqrt{n}$	2.00	1.41	1.15	1.00	0.89	0.82	0.76	0.71	0.67	0.63
$1.79/\sqrt{n}$	1.79	1.26	1.03	0.89	0.80	0.73	0.68	0.63	0.60	0.57

Two variable EVOP program calculation worksheet

Figure 5.40 Work sheet cycle No. 3.

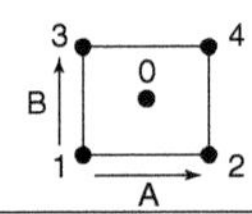

2^2 Factorial with added reference condition

Cycle n :
10
Response:YIELD

Project S-1
Phase III
Date

Calculation of Averages					
Operation condition	(0)	(1)	(2)	(3)	(4)
(i) Previous cycle sum	173.03	152.67	162.98	169.09	177.97
(ii) Previous cycle average	86.52	76.33	81.49	84.55	88.99
(iii) New observations	79.14	76.75	80.17	84.86	94.01
(iv) Differences (ii) less (iii)	7.38	-0.42	1.32	-0.32	-5.02
(v) New sums	165.66	153.08	161.66	169.41	182.99
(vi) New averages : $\bar{y}_i$	82.83	76.54	80.83	84.71	91.50

Calculation of Standard Deviation	
Prior estimate of σ	= 3.00
Previous sum **s**	= 27.05
Previous average **s**	= 3.38
New **s** = range × $f\delta,n$	= 5.08
Range	= 12.40
New sum **s**	= 32.13
New average **s** = $\frac{\text{new sum } s}{n-1}$	= 3.57

Calculation of Effects		
Phase mean	$= 1/5\,(\bar{y}_0 + \bar{y}_1 + \bar{y}_2 + \bar{y}_3 + \bar{y}_4)$	= 83.28
(A) effect	$= ½\,(\bar{y}_2 + \bar{y}_4 - \bar{y}_1 - \bar{y}_3)$	= 5.54
(B) effect	$= ½\,(\bar{y}_3 + \bar{y}_4 - \bar{y}_1 - \bar{y}_2)$	= 9.42
(AB) effect	$= ½\,(\bar{y}_1 + \bar{y}_4 - \bar{y}_3 - \bar{y}_2)$	= 1.25
Change in mean effect	$=$ phase mean $- \bar{y}_0$	= 0.45

Calculation of 2 S.E. Limits		
For new average **s** :	$\pm (2/\sqrt{n})s$	= 2.25
For new effects :	$\pm (2/\sqrt{n})s$	= 2.25
For change in mean :	$\pm (1.79/\sqrt{n})s$	= 2.03

Table of Multiplying Factors

n	1	2	3	4	5	6	7	8	9	10
$f\delta,n$		0.30	0.35	0.37	0.38	0.39	0.40	0.40	0.40	0.41
1 / n	1.00	0.50	0.33	0.25	0.20	0.17	0.14	0.12	0.11	0.10
1 / (n - 1)		1.00	0.50	0.33	0.25	0.20	0.17	0.14	0.12	0.11
$2/\sqrt{n}$	2.00	1.41	1.15	1.00	0.89	0.82	0.76	0.71	0.67	0.63
$1.79/\sqrt{n}$	1.79	1.26	1.03	0.89	0.80	0.73	0.68	0.63	0.60	0.57

Two variable EVOP program calculation worksheet

Figure 5.41 Work sheet cycle No. 10.

Table 5.31 2^2 BEVOP information board.

INFORMATION BOARD AFTER PHASE NO. III

Last Cycle Completed : n = 10

		RESPONSE No. 1	RESPONSE No. 2	RESPONSE No. 3
REQUIREMENT :		MAXIMIZE RESPONSE		
PROCESS AVERAGES :		84.71 (top left), 91.50 (top right), 82.83 (center), 76.54 (bottom left), 80.83 (bottom right); B (vertical axis), A (horizontal axis)		
2.S.E. LIMITS FOR AVERAGES		2.25		
EFFECTS WITH 2.S.E	PHASE MEAN	83.28		
	FACTOR A	5.54 ± 2.25		
	FACTOR B	9.42 ± 2.25		
	INTERACTION AB	1.25 ± 2.25		
	CHANGE IN MEAN	0.45 ± 2.03		
ESTIMATED STANDARD DEVIATION S		3.57		
PRIOR ESTIMATE OF STANDARD DEV.		3.00		

NOTE:

1. The Center of Experiment-Reference Conditions (The Current Best-Known Process Conditions)
2. The Change in Mean provides a measure of whether the conditions being run the current phase yield an average result better or worse than the reference conditions.
3. The Phase Mean is the Mean Response Over All Conditions Being Run in the present Phase.
4*. The change in Mean is a measure of the over-all Curvature of the Response Surface. It will be positive near a MINIMUM and negative near a MAXIMUM.
5. Sequencing of the Runs-Randomization, if runs can be randomized without too much difficulty, than this should be done.

*A Diagonally oriented saddle represented by equal coefficients of opposite sign could occure and yield a zero change in mean effect, but such a phenomenon would rarely be met in practice.

Conclusions

- Two factors are highly significant with positive sign. We are moving toward optimum (maximum).
- The *phase mean* is increased significantly.
- The estimated standard deviation is very close to the prior estimate of the standard deviation.

Recommendation for Phase IV

- It is recommended to move the center of the experiment ($A_0 = 30$; $B_0 = 30$) with increase in both factors ($A_0 = 50$; $B_0 = 50$).
- To keep the same factor's variation 5.0.
- To keep the same noise level STDEV = 3.0 (Table 5.32, Figures 5.42–5.44, Table 5.33).

Table 5.32 2^2 BEVOP simulation.

PROJECT : S-1
PHASE : IV
DATE :

BEVOP SIMULATION

S-1

I. CENTER OF EXPERIMENT : Ao = 50.00; Bo = 50.00 [Gen Random]

II. FACTOR'S INTERVAL OF VARIATION : Δ a = 5.00; Δ b = 5.00

III. PROCESS (PROPERTY-RESPONSE) NOISE : STDV = 3.00

EVOP WORK SHEET

TRIALS ORDER			FACTOR		RESPONSE									
			A	B	n = 1	n = 2	n = 3	n = 4	n = 5	n = 6	n = 7	n = 8	n = 9	n = 10
0		Block 1	50.00	50.00	100.41	96.30	97.25	102.15	96.72	100.77	95.63	101.69	97.41	98.78
1		Block 1	45.00	45.00	102.13	95.42	102.41	97.53	97.47	99.62	102.83	97.03	99.00	99.79
2		Block 1	55.00	45.00	94.79	94.91	99.43	101.29	98.35	99.19	102.84	101.91	101.54	97.21
3		Block 1	45.00	55.00	101.76	105.82	99.78	99.19	100.00	100.44	103.28	106.17	106.53	96.07
4		Block 1	55.00	55.00	100.04	101.43	103.17	92.16	94.31	96.39	100.35	104.04	97.75	101.83

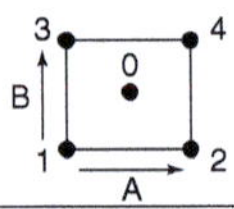

2^2 Factorial with added reference condition

Cycle n :
1
Response:YIELD

Project S-1
Phase IV
Date

Calculation of Averages						Calculation of Standard Deviation	
Operation condition	(0)	(1)	(2)	(3)	(4)	Prior estimate of σ	= 3.00
(i) Previous cycle sum						Previous sum **s**	=
(ii) Previous cycle average						Previous average **s**	=
(iii) New observations	100.41	102.13	94.79	101.76	100.04	New **s** = range × ***fδ,n***	=
(iv) Differences (ii) less (iii)						Range	=
(v) New sums	100.41	102.13	94.79	101.76	100.04	New sum **s**	=
(vi) New averages : $\bar{y}_i$	100.41	102.13	94.79	101.76	100.04	New average **s** = new sum **s** / (n - 1)	=

Calculation of Effects			Calculation of 2 S.E. Limits		
Phase mean	$= 1/5\,(\bar{y}_0 + \bar{y}_1 + \bar{y}_2 + \bar{y}_3 + \bar{y}_4)$	= 99.83			
(A) ———— effect	$= \frac{1}{2}(\bar{y}_2 + \bar{y}_4 - \bar{y}_1 - \bar{y}_3)$	= -4.53	For new average **s** :	$\pm(2/\sqrt{n})s$	= 6.00
(B) ———— effect	$= \frac{1}{2}(\bar{y}_3 + \bar{y}_4 - \bar{y}_1 - \bar{y}_2)$	= 2.44	For new effects :	$\pm(2/\sqrt{n})s$	= 6.00
(AB) ———— effect	$= \frac{1}{2}(\bar{y}_1 + \bar{y}_4 - \bar{y}_3 - \bar{y}_2)$	= 2.81	For change in mean :	$\pm(1.79/\sqrt{n})s$	= 5.37
Change in mean effect	= phase mean - $\bar{y}_0$	= -0.59			

Table of Multiplying Factors

n	1	2	3	4	5	6	7	8	9	10
fδ,n		0.30	0.35	0.37	0.38	0.39	0.40	0.40	0.40	0.41
1 / n	1.00	0.50	0.33	0.25	0.20	0.17	0.14	0.12	0.11	0.10
1 / (n - 1)		1.00	0.50	0.33	0.25	0.20	0.17	0.14	0.12	0.11
$2/\sqrt{n}$	2.00	1.41	1.15	1.00	0.89	0.82	0.76	0.71	0.67	0.63
$1.79/\sqrt{n}$	1.79	1.26	1.03	0.89	0.80	0.73	0.68	0.63	0.60	0.57

Two variable EVOP program calculation worksheet

Figure 5.42 Work sheet cycle No. 1.

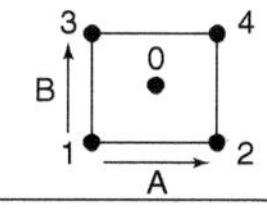

2^2 Factorial with added reference condition

Cycle n : *3*
Response:YIELD

Project S-1
Phase IV
Date

Calculation of Averages						Calculation of Standard Deviation	
Operation condition	(0)	(1)	(2)	(3)	(4)	Prior estimate of σ	= 3.00
(i) Previous cycle sum	196.72	197.55	189.70	207.58	201.47	Previous sum **s**	= 3.23
(ii) Previous cycle average	98.36	98.78	94.85	103.79	100.74	Previous average **s**	= 3.23
(iii) New observations	97.25	102.41	99.43	99.78	103.17	New **s** = range × $f\delta,n$	= 3.01
(iv) Differences (ii) less (iii)	1.11	-3.63	-4.58	4.01	-2.43	Range	= 8.59
(v) New sums	195.61	201.18	194.29	203.56	203.90	New sum **s**	= 6.24
(vi) New averages : $\bar{y}_i$	97.80	100.59	97.14	101.78	101.95	New average **s** = $\frac{\text{new sum s}}{n-1}$	= 3.12

Calculation of Effects			Calculation of 2 S.E. Limits	
Phase mean	= $1/5(\bar{y}_0 + \bar{y}_1 + \bar{y}_2 + \bar{y}_3 + \bar{y}_4)$	= 99.85		
(A) ———— effect	= $\frac{1}{2}(\bar{y}_2 + \bar{y}_4 - \bar{y}_1 - \bar{y}_3)$	= -1.64	For new average **s** : $\pm(2/\sqrt{n})s$	= 3.59
(B) ———— effect	= $\frac{1}{2}(\bar{y}_3 + \bar{y}_4 - \bar{y}_1 - \bar{y}_2)$	= 3.00	For new effects : $\pm(2/\sqrt{n})s$	= 3.59
(AB) ———— effect	= $\frac{1}{2}(\bar{y}_1 + \bar{y}_4 - \bar{y}_3 - \bar{y}_2)$	= 1.81	For change in mean : $\pm(1.79/\sqrt{n})s$	= 3.21
Change in mean effect	= phase mean - $\bar{y}_0$	= 2.05		

Table of Multiplying Factors

n	1	2	3	4	5	6	7	8	9	10
$f\delta,n$		0.30	0.35	0.37	0.38	0.39	0.40	0.40	0.40	0.41
1 / n	1.00	0.50	0.33	0.25	0.20	0.17	0.14	0.12	0.11	0.10
1 / (n - 1)		1.00	0.50	0.33	0.25	0.20	0.17	0.14	0.12	0.11
$2/\sqrt{n}$	2.00	1.41	1.15	1.00	0.89	0.82	0.76	0.71	0.67	0.63
$1.79/\sqrt{n}$	1.79	1.26	1.03	0.89	0.80	0.73	0.68	0.63	0.60	0.57

Two variable EVOP program calculation worksheet

Figure 5.43 Work sheet cycle No. 3.

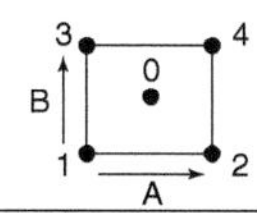

2^2 Factorial with added reference condition

Cycle n : *10*
Response:YIELD

Project S-1
Phase IV
Date

Calculation of Averages						Calculation of Standard Deviation	
Operation condition	(0)	(1)	(2)	(3)	(4)	Prior estimate of σ	= 3.00
(i) Previous cycle sum	197.05	197.96	202.95	210.52	198.87	Previous sum **s**	= 23.45
(ii) Previous cycle average	98.53	98.98	101.48	105.26	99.43	Previous average **s**	= 2.93
(iii) New observations	98.78	99.79	97.21	96.07	101.83	New **s** = range × $f\delta,n$	= 4.75
(iv) Differences (ii) less (iii)	–0.26	–0.81	4.27	9.20	–2.39	Range	= 11.59
(v) New sums	197.31	198.77	198.68	201.33	201.26	New sum **s**	= 28.20
(vi) New averages : $\bar{y}_i$	98.65	99.39	99.34	100.66	100.63	New average **s** = $\frac{\text{new sum s}}{n-1}$	= 3.13

Calculation of Effects			Calculation of 2 S.E. Limits	
Phase mean	= $1/5(\bar{y}_0 + \bar{y}_1 + \bar{y}_2 + \bar{y}_3 + \bar{y}_4)$	= 99.74		
(A) ———— effect	= $\frac{1}{2}(\bar{y}_2 + \bar{y}_4 - \bar{y}_1 - \bar{y}_3)$	= -0.04	For new average **s** : $\pm(2/\sqrt{n})s$	= 1.97
(B) ———— effect	= $\frac{1}{2}(\bar{y}_3 + \bar{y}_4 - \bar{y}_1 - \bar{y}_2)$	= 1.28	For new effects : $\pm(2/\sqrt{n})s$	= 1.97
(AB) ———— effect	= $\frac{1}{2}(\bar{y}_1 + \bar{y}_4 - \bar{y}_3 - \bar{y}_2)$	= 0.00	For change in mean : $\pm(1.79/\sqrt{n})s$	= 1.79
Change in mean effect	= phase mean – $\bar{y}_0$	= 1.08		

Table of Multiplying Factors

n	1	2	3	4	5	6	7	8	9	10
$f\delta,n$		0.30	0.35	0.37	0.38	0.39	0.40	0.40	0.40	0.41
1 / n	1.00	0.50	0.33	0.25	0.20	0.17	0.14	0.12	0.11	0.10
1 / (n – 1)		1.00	0.50	0.33	0.25	0.20	0.17	0.14	0.12	0.11
$2/\sqrt{n}$	2.00	1.41	1.15	1.00	0.89	0.82	0.76	0.71	0.67	0.63
$1.79/\sqrt{n}$	1.79	1.26	1.03	0.89	0.80	0.73	0.68	0.63	0.60	0.57

Two variable EVOP program calculation worksheet

Figure 5.44 Work sheet cycle No. 10.

Table 5.33 2^2 BEVOP information board.

INFORMATION BOARD AFTER PHASE NO. IV Last Cycle Completed : n = 10			
	RESPONSE No. 1	RESPONSE No. 2	RESPONSE No. 3
REQUIREMENT :	MAXIMIZE RESPONSE		
PROCESS AVERAGES :	100.66 100.63 B 98.65 99.39 99.34 A		
2.S.E. LIMITS FOR AVERAGES	1.97		
EFFECTS WITH 2.S.E — PHASE MEAN	99.74		
FACTOR A	-0.04 ± 1.97		
FACTOR B	1.28 ± 1.97		
INTERACTION AB	0.00 ± 1.97		
CHANGE IN MEAN	1.08 ± 1.79		
ESTIMATED STANDARD DEVIATION S	3.13		
PRIOR ESTIMATE OF STANDARD DEV.	3.00		

NOTE:

1. The Center of Experiment-Reference Conditions (The Current Best-Known Process Conditions)
2. The Change in Mean provides a measure of whether the conditions being run the current phase yield an average result better or worse than the reference conditions.
3. The Phase Mean is the Mean Response Over All Conditions Being Run in the present Phase.
4*. The change in Mean is a measure of the over-all Curvature of the Response Surface. It will be positive near a MINIMUM and negative near a MAXIMUM.
5. Sequencing of the Runs-Randomization, if runs can be randomized without too much difficulty, than this should be done.

*A Diagonally oriented saddle represented by equal coefficients of opposite sign could occure and yield a zero change in mean effect, but such a phenomenon would rarely be met in practice.

Conclusions

- Two factors are not significant. We have reached optimum (Plato).
- The *phase mean* is increased significantly.
- The estimated standard deviation is very close to the prior estimate of the standard deviation.

Recommendation for Phase V

- It is recommended to move the center beyond optimum ($A_0 = 50$; $B_0 = 50$) with increase in both factors ($A_0 = 80$; $B_0 = 80$).
- To keep the same factor's variation 5.0.
- To keep the same noise level STDEV = 3.0 (Table 5.34, Figures 5.45–5.47, Table 5.35).

Table 5.34 2^2 BEVOP simulation.

PROJECT : S-1
PHASE : V
DATE :

BEVOP SIMULATION

S-1

I. CENTER OF EXPERIMENT : Ao = 80.00, Bo = 80.00 [Gen Random]

II. FACTOR'S INTERVAL OF VARIATION : Δa = 5.00, Δb = 5.00

III. PROCESS (PROPERTY-RESPONSE) NOISE : STDV = 3.00

EVOP WORK SHEET

TRIALS ORDER			FACTOR		RESPONSE									
			A	B	n = 1	n = 2	n = 3	n = 4	n = 5	n = 6	n = 7	n = 8	n = 9	n = 10
0		Block 1	80.00	80.00	64.41	60.30	61.25	66.15	60.72	64.77	59.63	65.69	61.41	62.78
1		Block 1	75.00	75.00	78.13	71.42	78.41	73.53	73.47	75.62	78.83	73.03	75.00	75.79
2		Block 1	85.00	75.00	58.79	58.91	63.43	65.29	62.35	63.19	66.84	65.91	65.54	61.21
3		Block 1	75.00	85.00	65.76	69.82	63.78	63.19	64.00	64.44	67.28	70.17	70.53	60.07
4		Block 1	85.00	85.00	52.04	53.43	55.17	44.16	46.31	48.39	52.35	56.04	49.75	53.83

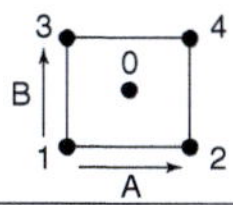

2^2 Factorial with added reference condition

Cycle n : *1* *Response:YIELD*

Project S-1
Phase V
Date

Calculation of Averages						Calculation of Standart Deviation	
Operation condition	(0)	(1)	(2)	(3)	(4)	Prior estimate of σ	= 3.00
(i) Previous cycle sum						Previous sum **s**	=
(ii) Previous cycle average						Previous average **s**	=
(iii) New observations	64.41	78.13	58.79	65.76	52.04	New **s** = range × $f\delta,n$	=
(iv) Differences (ii) less (iii)						Range	=
(v) New sums	64.41	78.13	58.79	65.76	52.04	New sum **s**	=
(vi) New averages : $\bar{y}i$	64.41	78.13	58.79	65.76	52.04	New average **s** = new sum **s** / (n - 1)	=

Calculation of Effects			Calculation of 2 S.E. Limits		
Phase mean	= 1/5 ($\bar{y}o + \bar{y}1 + \bar{y}2 + \bar{y}3 + \bar{y}4$)	= 63.83			
(A) effect	= ½ ($\bar{y}2 + \bar{y}4 - \bar{y}1 - \bar{y}3$)	= -16.53	For new average **s** :	± ($2/\sqrt{n}$)s	= 6.00
(B) effect	= ½ ($\bar{y}3 + \bar{y}4 - \bar{y}1 - \bar{y}2$)	= -9.56	For new effects :	± ($2/\sqrt{n}$)s	= 6.00
(AB) effect	= ½ ($\bar{y}1 + \bar{y}4 - \bar{y}3 - \bar{y}2$)	= 2.81	For change in mean :	± ($1.79/\sqrt{n}$)s	= 5.37
Change in mean effect	= phase mean - $\bar{y}o$	= -0.59			

Table of Multiplying Factors

n	1	2	3	4	5	6	7	8	9	10
$f\delta,n$		0.30	0.35	0.37	0.38	0.39	0.40	0.40	0.40	0.41
1 / n	1.00	0.50	0.33	0.25	0.20	0.17	0.14	0.12	0.11	0.10
1 / (n - 1)		1.00	0.50	0.33	0.25	0.20	0.17	0.14	0.12	0.11
$2/\sqrt{n}$	2.00	1.41	1.15	1.00	0.89	0.82	0.76	0.71	0.67	0.63
$1.79/\sqrt{n}$	1.79	1.26	1.03	0.89	0.80	0.73	0.68	0.63	0.60	0.57

Two variable EVOP program calculation worksheet

Figure 5.45 Work sheet cycle No. 1.

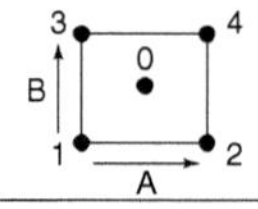

2^2 *Factorial with added reference condition*

Cycle n : 3
Response:YIELD

Project S-1
Phase IV
Date

Calculation of Averages						Calculation of Standard Deviation	
Operation condition	(0)	(1)	(2)	(3)	(4)	Prior estimate of σ	= 3.00
(i) Previous cycle sum	124.72	149.55	117.70	135.58	105.47	Previous sum **s**	= 3.23
(ii) Previous cycle average	62.36	74.78	58.85	67.79	52.74	Previous average **s**	= 3.23
(iii) New observations	61.25	78.41	63.43	63.78	55.17	New **s** = range × ***fδ,n***	= 3.01
(iv) Differences (ii) less (iii)	1.11	-3.63	-4.58	4.01	-2.43	Range	= 8.59
(v) New sums	123.61	153.18	122.29	131.56	107.90	New sum **s**	= 6.24
(vi) New averages : $\bar{y}_i$	61.80	76.59	61.14	65.78	53.95	New average **s** = new sum **s** / (n - 1)	= 3.12

Calculation of Effects

Phase mean	$= 1/5\,(\bar{y}_0 + \bar{y}_1 + \bar{y}_2 + \bar{y}_3 + \bar{y}_4)$	= 63.85
(A) effect	$= ½\,(\bar{y}_2 + \bar{y}_4 - \bar{y}_1 - \bar{y}_3)$	= -13.64
(B) effect	$= ½\,(\bar{y}_3 + \bar{y}_4 - \bar{y}_1 - \bar{y}_2)$	= -9.00
(AB) effect	$= ½\,(\bar{y}_1 + \bar{y}_4 - \bar{y}_3 - \bar{y}_2)$	= 1.81
Change in mean effect	= phase mean - $\bar{y}_0$	= 2.05

Calculation of 2 S.E. Limits

For new average **s** :	$\pm (2/\sqrt{n})s$	= 3.59
For new effects :	$\pm (2/\sqrt{n})s$	= 3.59
For change in mean :	$\pm (1.79/\sqrt{n})s$	= 3.21

Table of Multiplying Factors

n	1	2	3	4	5	6	7	8	9	10
fδ,n		0.30	0.35	0.37	0.38	0.39	0.40	0.40	0.40	0.41
1 / n	1.00	0.50	0.33	0.25	0.20	0.17	0.14	0.12	0.11	0.10
1 / (n - 1)		1.00	0.50	0.33	0.25	0.20	0.17	0.14	0.12	0.11
$2/\sqrt{n}$	2.00	1.41	1.15	1.00	0.89	0.82	0.76	0.71	0.67	0.63
$1.79/\sqrt{n}$	1.79	1.26	1.03	0.89	0.80	0.73	0.68	0.63	0.60	0.57

Two variable EVOP program calculation worksheet

Figure 5.46 Work sheet cycle No. 3.

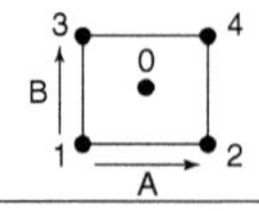

2^2 *Factorial with added reference condition*

Cycle n : 10
Response:YIELD

Project S-1
Phase V
Date

Calculation of Averages						Calculation of Standard Deviation	
Operation condition	(0)	(1)	(2)	(3)	(4)	Prior estimate of σ	= 3.00
(i) Previous cycle sum	125.05	149.96	130.95	138.52	102.87	Previous sum **s**	= 23.45
(ii) Previous cycle average	62.53	74.98	65.48	69.26	51.43	Previous average **s**	= 2.93
(iii) New observations	62.78	75.79	61.21	60.07	53.83	New **s** = range × ***fδ,n***	= 4.75
(iv) Differences (ii) less (iii)	-0.26	-0.81	4.27	9.20	-2.39	Range	= 11.59
(v) New sums	125.31	150.77	126.68	129.33	105.26	New sum **s**	= 28.20
(vi) New averages : $\bar{y}_i$	62.65	75.39	63.34	64.66	52.63	New average **s** = new sum **s** / (n - 1)	= 3.13

Calculation of Effects

Phase mean	$= 1/5\,(\bar{y}_0 + \bar{y}_1 + \bar{y}_2 + \bar{y}_3 + \bar{y}_4)$	= 63.74
(A) effect	$= ½\,(\bar{y}_2 + \bar{y}_4 - \bar{y}_1 - \bar{y}_3)$	= -12.04
(B) effect	$= ½\,(\bar{y}_3 + \bar{y}_4 - \bar{y}_1 - \bar{y}_2)$	= -10.72
(AB) effect	$= ½\,(\bar{y}_1 + \bar{y}_4 - \bar{y}_3 - \bar{y}_2)$	= 0.00
Change in mean effect	= phase mean - $\bar{y}_0$	= 1.08

Calculation of 2 S.E. Limits

For new average **s** :	$\pm (2/\sqrt{n})s$	= 1.97
For new effects :	$\pm (2/\sqrt{n})s$	= 1.97
For change in mean :	$\pm (1.79/\sqrt{n})s$	= 1.79

Table of Multiplying Factors

n	1	2	3	4	5	6	7	8	9	10
fδ,n		0.30	0.35	0.37	0.38	0.39	0.40	0.40	0.40	0.41
1 / n	1.00	0.50	0.33	0.25	0.20	0.17	0.14	0.12	0.11	0.10
1 / (n - 1)		1.00	0.50	0.33	0.25	0.20	0.17	0.14	0.12	0.11
$2/\sqrt{n}$	2.00	1.41	1.15	1.00	0.89	0.82	0.76	0.71	0.67	0.63
$1.79/\sqrt{n}$	1.79	1.26	1.03	0.89	0.80	0.73	0.68	0.63	0.60	0.57

Two variable EVOP program calculation worksheet

Figure 5.47 Work sheet cycle No. 10.

Table 5.35 2^2 BEVOP information board.

INFORMATION BOARD AFTER PHASE NO. V Last Cycle Completed : n = 10				
		RESPONSE No. 1	RESPONSE No. 2	RESPONSE No. 3
REQUIREMENT :		MAXIMIZE RESPONSE		
PROCESS AVERAGES :		64.66 (top left), 52.63 (top right), 62.65 (center), 75.39 (bottom left), 63.34 (bottom right); axes: B, A		
2.S.E. LIMITS FOR AVERAGES		1.97		
EFFECTS WITH 2.S.E	PHASE MEAN FACTOR A FACTOR B INTERACTION AB CHANGE IN MEAN	63.74 -12.04 ± 1.97 -10.72 ± 1.97 0.00 ± 1.97 1.08 ± 1.79		
ESTIMATED STANDARD DEVIATION S		3.13		
PRIOR ESTIMATE OF STANDARD DEV.		3.00		

NOTE:

1. The Center of Experiment-Reference Conditions (The Current Best-Known Process Conditions)
2. The Change in Mean provides a measure of whether the conditions being run the current phase yield an average result better or worse than the reference conditions.
3. The Phase Mean is the Mean Response Over All Conditions Being Run in the present Phase.
4*. The change in Mean is a measure of the over-all Curvature of the Response Surface. It will be positive near a MINIMUM and negative near a MAXIMUM.
5. Sequencing of the Runs-Randomization, if runs can be randomized without too much difficulty, than this should be done.

*A Diagonally oriented saddle represented by equal coefficients of opposite sign could occure and yield a zero change in mean effect, but such a phenomenon would rarely be met in practice.

Conclusions

- Two factors are highly significant with negative sign. We are moving away from optimum.
- The *phase mean* is reduced significantly.
- The estimated standard deviation is very close to the prior estimate of the standard deviation.

Recommendation

- It is recommended to move the center ($A_0 = 80$; $B_0 = 80$) back to optimum position ($A_0 = 50$; $B_0 = 50$).
- To keep the same factor's variation 5.0.
- To keep the same noise level STDEV = 3.0.

Solution

It is easy to see from the response surface and contour diagram that optimum (maximum) yield is 100[%] for factor's settings $A_0 = 50$; $B_0 = 50$ (Figures 5.48 and 5.49).

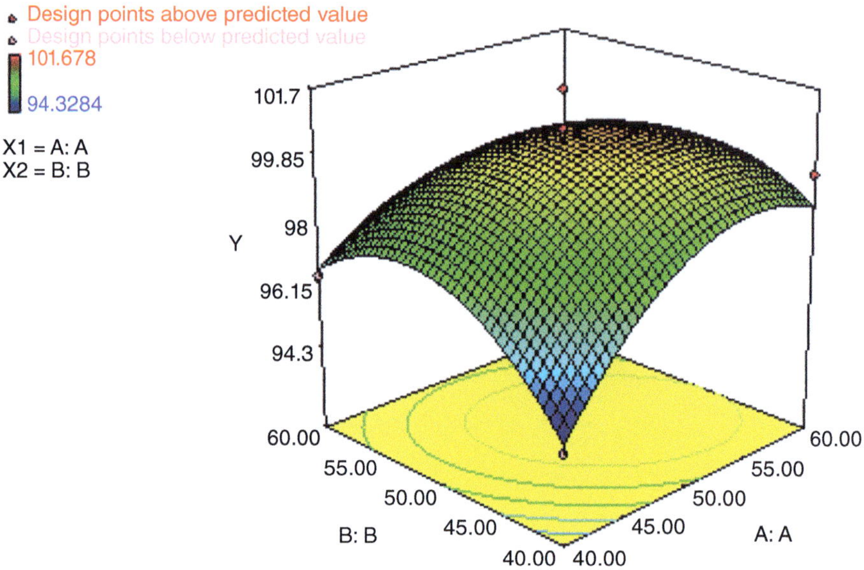

Figure 5.48 Response surface.

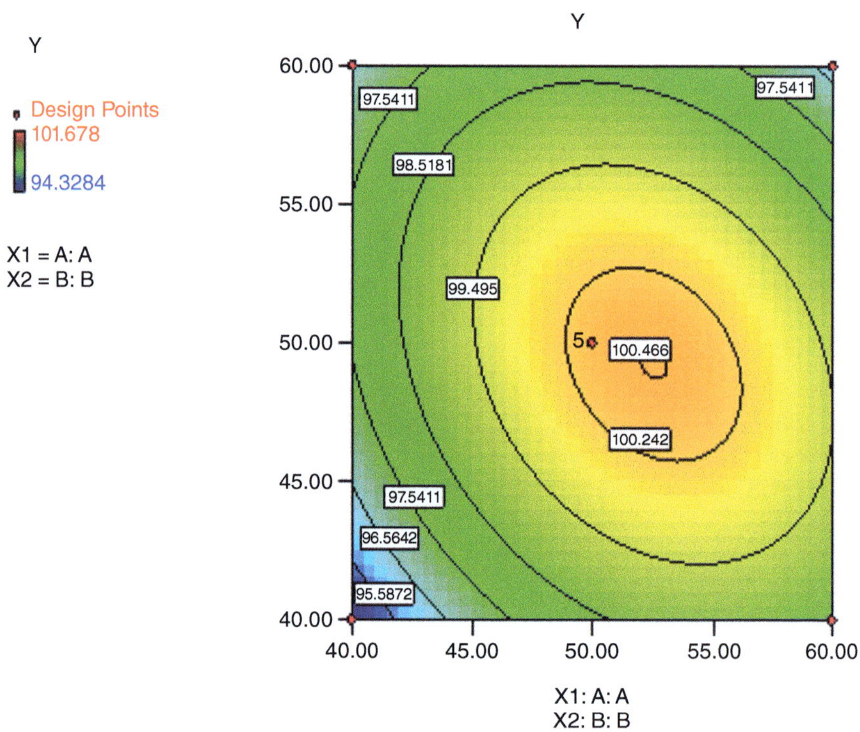

Figure 5.49 Contour diagram.

5.7.1.2 Simulation No. 2: 2^2 BEVOP

In the fine tuning of the batch process, two factors were being studied. These were the mixing speed (A) and mixing time (B) [3]. The current best process conditions are: $A_0 = 60$ [rpm], $B_0 = 180$ [min], and corresponding product viscosity was around $y = 2{,}406{,}000$ [cp]. Practically, viscosity range is: $1{,}000{,}000 < y < 2{,}500{,}000$ [cp]. The experimental limits for mixing speed are $40 < A > 80$ [rpm] and for the mixing time $40 < B < 180$ [min]. Start process simulation with next factor's intervals: $\Delta a = 5$; $\Delta b = 5$. Process noise measured by STDEV = 10,000 [cp] can be set on different value to test efficiency of the EVOP optimization. The objective of this study is to *minimize* product viscosity using EVOP. The appropriate mathematical function was used to calculate product response numbers plus noise (STDEV multiplied by Random Normal Deviates) to mask response and simulate a real plant viscosity data.

Phase I: The settings (Center Point , Factor's Interval, and Process Noise) and viscosity for different process conditions were calculated using Microsoft Excel simulation in Table 5.36. The corresponding calculations for ten cycles, phase I, are shown in work sheets in Figure 5.50–5.52. The information board that would appear at the end of cycle 10, after transfer of the various calculated quantities from the appropriate worksheet, is shown in Table 5.37.

Table 5.36 2^2 BEVOP simulation.

PROJECT : S-2
PHASE : I
DATE :

BEVOP SIMULATION

S-2

I. CENTER OF EXPERIMENT :	Ao =	60.00	Gen Random
	Bo =	180.00	
II. FACTOR'S INTERVAL OF VARIATION :	Δ a =	5.00	
	Δ b =	5.00	
III. PROCESS (PROPERTY-RESPONSE) NOISE :	STDV =	10000.00	

EVOP WORK SHEET

TRIALS ORDER			FACTOR A	FACTOR B	RESPONSE n = 1	n = 2	n = 3	n = 4	n = 5	n = 6	n = 7	n = 8	n = 9	n = 10
0		Block 1	60.00	180.00	2424276.2	2399360.8	2421389.7	2412143.8	2426532.8	2417034.7	2449179.8	2418273.6	2445945.3	2402599.3
1		Block 1	55.00	175.00	2284671.7	2284839.6	2279656.3	2289573.4	2271459.9	2280589.8	2291295.0	2302009.4	2291084.9	2275409.7
2		Block 1	65.00	175.00	2289180.0	2305354.2	2286012.9	2294015.5	2310617.3	2304808.0	2297561.8	2302008.0	2304925.2	2306013.2
3		Block 1	55.00	185.00	2543275.2	2549096.6	2552263.7	2558524.7	2542177.2	2562422.8	2548091.8	2555993.2	2562002.9	2547956.5
4		Block 1	65.00	185.00	2567730.4	2551279.6	2565773.9	2558412.9	2556054.4	2561014.8	2557558.1	2556294.2	2568590.7	2579763.4

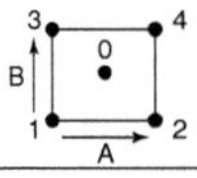

2^2 Factorial with added reference condition

Cycle n :
1
Response:YIELD

Project S-2
Phase I
Date

Calculation of Averages					
Operation condition	(0)	(1)	(2)	(3)	(4)
(i) Previous cycle sum					
(ii) Previous cycle average					
(iii) New observations	2424276.24	2284671.70	2289180.02	2543275.17	2567730.39
(iv) Differences (ii) less (iii)					
(v) New sums	2424276.24	2284671.70	2289180.02	2543275.17	2567730.39
(vi) New averages : ÿi	2424276.24	2284671.70	2289180.02	2543275.17	2567730.39

Calculation of Standard Deviation	
Prior estimate of σ	= 10000.00
Previous sum **s**	=
Previous average **s**	=
New **s** = range × **fδ,n**	=
Range	=
New sum **s**	=
New average **s** = new sum **s** / n - 1	=

Calculation of Effects		
Phase mean	= 1/5 (ÿo + ÿ1 + ÿ2 + ÿ3 + ÿ4) =	2421826.71
(A) ——— effect	= ½ (ÿ2 + ÿ4 - ÿ1 - ÿ3) =	14481.77
(B) ——— effect	= ½ (ÿ3 + ÿ4 - ÿ1 - ÿ2) =	268576.92
(AB) ——— effect	= ½ (ÿ1 + ÿ4 - ÿ3 - ÿ2) =	9973.45
Change in mean effect	= phase mean - ÿo =	-2449.54

Calculation of 2 S.E. Limits		
For new average **s** :	$\pm (2/\sqrt{n})s$	= 20000.00
For new effects :	$\pm (2/\sqrt{n})s$	= 20000.00
For change in mean :	$\pm (1.79/\sqrt{n})s$	= 17900.00

Table of Multiplying Factors

n	1	2	3	4	5	6	7	8	9	10
fδ,n		0.30	0.35	0.37	0.38	0.39	0.40	0.40	0.40	0.41
1 / n	1.00	0.50	0.33	0.25	0.20	0.17	0.14	0.12	0.11	0.10
1 / (n - 1)		1.00	0.50	0.33	0.25	0.20	0.17	0.14	0.12	0.11
$2/\sqrt{n}$	2.00	1.41	1.15	1.00	0.89	0.82	0.76	0.71	0.67	0.63
$1.79/\sqrt{n}$	1.79	1.26	1.03	0.89	0.80	0.73	0.68	0.63	0.60	0.57

Two variable EVOP program calculation worksheet

Figure 5.50 Work sheet cycle No. 1.

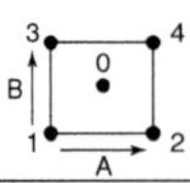

2^2 Factorial with added reference condition

Cycle n :
3
Response:YIELD

Project S-2
Phase I
Date

Calculation of Averages					
Operation condition	(0)	(1)	(2)	(3)	(4)
(i) Previous cycle sum	4823637.06	4569511.28	4594534.23	5092371.78	5119010.01
(ii) Previous cycle average	2411818.53	2284755.64	2297267.12	2546185.89	2559505.00
(iii) New observations	2421389.67	2279656.25	2286012.88	2552263.66	2565773.94
(iv) Differences (ii) less (iii)	-9571.14	5099.39	11254.24	-6077.77	-6268.94
(v) New sums	4833208.20	4564411.89	4583279.99	5098449.56	5125278.95
(vi) New averages : ÿi	2416604.10	2282205.94	2291640.00	2549224.78	2562639.47

Calculation of Standard Deviation	
Prior estimate of σ	= 10000.00
Previous sum **s**	= 12326.89
Previous average **s**	= 12326.89
New **s** = range × **fδ,n**	= 7288.88
Range	= 20825.38
New sum **s**	= 19615.77
New average **s** = new sum **s** / n - 1	= 9807.88

Calculation of Effects		
Phase mean	= 1/5 (ÿo + ÿ1 + ÿ2 + ÿ3 + ÿ4) =	2420462.86
(A) ——— effect	= ½ (ÿ2 + ÿ4 - ÿ1 - ÿ3) =	11424.37
(B) ——— effect	= ½ (ÿ3 + ÿ4 - ÿ1 - ÿ2) =	269009.16
(AB) ——— effect	= ½ (ÿ1 + ÿ4 - ÿ3 - ÿ2) =	1990.32
Change in mean effect	= phase mean - ÿo =	3858.76

Calculation of 2 S.E. Limits		
For new average **s** :	$\pm (2/\sqrt{n})s$	= 11279.07
For new effects :	$\pm (2/\sqrt{n})s$	= 11279.07
For change in mean :	$\pm (1.79/\sqrt{n})s$	= 10102.12

Table of Multiplying Factors

n	1	2	3	4	5	6	7	8	9	10
fδ,n		0.30	0.35	0.37	0.38	0.39	0.40	0.40	0.40	0.41
1 / n	1.00	0.50	0.33	0.25	0.20	0.17	0.14	0.12	0.11	0.10
1 / (n - 1)		1.00	0.50	0.33	0.25	0.20	0.17	0.14	0.12	0.11
$2/\sqrt{n}$	2.00	1.41	1.15	1.00	0.89	0.82	0.76	0.71	0.67	0.63
$1.79/\sqrt{n}$	1.79	1.26	1.03	0.89	0.80	0.73	0.68	0.63	0.60	0.57

Two variable EVOP program calculation worksheet

Figure 5.51 Work sheet cycle No. 3.

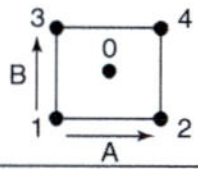

2^2 Factorial with added reference condition

Cycle n : *10*
Response:YIELD

Project S-2
Phase I
Date

Calculation of Averages						Calculation of Standard Deviation	
Operation condition	(0)	(1)	(2)	(3)	(4)	Prior estimate of σ	= 10000.00
(i) Previous cycle sum	4872063.09	4584821.43	4606135.99	5115828.50	5126040.42	Previous sum **s**	= 80811.05
(ii) Previous cycle average	2436031.54	2292410.72	2303068.00	2557914.25	2563020.21	Previous average **s**	= 10101.38
(iii) New observations	2402599.30	2275409.73	2306013.21	2547956.49	2579763.44	New **s** = range × $f\delta,n$	= 20571.94
(iv) Differences (ii) less (iii)	33432.24	17000.99	-2945.21	9957.76	-16743.23	Range	= 50175.47
(v) New sums	4838630.84	4567820.45	4609081.21	5105870.74	5142783.65	New sum **s**	= 101382.99
(vi) New averages : $\bar{y}i$	2419315.42	2283910.22	2304540.60	2552935.37	2571391.83	New average **s** = new sum **s** / n - 1	= 11264.78

Calculation of Effects			Calculation of 2 S.E. Limits
Phase mean	$= 1/5\,(\bar{y}o + \bar{y}1 + \bar{y}2 + \bar{y}3 + \bar{y}4)$ =	2426418.69	
(A) ———— effect	$= ½\,(\bar{y}2 + \bar{y}4 - \bar{y}1 - \bar{y}3)$ =	19543.42	For new average **s** : $\pm(2/\sqrt{n})s$ = 7096.81
(B) ———— effect	$= ½\,(\bar{y}3 + \bar{y}4 - \bar{y}1 - \bar{y}2)$ =	267938.19	For new effects : $\pm(2/\sqrt{n})s$ = 7096.81
(AB) ———— effect	$= ½\,(\bar{y}1 + \bar{y}4 - \bar{y}3 - \bar{y}2)$ =	-1086.96	For change in mean : $\pm(1.79/\sqrt{n})s$ = 6420.92
Change in mean effect	= phase mean - $\bar{y}o$ =	7103.27	

Table of Multiplying Factors

n	1	2	3	4	5	6	7	8	9	10
$f\delta,n$		0.30	0.35	0.37	0.38	0.39	0.40	0.40	0.40	0.41
1 / n	1.00	0.50	0.33	0.25	0.20	0.17	0.14	0.12	0.11	0.10
1 / (n - 1)		1.00	0.50	0.33	0.25	0.20	0.17	0.14	0.12	0.11
$2/\sqrt{n}$	2.00	1.41	1.15	1.00	0.89	0.82	0.76	0.71	0.67	0.63
$1.79/\sqrt{n}$	1.79	1.26	1.03	0.89	0.80	0.73	0.68	0.63	0.60	0.57

Two variable EVOP program calculation worksheet

Figure 5.52 Work sheet cycle No. 10.

Table 5.37 2^2 BEVOP information board.

INFORMATION BOARD AFTER PHASE NO. I

Last Cycle Completed : n = 10

	RESPONSE No. 1	RESPONSE No. 2	RESPONSE No. 3
REQUIREMENT :	MINIMIZE RESPONSE		
PROCESS AVERAGES :	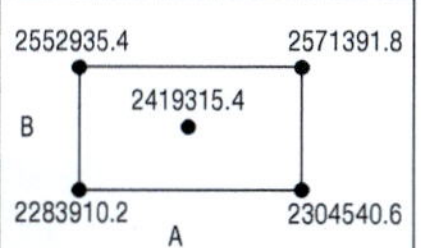		
2.S.E. LIMITS FOR AVERAGES	7096.81		
EFFECTS WITH 2.S.E — PHASE MEAN	2426418.69		
FACTOR A	19543.42 ± 7096.81		
FACTOR B	267938.19 ± 7096.81		
INTERACTION AB	-1086.96 ± 7096.81		
CHANGE IN MEAN	7103.27 ± 6420.92		
ESTIMATED STANDARD DEVIATION S	11264.78		
PRIOR ESTIMATE OF STANDARD DEV.	10000.00		

NOTE:

1. The Center of Experiment-Reference Conditions (The Current Best-Known Process Conditions)
2. The Change in Mean provides a measure of whether the conditions being run the current phase yield an average result better or worse than the reference conditions.
3. The Phase Mean is the Mean Response Over All Conditions Being Run in the present Phase.
4*. The change in Mean is a measure of the over-all Curvature of the Response Surface. It will be positive near a MINIMUM and negative near a MAXIMUM.
5. Sequencing of the Runs-Randomization, if runs can be randomized without too much difficulty, than this should be done.

*A Diagonally oriented saddle represented by equal coefficients of opposite sign could occure and yield a zero change in mean effect, but such a phenomenon would rarely be met in practice.

Conclusions

- Two factors are highly significant with positive sign. It means higher factor's values will increase product viscosity. To reduce viscosity (minimize) means to decrease factor's settings.
- The *change in mean* is also significant. Maybe some slight curvature in viscosity response surface.
- The estimated standard deviation is very close to the prior estimate of the standard deviation.

Recommendation for Phase II

- It is recommended to move the center of the experi ment ($A_0 = 60$; $B_0 = 180$) to the new position with significantly decreased both factors ($A_0 = 50$; $B_0 = 120$).
- To keep the same factor's variation 5.0.
- To keep the same noise level STDEV = 10,000.0 (Table 5.38, Figures 5.53–5.55, Table 5.39).

Table 5.38 2^2 BEVOP simulation.

PROJECT : S-2
PHASE : II
DATE :

BEVOP SIMULATION

S-2

I. CENTER OF EXPERIMENT : Ao = 50.00, Bo = 120.00 [Gen Random]

II. FACTOR'S INTERVAL OF VARIATION : Δa = 5.00, Δb = 5.00

III. PROCESS (PROPERTY-RESPONSE) NOISE : STDV = 10000.00

EVOP WORK SHEET

TRIALS ORDER			FACTOR A	FACTOR B	RESPONSE n = 1	n = 2	n = 3	n = 4	n = 5	n = 6	n = 7	n = 8	n = 9	n = 10
0		Block 1	50.00	120.00	1086776.6	1061861.2	1083890.1	1074644.2	1089033.2	1079535.1	1111680.2	1080774.0	1108445.7	1065099.7
1		Block 1	45.00	115.00	991644.1	991812.0	986628.7	996545.8	978432.3	987562.2	998267.4	1008981.8	998057.3	982382.1
2		Block 1	55.00	115.00	997012.4	1013186.6	993845.3	1001847.9	1018449.7	1012640.4	1005394.2	1009840.4	1012757.6	1013845.6
3		Block 1	45.00	125.00	1160443.6	1166265.0	1169432.1	1175693.1	1159345.6	1179591.2	1165260.2	1173161.6	1179171.3	1165124.9
4		Block 1	55.00	125.00	1185758.8	1169308.0	1183802.3	1176441.3	1174082.8	1179043.2	1175586.5	1174322.6	1186619.1	1197791.8

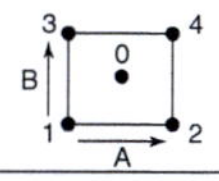

2^2 Factorial with added reference condition

Cycle n : 1
Response:YIELD

Project S-2
Phase II
Date

Calculation of Averages					
Operation condition	(0)	(1)	(2)	(3)	(4)
(i) Previous cycle sum					
(ii) Previous cycle average					
(iii) New observations	1086776.64	991644.10	997012.42	1160443.57	1185758.79
(iv) Differences (ii) less (iii)					
(v) New sums	1086776.64	991644.10	997012.42	1160443.57	1185758.79
(vi) New averages : $\bar{y}i$	1086776.64	991644.10	997012.42	1160443.57	1185758.79

Calculation of Standard Deviation	
Prior estimate of σ	= 10000.00
Previous sum **s**	=
Previous average **s**	=
New **s** = range × $f\delta,n$	=
Range	=
New sum **s**	=
New average **s** = new sum **s** / n - 1	=

Calculation of Effects		
Phase mean	= 1/5 ($\bar{y}0 + \bar{y}1 + \bar{y}2 + \bar{y}3 + \bar{y}4$) =	1084327.11
(A) effect	= ½ ($\bar{y}2 + \bar{y}4 - \bar{y}1 - \bar{y}3$) =	15341.77
(B) effect	= ½ ($\bar{y}3 + \bar{y}4 - \bar{y}1 - \bar{y}2$) =	178772.92
(AB) effect	= ½ ($\bar{y}1 + \bar{y}4 - \bar{y}3 - \bar{y}2$) =	9973.45
Change in mean effect	= phase mean - $\bar{y}0$ =	-2449.54

Calculation of 2 S.E. Limits		
For new average **s** :	± (2/√n)s	= 20000.00
For new effects :	± (2/√n)s	= 20000.00
For change in mean :	± (1.79/√n)s	= 17900.00

Table of Multiplying Factors

n	1	2	3	4	5	6	7	8	9	10
$f\delta,n$		0.30	0.35	0.37	0.38	0.39	0.40	0.40	0.40	0.41
1 / n	1.00	0.50	0.33	0.25	0.20	0.17	0.14	0.12	0.11	0.10
1 / (n - 1)		1.00	0.50	0.33	0.25	0.20	0.17	0.14	0.12	0.11
2 / √n	2.00	1.41	1.15	1.00	0.89	0.82	0.76	0.71	0.67	0.63
1.79 / √n	1.79	1.26	1.03	0.89	0.80	0.73	0.68	0.63	0.60	0.57

Two variable EVOP program calculation worksheet

Figure 5.53 Work sheet cycle No. 1.

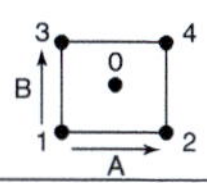

2^2 Factorial with added reference condition

Cycle n : 3
Response:YIELD

Project S-2
Phase II
Date

Calculation of Averages					
Operation condition	(0)	(1)	(2)	(3)	(4)
(i) Previous cycle sum	2148637.86	1983456.08	2010199.03	2326708.58	2355066.81
(ii) Previous cycle average	1074318.93	991728.04	1005099.52	1163354.29	1177533.40
(iii) New observations	1083890.07	986628.65	993845.28	1169432.06	1183802.34
(iv) Differences (ii) less (iii)	-9571.14	5099.39	11254.24	-6077.77	-6268.94
(v) New sums	2158209.00	1978356.69	1998944.79	2332786.36	2361335.75
(vi) New averages : $\bar{y}i$	1079104.50	989178.34	999472.40	1166393.18	1180667.87

Calculation of Standard Deviation	
Prior estimate of σ	= 10000.00
Previous sum **s**	= 12326.89
Previous average **s**	= 12326.89
New **s** = range × $f\delta,n$	= 7288.88
Range	= 20825.38
New sum **s**	= 19615.77
New average **s** = new sum **s** / n - 1	= 9807.88

Calculation of Effects		
Phase mean	= 1/5 ($\bar{y}0 + \bar{y}1 + \bar{y}2 + \bar{y}3 + \bar{y}4$) =	1082963.26
(A) effect	= ½ ($\bar{y}2 + \bar{y}4 - \bar{y}1 - \bar{y}3$) =	12284.37
(B) effect	= ½ ($\bar{y}3 + \bar{y}4 - \bar{y}1 - \bar{y}2$) =	179205.16
(AB) effect	= ½ ($\bar{y}1 + \bar{y}4 - \bar{y}3 - \bar{y}2$) =	1990.32
Change in mean effect	= phase mean - $\bar{y}0$ =	3858.76

Calculation of 2 S.E. Limits		
For new average **s** :	± (2/√n)s	= 11279.07
For new effects :	± (2/√n)s	= 11279.07
For change in mean :	± (1.79/√n)s	= 10102.12

Table of Multiplying Factors

n	1	2	3	4	5	6	7	8	9	10
$f\delta,n$		0.30	0.35	0.37	0.38	0.39	0.40	0.40	0.40	0.41
1 / n	1.00	0.50	0.33	0.25	0.20	0.17	0.14	0.12	0.11	0.10
1 / (n - 1)		1.00	0.50	0.33	0.25	0.20	0.17	0.14	0.12	0.11
2 / √n	2.00	1.41	1.15	1.00	0.89	0.82	0.76	0.71	0.67	0.63
1.79 / √n	1.79	1.26	1.03	0.89	0.80	0.73	0.68	0.63	0.60	0.57

Two variable EVOP program calculation worksheet

Figure 5.54 Work sheet cycle No. 3.

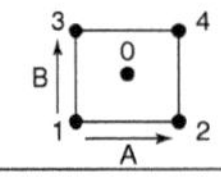

2^2 Factorial with added reference condition

Cycle n : 10
Response: YIELD

Project S-2
Phase II
Date

Calculation of Averages

Operation condition	(0)	(1)	(2)	(3)	(4)
(i) Previous cycle sum	2197063.89	1998766.23	2021800.79	2350165.30	2362097.22
(ii) Previous cycle average	1098531.94	999383.12	1010900.40	1175082.65	1181048.61
(iii) New observations	1065099.70	982382.13	1013845.61	1165124.89	1197791.84
(iv) Differences (ii) less (iii)	33432.24	17000.99	−2945.21	9957.76	−16743.23
(v) New sums	2163631.64	1981765.25	2024746.01	2340207.54	2378840.45
(vi) New averages : $\bar{y}_i$	1081815.82	990882.62	1012373.00	1170103.77	1189420.23

Calculation of Standard Deviation

Prior estimate of σ	= 10000.00
Previous sum **s**	= 80811.05
Previous average **s**	= 10101.38
New **s** = range × $f\delta,n$	= 20571.94
Range	= 50175.47
New sum **s**	= 101382.99
New average **s** = $\frac{\text{new sum } s}{n-1}$	= 11264.78

Calculation of Effects

Phase mean	$= 1/5\,(\bar{y}_0 + \bar{y}_1 + \bar{y}_2 + \bar{y}_3 + \bar{y}_4)$ =	1088919.00
(A) effect	$= \frac{1}{2}(\bar{y}_2 + \bar{y}_4 - \bar{y}_1 - \bar{y}_3)$ =	20403.42
(B) effect	$= \frac{1}{2}(\bar{y}_3 + \bar{y}_4 - \bar{y}_1 - \bar{y}_2)$ =	178134.19
(AB) effect	$= \frac{1}{2}(\bar{y}_1 + \bar{y}_4 - \bar{y}_3 - \bar{y}_2)$ =	-1086.96
Change in mean effect	= phase mean - $\bar{y}_0$ =	7103.27

Calculation of 2 S.E. Limits

For new average **s** :	$\pm (2/\sqrt{n})s$	= 7096.81
For new effects :	$\pm (2/\sqrt{n})s$	= 7096.81
For change in mean :	$\pm (1.79/\sqrt{n})s$	= 6420.92

Table of Multiplying Factors

n	1	2	3	4	5	6	7	8	9	10
$f\delta,n$		0.30	0.35	0.37	0.38	0.39	0.40	0.40	0.40	0.41
$1/n$	1.00	0.50	0.33	0.25	0.20	0.17	0.14	0.12	0.11	0.10
$1/(n-1)$		1.00	0.50	0.33	0.25	0.20	0.17	0.14	0.12	0.11
$2/\sqrt{n}$	2.00	1.41	1.15	1.00	0.89	0.82	0.76	0.71	0.67	0.63
$1.79/\sqrt{n}$	1.79	1.26	1.03	0.89	0.80	0.73	0.68	0.63	0.60	0.57

Two variable EVOP program calculation worksheet

Figure 5.55 Work sheet cycle No. 10.

Table 5.39 2^2 BEVOP information board.

INFORMATION BOARD AFTER PHASE NO. II

Last Cycle Completed : n = 10

<table>
<tr><td></td><td>RESPONSE No. 1</td><td>RESPONSE No. 2</td><td>RESPONSE No. 3</td></tr>
<tr><td>REQUIREMENT :</td><td>MINIMIZE RESPONSE</td><td></td><td></td></tr>
<tr><td>PROCESS AVERAGES :</td><td>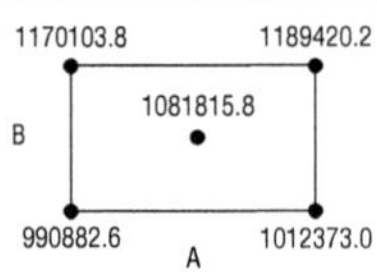
</td><td></td><td></td></tr>
<tr><td>2.S.E. LIMITS FOR AVERAGES</td><td>7096.81</td><td></td><td></td></tr>
<tr><td>EFFECTS WITH 2.S.E
PHASE MEAN
FACTOR A
FACTOR B
INTERACTION AB
CHANGE IN MEAN</td><td>
1088919.09
20403.42 ± 7096.81
178134.19 ± 7096.81
-1086.96 ± 7096.81
7103.27 ± 6420.92</td><td></td><td></td></tr>
<tr><td>ESTIMATED STANDARD DEVIATION S
PRIOR ESTIMATE OF STANDARD DEV.</td><td>11264.78
10000.00</td><td></td><td></td></tr>
</table>

NOTE:

1. The Center of Experiment-Reference Conditions (The Current Best-Known Process Conditions)
2. The Change in Mean provides a measure of whether the conditions being run the current phase yield an average result better or worse than the reference conditions.
3. The Phase Mean is the Mean Response Over All Conditions Being Run in the present Phase.
4*. The change in Mean is a measure of the over-all Curvature of the Response Surface. It will be positive near a MINIMUM and negative near a MAXIMUM.
5. Sequencing of the Runs-Randomization, if runs can be randomized without too much difficulty, than this should be done.

*A Diagonally oriented saddle represented by equal coefficients of opposite sign could occure and yield a zero change in mean effect, but such a phenomenon would rarely be met in practice.

Conclusions

- Two factors are highly significant with positive sign. It means higher factor's values will increase product viscosity. To reduce viscosity (minimize) means to decrease factors significantly.
- The *phase mean* is reduced significantly.
- The *change in mean* is significant.
- The estimated standard deviation is very close to the prior estimate of the standard deviation.

Recommendation for Phase III

- It is recommended to move the center of the experiment ($A_0 = 50$; $B_0 = 120$) into new center point ($A_0 = 40$; $B_0 = 40$).
- To keep the same factor's variation 5.0.
- To keep the same noise level STDEV = 10,000.0 (Table 5.40, Figures 5.56–5.58, Table 5.41).

Table 5.40 2^2 BEVOP simulation.

PROJECT : S-2
PHASE : III
DATE :

BEVOP SIMULATION

S-2

I. CENTER OF EXPERIMENT :	Ao =	40.00	Gen Random
	Bo =	40.00	
II. FACTOR'S INTERVAL OF VARIATION :	Δ a =	5.00	
	Δ b =	5.00	
III. PROCESS (PROPERTY-RESPONSE) NOISE :	STDV =	10000.00	

EVOP WORK SHEET

TRIALS ORDER			FACTOR A	FACTOR B	RESPONSE n = 1	n = 2	n = 3	n = 4	n = 5	n = 6	n = 7	n = 8	n = 9	n = 10
0		Block 1	40.00	40.00	145289.8	120374.4	142403.3	133157.4	147546.4	138048.3	170193.4	139287.2	166958.9	123612.9
1		Block 1	35.00	35.00	108763.3	108931.2	103747.9	113665.0	95551.5	104681.4	115386.6	126101.0	115176.5	99501.3
2		Block 1	45.00	35.00	116991.6	133165.8	113824.5	121827.1	138428.9	132619.6	125373.4	129819.6	132736.8	133824.8
3		Block 1	35.00	45.00	157490.8	163312.2	166479.3	172740.3	156392.8	176638.4	162307.4	170208.8	176218.5	162172.1
4		Block 1	45.00	45.00	185666.0	169215.2	183709.5	176348.5	173990.0	178950.4	175493.7	174229.8	186526.3	197699.0

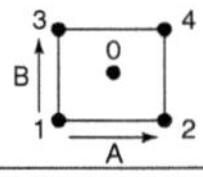

2^2 Factorial with added reference condition

Cycle n :
1
Response:YIELD

Project S-2
Phase III
Date

Calculation of Averages						Calculation of Standard Deviation	
Operation condition	(0)	(1)	(2)	(3)	(4)	Prior estimate of σ	= 10000.00
(i) Previous cycle sum						Previous sum **s**	=
(ii) Previous cycle average						Previous average **s**	=
(iii) New observations	145289.84	108763.30	116991.62	157490.77	185665.99	New **s** = range × **fδ,n**	=
(iv) Differences (ii) less (iii)						Range	=
(v) New sums	145289.84	108763.30	116991.62	157490.77	185665.99	New sum **s**	=
(vi) New averages : $\bar{y}i$	145289.84	108763.30	116991.62	157490.77	185665.99	New average **s** = new sum **s** / n - 1	=

Calculation of Effects			Calculation of 2 S.E. Limits	
Phase mean	= 1/5 ($\bar{y}$o + $\bar{y}$1 + $\bar{y}$2 + $\bar{y}$3 + $\bar{y}$4) =	142840.31		
(A) ——— effect	= ½ ($\bar{y}$2 + $\bar{y}$4 - $\bar{y}$1 - $\bar{y}$3) =	18201.77	For new average **s** : ± (2/√n)s	= 20000.00
(B) ——— effect	= ½ ($\bar{y}$3 + $\bar{y}$4 - $\bar{y}$1 - $\bar{y}$2) =	58700.92	For new effects : ± (2/√n)s	= 20000.00
(AB) ——— effect	= ½ ($\bar{y}$1 + $\bar{y}$4 - $\bar{y}$3 - $\bar{y}$2) =	9973.45	For change in mean : ± (1.79/√n)s	= 17900.00
Change in mean effect	= phase mean - $\bar{y}$o =	-2449.54		

Table of Multiplying Factors

n	1	2	3	4	5	6	7	8	9	10
fδ,n		0.30	0.35	0.37	0.38	0.39	0.40	0.40	0.40	0.41
1 / n	1.00	0.50	0.33	0.25	0.20	0.17	0.14	0.12	0.11	0.10
1 / (n - 1)		1.00	0.50	0.33	0.25	0.20	0.17	0.14	0.12	0.11
2 / √n	2.00	1.41	1.15	1.00	0.89	0.82	0.76	0.71	0.67	0.63
1.79 / √n	1.79	1.26	1.03	0.89	0.80	0.73	0.68	0.63	0.60	0.57

Two variable EVOP program calculation worksheet

Figure 5.56 Work sheet cycle No. 1.

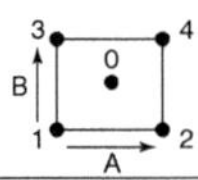

2^2 Factorial with added reference condition

Cycle n :
3
Response:YIELD

Project S-2
Phase III
Date

Calculation of Averages						Calculation of Standard Deviation	
Operation condition	(0)	(1)	(2)	(3)	(4)	Prior estimate of σ	= 10000.00
(i) Previous cycle sum	265664.26	217694.48	250157.43	320802.98	354881.21	Previous sum **s**	= 12326.89
(ii) Previous cycle average	132832.13	108847.24	125078.72	160401.49	177440.60	Previous average **s**	= 12326.89
(iii) New observations	142403.27	103747.85	113824.48	166479.26	183709.54	New **s** = range × **fδ,n**	= 7288.88
(iv) Differences (ii) less (iii)	-9571.14	5099.39	11254.24	-6077.77	-6268.94	Range	= 20825.38
(v) New sums	275235.40	212595.09	238903.19	326880.76	361150.15	New sum **s**	= 19615.77
(vi) New averages : $\bar{y}i$	137617.70	106297.54	119451.60	163440.38	180575.07	New average **s** = new sum **s** / n - 1	= 9807.88

Calculation of Effects			Calculation of 2 S.E. Limits	
Phase mean	= 1/5 ($\bar{y}$o + $\bar{y}$1 + $\bar{y}$2 + $\bar{y}$3 + $\bar{y}$4) =	141476.46		
(A) ——— effect	= ½ ($\bar{y}$2 + $\bar{y}$4 - $\bar{y}$1 - $\bar{y}$3) =	15144.37	For new average **s** : ± (2/√n)s	= 11279.07
(B) ——— effect	= ½ ($\bar{y}$3 + $\bar{y}$4 - $\bar{y}$1 - $\bar{y}$2) =	59133.16	For new effects : ± (2/√n)s	= 11279.07
(AB) ——— effect	= ½ ($\bar{y}$1 + $\bar{y}$4 - $\bar{y}$3 - $\bar{y}$2) =	1990.32	For change in mean : ± (1.79/√n)s	= 10102.12
Change in mean effect	= phase mean - $\bar{y}$o =	3858.76		

Table of Multiplying Factors

n	1	2	3	4	5	6	7	8	9	10
fδ,n		0.30	0.35	0.37	0.38	0.39	0.40	0.40	0.40	0.41
1 / n	1.00	0.50	0.33	0.25	0.20	0.17	0.14	0.12	0.11	0.10
1 / (n - 1)		1.00	0.50	0.33	0.25	0.20	0.17	0.14	0.12	0.11
2 / √n	2.00	1.41	1.15	1.00	0.89	0.82	0.76	0.71	0.67	0.63
1.79 / √n	1.79	1.26	1.03	0.89	0.80	0.73	0.68	0.63	0.60	0.57

Two variable EVOP program calculation worksheet

Figure 5.57 Work sheet cycle No. 3.

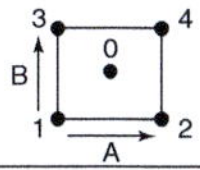

2^2 Factorial with added reference condition

Cycle n :
10
Response:YIELD

Project S-2
Phase III
Date

Calculation of Averages

Operation condition	(0)	(1)	(2)	(3)	(4)
(i) Previous cycle sum	314090.29	233004.63	261759.19	344259.70	361911.62
(ii) Previous cycle average	157045.14	116502.32	130879.60	172129.85	180955.81
(iii) New observations	123612.90	99501.33	133824.81	162172.09	197699.04
(iv) Differences (ii) less (iii)	33432.24	17000.99	-2945.21	9957.76	-16743.23
(v) New sums	280658.04	216003.65	264704.41	334301.94	378654.85
(vi) New averages : $\bar{y}_i$	140329.02	108001.82	132352.20	167150.97	189327.43

Calculation of Standard Deviation

Prior estimate of σ	= 10000.00
Previous sum **s**	= 80811.05
Previous average **s**	= 10101.38
New **s** = range × $f\delta,n$	= 20571.94
Range	= 50175.47
New sum **s**	= 101382.99
New average **s** = $\frac{\text{new sum } s}{n-1}$	= 11264.78

Calculation of Effects

Phase mean	= $1/5\,(\bar{y}_0 + \bar{y}_1 + \bar{y}_2 + \bar{y}_3 + \bar{y}_4)$ =	147432.29
(A) ———— effect	= $\frac{1}{2}\,(\bar{y}_2 + \bar{y}_4 - \bar{y}_1 - \bar{y}_3)$ =	23263.42
(B) ———— effect	= $\frac{1}{2}\,(\bar{y}_3 + \bar{y}_4 - \bar{y}_1 - \bar{y}_2)$ =	58062.19
(AB) ———— effect	= $\frac{1}{2}\,(\bar{y}_1 + \bar{y}_4 - \bar{y}_3 - \bar{y}_2)$ =	-1086.96
Change in mean effect	= phase mean - $\bar{y}_0$ =	7103.27

Calculation of 2 S.E. Limits

For new average **s** : $\pm(2/\sqrt{n})s$	= 7096.81
For new effects : $\pm(2/\sqrt{n})s$	= 7096.81
For change in mean : $\pm(1.79/\sqrt{n})s$	= 6420.92

Table of Multiplying Factors

n	1	2	3	4	5	6	7	8	9	10
$f\delta,n$		0.30	0.35	0.37	0.38	0.39	0.40	0.40	0.40	0.41
1 / n	1.00	0.50	0.33	0.25	0.20	0.17	0.14	0.12	0.11	0.10
1 / (n - 1)		1.00	0.50	0.33	0.25	0.20	0.17	0.14	0.12	0.11
$2/\sqrt{n}$	2.00	1.41	1.15	1.00	0.89	0.82	0.76	0.71	0.67	0.63
$1.79/\sqrt{n}$	1.79	1.26	1.03	0.89	0.80	0.73	0.68	0.63	0.60	0.57

Two variable EVOP program calculation worksheet

Figure 5.58 Work sheet cycle No. 10.

Table 5.41 2^2 BEVOP information board.

INFORMATION BOARD AFTER PHASE NO. III

Last Cycle Completed : n = 10

	RESPONSE No. 1	RESPONSE No. 2	RESPONSE No. 3
REQUIREMENT :	MINIMIZE RESPONSE		
PROCESS AVERAGES :	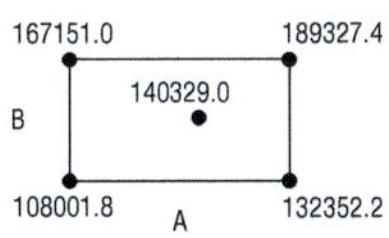		
2.S.E. LIMITS FOR AVERAGES	7096.81		
EFFECTS WITH 2.S.E — PHASE MEAN	147432.29		
FACTOR A	23263.42 ± 7096.81		
FACTOR B	58062.19 ± 7096.81		
INTERACTION AB	-1086.96 ± 7096.81		
CHANGE IN MEAN	7103.27 ± 6420.92		
ESTIMATED STANDARD DEVIATION S	11264.78		
PRIOR ESTIMATE OF STANDARD DEV.	10000.00		

NOTE:

1. The Center of Experiment-Reference Conditions (The Current Best-Known Process Conditions)
2. The Change in Mean provides a measure of whether the conditions being run the current phase yield an average result better or worse than the reference conditions.
3. The Phase Mean is the Mean Response Over All Conditions Being Run in the present Phase.
4*. The change in Mean is a measure of the over-all Curvature of the Response Surface. It will be positive near a MINIMUM and negative near a MAXIMUM.
5. Sequencing of the Runs-Randomization, if runs can be randomized without too much difficulty, than this should be done.

*A Diagonally oriented saddle represented by equal coefficients of opposite sign could occure and yield a zero change in mean effect, but such a phenomenon would rarely be met in practice.

Solutions

From the response surface and contour diagram that optimum (minimum) viscosity was achieved for the factors settings $A_0 = 40$; $B_0 = 40$ (Figures 5.59 and 5.60).

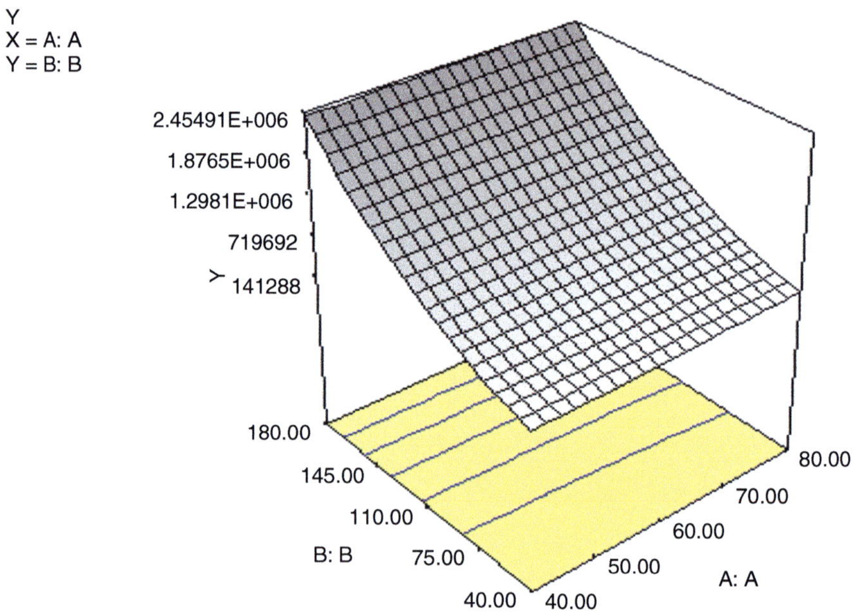

Figure 5.59 Response surface.

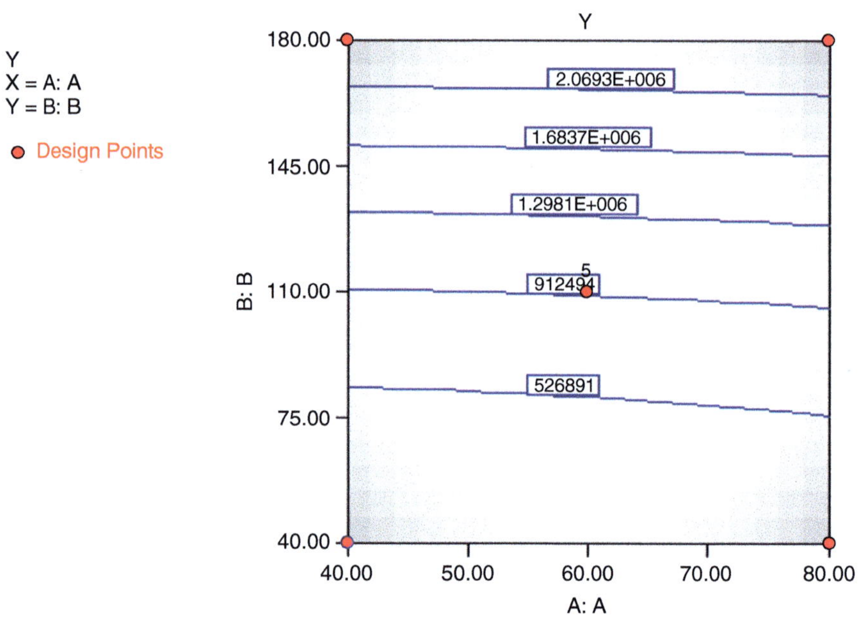

Figure 5.60 Contour diagram.

5.7.1.3 Simulation No. 3: 2^2 BEVOP

Two process factors were being studied in product development. These were factors (A) and (B) at which the process was conducted. The current best process conditions were: $A_0 = 2.5$; $B_0 = 2.5$ and corresponding product property was around $y = 7$. The experimental limits are for factors A [−5; +5] and B [−5; +5]. Theoretically, response range is: $0.0 < y < 100$ [%]. Process noise measured by STDEV = 1.0. Start simulation with next factor's intervals: $\Delta a = 1.0$ and $\Delta b = 1.0$. The objective is to *maximize* response using EVOP. The appropriate mathematical function was used to calculate product response numbers plus noise (STDEV multiplied by Random Normal Deviates) to mask response and simulate real plant data. It is important to mention that this type of optimization is very complex and difficult.

Phase I: The settings (Center Point , Factor's Interval, and Process Noise) and product response for different process conditions have been completed using Excel simulation in Table 5.42. The corresponding calculations for ten cycles, phase I, have been shown in work sheets in Figure 5.61–5.63. The information board that would appear at the end of cycle 10, after transfer of the various calculated quantities from the appropriate worksheet, is shown in Table 5.43 (Figures 5.61–5.63).

Table 5.42 2^2 BEVOP simulation.

PROJECT : S-3
PHASE : I
DATE :

BEVOP SIMULATION

S-3

I. CENTER OF EXPERIMENT : Ao = 2.50, Bo = 2.50 [Gen Random]

II. FACTOR'S INTERVAL OF VARIATION : Δ a = 1.00, Δ b = 1.00

III. PROCESS (PROPERTY-RESPONSE) NOISE : STDV = 2.00

EVOP WORK SHEET

TRIALS ORDER			FACTOR		RESPONSE									
			A	B	n = 1	n = 2	n = 3	n = 4	n = 5	n = 6	n = 7	n = 8	n = 9	n = 10
0		Block 1	2.50	2.50	5.50	4.97	2.77	5.25	8.34	4.56	7.18	8.09	7.54	4.33
1		Block 1	1.50	1.50	2.81	-0.67	3.01	-1.68	-0.04	6.87	4.18	3.27	1.49	4.87
2		Block 1	3.50	1.50	22.40	22.56	21.94	21.47	20.12	20.85	24.15	25.22	18.05	19.95
3		Block 1	1.50	3.50	-9.91	-9.07	-9.36	-6.28	-8.58	-6.03	-7.05	-6.86	-5.97	-7.02
4		Block 1	3.50	3.50	12.22	12.60	11.89	13.43	11.61	14.44	11.83	10.19	14.79	14.57

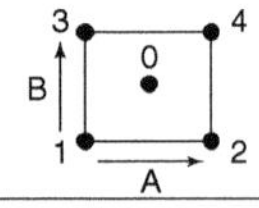

2^2 Factorial with added reference condition

Cycle n : *1*

Response:YIELD

Project S-3
Phase I
Date ______

Calculation of Averages						Calculation of Standard Deviation	
Operation condition	(0)	(1)	(2)	(3)	(4)	Prior estimate of **σ**	= 2.00
(i) Previous cycle sum						Previous sum **s**	=
(ii) Previous cycle average						Previous average **s**	=
(iii) New observations	5.50	2.81	22.40	-9.91	12.22	New **s** = range × ***f δ,n***	=
(iv) Differences (ii) less (iii)						Range	=
(v) New sums	5.50	2.81	22.40	-9.91	12.22	New sum **s**	=
(vi) New averages : $\bar{y}i$	5.50	2.81	22.40	-9.91	12.22	New average **s** = new sum **s** / n - 1	=

Calculation of Effects			Calculation of 2 S.E. Limits		
Phase mean	= 1/5 ($\bar{y}o$ + $\bar{y}1$ + $\bar{y}2$ + $\bar{y}3$ + $\bar{y}4$)	= 6.61			
(A) effect	= ½ ($\bar{y}2$ + $\bar{y}4$ - $\bar{y}1$ - $\bar{y}3$)	= 20.86	For new average **s** :	± $(2/\sqrt{n})$s	= 4.00
(B) effect	= ½ ($\bar{y}3$ + $\bar{y}4$ - $\bar{y}1$ - $\bar{y}2$)	= -11.45	For new effects :	± $(2/\sqrt{n})$s	= 4.00
(AB) effect	= ½ ($\bar{y}1$ + $\bar{y}4$ - $\bar{y}3$ - $\bar{y}2$)	= 1.27	For change in mean :	± $(1.79/\sqrt{n})$s	= 3.58
Change in mean effect	= phase mean - $\bar{y}o$	= 1.11			

Table of Multiplying Factors

n	1	2	3	4	5	6	7	8	9	10
f δ,n		0.30	0.35	0.37	0.38	0.39	0.40	0.40	0.40	0.41
1 / n	1.00	0.50	0.33	0.25	0.20	0.17	0.14	0.12	0.11	0.10
1 / (n - 1)		1.00	0.50	0.33	0.25	0.20	0.17	0.14	0.12	0.11
$2/\sqrt{n}$	2.00	1.41	1.15	1.00	0.89	0.82	0.76	0.71	0.67	0.63
$1.79/\sqrt{n}$	1.79	1.26	1.03	0.89	0.80	0.73	0.68	0.63	0.60	0.57

Two variable EVOP program calculation worksheet

Figure 5.61 Work sheet cycle No. 1.

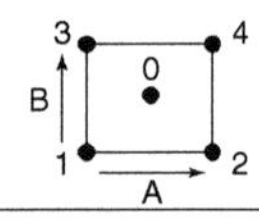

2^2 Factorial with added reference condition

Cycle n : *3*

Response:YIELD

Project S-3
Phase I
Date ______

Calculation of Averages						Calculation of Standard Deviation	
Operation condition	(0)	(1)	(2)	(3)	(4)	Prior estimate of **σ**	= 2.00
(i) Previous cycle sum	10.47	2.14	44.97	-18.98	24.82	Previous sum **s**	= 1.30
(ii) Previous cycle average	5.24	1.07	22.48	-9.49	12.41	Previous average **s**	= 1.30
(iii) New observations	2.77	3.01	21.94	-9.36	11.89	New **s** = range × ***f δ,n***	= 1.54
(iv) Differences (ii) less (iii)	2.46	-1.94	0.54	-0.13	0.52	Range	= 4.41
(v) New sums	8.01	4.08	44.43	-18.85	24.30	New sum **s**	= 2.84
(vi) New averages : $\bar{y}i$	4.01	2.04	22.21	-9.43	12.15	New average **s** = new sum **s** / n - 1	= 1.42

Calculation of Effects			Calculation of 2 S.E. Limits		
Phase mean	= 1/5 ($\bar{y}o$ + $\bar{y}1$ + $\bar{y}2$ + $\bar{y}3$ + $\bar{y}4$)	= 6.20			
(A) effect	= ½ ($\bar{y}2$ + $\bar{y}4$ - $\bar{y}1$ - $\bar{y}3$)	= 20.87	For new average **s** :	± $(2/\sqrt{n})$s	= 1.63
(B) effect	= ½ ($\bar{y}3$ + $\bar{y}4$ - $\bar{y}1$ - $\bar{y}2$)	= -10.77	For new effects :	± $(2/\sqrt{n})$s	= 1.63
(AB) effect	= ½ ($\bar{y}1$ + $\bar{y}4$ - $\bar{y}3$ - $\bar{y}2$)	= 0.70	For change in mean :	± $(1.79/\sqrt{n})$s	= 1.46
Change in mean effect	= phase mean - $\bar{y}o$	= 2.19			

Table of Multiplying Factors

n	1	2	3	4	5	6	7	8	9	10
f δ,n		0.30	0.35	0.37	0.38	0.39	0.40	0.40	0.40	0.41
1 / n	1.00	0.50	0.33	0.25	0.20	0.17	0.14	0.12	0.11	0.10
1 / (n - 1)		1.00	0.50	0.33	0.25	0.20	0.17	0.14	0.12	0.11
$2/\sqrt{n}$	2.00	1.41	1.15	1.00	0.89	0.82	0.76	0.71	0.67	0.63
$1.79/\sqrt{n}$	1.79	1.26	1.03	0.89	0.80	0.73	0.68	0.63	0.60	0.57

Two variable EVOP program calculation worksheet

Figure 5.62 Work sheet cycle No. 3.

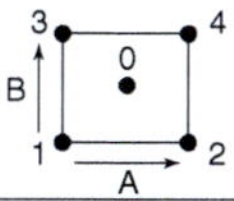

2^2 Factorial with added reference condition

Cycle n : 10
Response:YIELD

Project S-3
Phase I
Date

Calculation of Averages

Operation condition	(0)	(1)	(2)	(3)	(4)
(i) Previous cycle sum	14.75	5.04	41.92	-12.94	26.17
(ii) Previous cycle average	7.38	2.52	20.96	-6.47	13.08
(iii) New observations	4.33	4.87	19.95	-7.02	14.57
(iv) Differences (ii) less (iii)	3.05	-2.35	1.01	0.54	-1.48
(v) New sums	11.70	7.38	40.91	-13.49	27.65
(vi) New averages : $\bar{y}i$	5.85	3.69	20.46	-6.74	13.83

Calculation of Standard Deviation

Prior estimate of σ	= 2.00
Previous sum **s**	= 18.46
Previous average **s**	= 2.31
New **s** = range × $f\delta,n$	= 2.21
Range	= 5.40
New sum **s**	= 20.67
New average **s** = $\frac{\text{new sum } s}{n-1}$	= 2.30

Calculation of Effects

Phase mean	= $1/5\,(\bar{y}o + \bar{y}1 + \bar{y}2 + \bar{y}3 + \bar{y}4)$	= 7.42
(A) ———— effect	= $½\,(\bar{y}2 + \bar{y}4 - \bar{y}1 - \bar{y}3)$	= 18.67
(B) ———— effect	= $½\,(\bar{y}3 + \bar{y}4 - \bar{y}1 - \bar{y}2)$	= -8.53
(AB) ———— effect	= $½\,(\bar{y}1 + \bar{y}4 - \bar{y}3 - \bar{y}2)$	= 1.90
Change in mean effect	= phase mean - $\bar{y}o$	= 1.56

Calculation of 2 S.E. Limits

For new average **s** :	$\pm (2/\sqrt{n})s$	= 1.45
For new effects :	$\pm (2/\sqrt{n})s$	= 1.45
For change in mean :	$\pm (1.79/\sqrt{n})s$	= 1.31

Table of Multiplying Factors

n	1	2	3	4	5	6	7	8	9	10
$f\delta,n$		0.30	0.35	0.37	0.38	0.39	0.40	0.40	0.40	0.41
1 / n	1.00	0.50	0.33	0.25	0.20	0.17	0.14	0.12	0.11	0.10
1 / (n - 1)		1.00	0.50	0.33	0.25	0.20	0.17	0.14	0.12	0.11
$2/\sqrt{n}$	2.00	1.41	1.15	1.00	0.89	0.82	0.76	0.71	0.67	0.63
$1.79/\sqrt{n}$	1.79	1.26	1.03	0.89	0.80	0.73	0.68	0.63	0.60	0.57

Two variable EVOP program calculation worksheet

Figure 5.63 Work sheet cycle No. 10.

Table 5.43 2^2 BEVOP information board.

INFORMATION BOARD AFTER PHASE NO. I

Last Cycle Completed : n = 10

	RESPONSE No. 1	RESPONSE No. 2	RESPONSE No. 3
REQUIREMENT :	MAXIMIZE RESPONSE		
PROCESS AVERAGES :	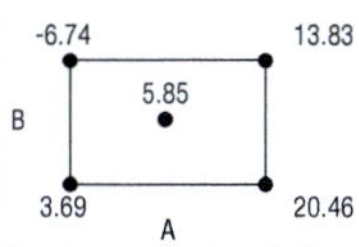		
2.S.E. LIMITS FOR AVERAGES	1.45		
EFFECTS WITH 2.S.E — PHASE MEAN	7.42		
FACTOR A	18.67 ± 1.45		
FACTOR B	-8.53 ± 1.45		
INTERACTION AB	1.90 ± 1.45		
CHANGE IN MEAN	1.56 ± 1.31		
ESTIMATED STANDARD DEVIATION S	2.30		
PRIOR ESTIMATE OF STANDARD DEV.	2.00		

NOTE:

1. The Center of Experiment-Reference Conditions (The Current Best-Known Process Conditions)
2. The Change in Mean provides a measure of whether the conditions being run the current phase yield an average result better or worse than the reference conditions.
3. The Phase Mean is the Mean Response Over All Conditions Being Run in the present Phase.
4*. The change in Mean is a measure of the over-all Curvature of the Response Surface. It will be positive near a MINIMUM and negative near a MAXIMUM.
5. Sequencing of the Runs-Randomization, if runs can be randomized without too much difficulty, than this should be done.

*A Diagonally oriented saddle represented by equal coefficients of opposite sign could occure and yield a zero change in mean effect, but such a phenomenon would rarely be met in practice.

Conclusions

- Factor A is highly significant with positive sign. It means higher factor's value will increase product yield. Factor B is also highly significant but with negative sign. Higher settings for factor B mean lower product yield. To maximize product yield, we need to increase factor A and reduce factor B.
- The *change in mean* is also significant. Maybe some slight curvature in viscosity response surface.

Recommendation for Phase II

- It is recommended to move the center of the experiment ($A_0 = 2.5$; $B_0 = 2.5$) to a new position with increase in factor A and lower factor B ($A_0 = 5.0$; $B_0 = 0.0$).
- To keep the same factor's variation 1.0.
- To keep the same noise level STDEV = 1.0 (Table 5.44, Figures 5.64–5.66, Table 5.45).

Table 5.44 2^2 BEVOP simulation.

PROJECT : S-3
PHASE : II
DATE :

BEVOP SIMULATION

S-3

I. CENTER OF EXPERIMENT : Ao = 5.00; Bo = 0.00 [Gen Random]

II. FACTOR'S INTERVAL OF VARIATION : Δ a = 1.00; Δ b = 1.00

III. PROCESS (PROPERTY-RESPONSE) NOISE : STDV = 1.00

EVOP WORK SHEET

TRIALS ORDER			FACTOR		RESPONSE									
			A	B	n = 1	n = 2	n = 3	n = 4	n = 5	n = 6	n = 7	n = 8	n = 9	n = 10
0		Block 1	5.00	0.00	50.40	47.91	50.11	49.19	50.63	49.68	52.89	49.80	52.57	48.23
1		Block 1	4.00	-1.00	31.00	31.00	31.00	31.00	31.00	31.00	31.00	31.00	31.00	31.00
2		Block 1	6.00	-1.00	71.00	71.00	71.00	71.00	71.00	71.00	71.00	71.00	71.00	71.00
3		Block 1	4.00	1.00	31.00	31.00	31.00	31.00	31.00	31.00	31.00	31.00	31.00	31.00
4		Block 1	6.00	1.00	71.00	71.00	71.00	71.00	71.00	71.00	71.00	71.00	71.00	71.00

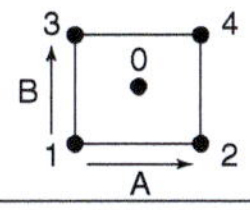

2^2 Factorial with added reference condition

Cycle n : *1*
Response:YIELD

Project S-3
Phase II
Date

Calculation of Averages

Operation condition	(0)	(1)	(2)	(3)	(4)
(i) Previous cycle sum					
(ii) Previous cycle average					
(iii) New observations	50.40	31.00	71.00	31.00	71.00
(iv) Differences (ii) less (iii)					
(v) New sums	50.40	31.00	71.00	31.00	71.00
(vi) New averages : $\bar{y}_i$	50.40	31.00	71.00	31.00	71.00

Calculation of Standard Deviation

Prior estimate of σ	= 1.00
Previous sum **s**	=
Previous average **s**	=
New **s** = range × $f\delta,n$	=
Range	=
New sum **s**	=
New average **s** = new sum **s** / (n - 1)	=

Calculation of Effects

Phase mean	$= 1/5\,(\bar{y}_0 + \bar{y}_1 + \bar{y}_2 + \bar{y}_3 + \bar{y}_4)$	= 50.88
(A) effect	$= ½\,(\bar{y}_2 + \bar{y}_4 - \bar{y}_1 - \bar{y}_3)$	= 40.00
(B) effect	$= ½\,(\bar{y}_3 + \bar{y}_4 - \bar{y}_1 - \bar{y}_2)$	= 0.00
(AB) effect	$= ½\,(\bar{y}_1 + \bar{y}_4 - \bar{y}_3 - \bar{y}_2)$	= 0.00
Change in mean effect	= phase mean - $\bar{y}_0$	= 0.48

Calculation of 2 S.E. Limits

For new average **s** :	$\pm(2/\sqrt{n})s$	= 2.00
For new effects :	$\pm(2/\sqrt{n})s$	= 2.00
For change in mean :	$\pm(1.79/\sqrt{n})s$	= 1.79

Table of Multiplying Factors

n	1	2	3	4	5	6	7	8	9	10
$f\delta,n$		0.30	0.35	0.37	0.38	0.39	0.40	0.40	0.40	0.41
1 / n	1.00	0.50	0.33	0.25	0.20	0.17	0.14	0.12	0.11	0.10
1 / (n - 1)		1.00	0.50	0.33	0.25	0.20	0.17	0.14	0.12	0.11
$2/\sqrt{n}$	2.00	1.41	1.15	1.00	0.89	0.82	0.76	0.71	0.67	0.63
$1.79/\sqrt{n}$	1.79	1.26	1.03	0.89	0.80	0.73	0.68	0.63	0.60	0.57

Two variable EVOP program calculation worksheet

Figure 5.64 Work sheet cycle No. 1.

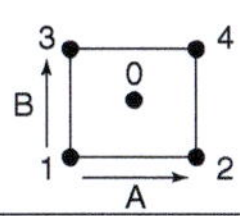

2^2 Factorial with added reference condition

Cycle n : *3*
Response:YIELD

Project S-3
Phase II
Date

Calculation of Averages

Operation condition	(0)	(1)	(2)	(3)	(4)
(i) Previous cycle sum	98.31	62.00	142.00	62.00	142.00
(ii) Previous cycle average	49.15	31.00	71.00	31.00	71.00
(iii) New observations	50.11	31.00	71.00	31.00	71.00
(iv) Differences (ii) less (iii)	-0.96	0.00	0.00	0.00	0.00
(v) New sums	99.27	62.00	142.00	62.00	142.00
(vi) New averages : $\bar{y}_i$	49.63	31.00	71.00	31.00	71.00

Calculation of Standard Deviation

Prior estimate of σ	= 1.00
Previous sum **s**	= 0.75
Previous average **s**	= 0.75
New **s** = range × $f\delta,n$	= 0.33
Range	= 0.96
New sum **s**	= 1.08
New average **s** = new sum **s** / (n - 1)	= 0.54

Calculation of Effects

Phase mean	$= 1/5\,(\bar{y}_0 + \bar{y}_1 + \bar{y}_2 + \bar{y}_3 + \bar{y}_4)$	= 50.73
(A) effect	$= ½\,(\bar{y}_2 + \bar{y}_4 - \bar{y}_1 - \bar{y}_3)$	= 40.00
(B) effect	$= ½\,(\bar{y}_3 + \bar{y}_4 - \bar{y}_1 - \bar{y}_2)$	= 0.00
(AB) effect	$= ½\,(\bar{y}_1 + \bar{y}_4 - \bar{y}_3 - \bar{y}_2)$	= 0.00
Change in mean effect	= phase mean - $\bar{y}_0$	= 1.09

Calculation of 2 S.E. Limits

For new average **s** :	$\pm(2/\sqrt{n})s$	= 0.62
For new effects :	$\pm(2/\sqrt{n})s$	= 0.62
For change in mean :	$\pm(1.79/\sqrt{n})s$	= 0.56

Table of Multiplying Factors

n	1	2	3	4	5	6	7	8	9	10
$f\delta,n$		0.30	0.35	0.37	0.38	0.39	0.40	0.40	0.40	0.41
1 / n	1.00	0.50	0.33	0.25	0.20	0.17	0.14	0.12	0.11	0.10
1 / (n - 1)		1.00	0.50	0.33	0.25	0.20	0.17	0.14	0.12	0.11
$2/\sqrt{n}$	2.00	1.41	1.15	1.00	0.89	0.82	0.76	0.71	0.67	0.63
$1.79/\sqrt{n}$	1.79	1.26	1.03	0.89	0.80	0.73	0.68	0.63	0.60	0.57

Two variable EVOP program calculation worksheet

Figure 5.65 Work sheet cycle No. 3.

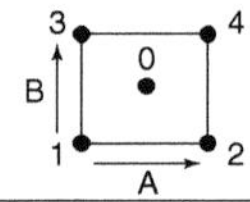

2^2 Factorial with added reference condition

Cycle n : 10

Response: YIELD

Project S-3
Phase II
Date

Calculation of Averages						Calculation of Standard Deviation	
Operation condition	(0)	(1)	(2)	(3)	(4)	Prior estimate of σ	= 1.00
(i) Previous cycle sum	103.15	62.00	142.00	62.00	142.00	Previous sum **s**	= 4.48
(ii) Previous cycle average	51.58	31.00	71.00	31.00	71.00	Previous average **s**	= 0.56
(iii) New observations	48.23	31.00	71.00	31.00	71.00	New **s** = range × $f\delta,n$	= 1.37
(iv) Differences (ii) less (iii)	3.34	0.00	0.00	0.00	0.00	Range	= 3.34
(v) New sums	99.81	62.00	142.00	62.00	142.00	New sum **s**	= 5.85
(vi) New averages : $\bar{y}_i$	49.90	31.00	71.00	31.00	71.00	New average **s** = $\frac{\text{new sum } s}{n-1}$	= 0.65

Calculation of Effects			Calculation of 2 S.E. Limits		
Phase mean	$= 1/5(\bar{y}_0 + \bar{y}_1 + \bar{y}_2 + \bar{y}_3 + \bar{y}_4)$	= 50.78			
(A) effect	$= \frac{1}{2}(\bar{y}_2 + \bar{y}_4 - \bar{y}_1 - \bar{y}_3)$	= 40.00	For new average **s** :	$\pm(2/\sqrt{n})s$	= 0.41
(B) effect	$= \frac{1}{2}(\bar{y}_3 + \bar{y}_4 - \bar{y}_1 - \bar{y}_2)$	= 0.00	For new effects :	$\pm(2/\sqrt{n})s$	= 0.41
(AB) effect	$= \frac{1}{2}(\bar{y}_1 + \bar{y}_4 - \bar{y}_3 - \bar{y}_2)$	= 0.00	For change in mean :	$\pm(1.79/\sqrt{n})s$	= 0.37
Change in mean effect	= phase mean $- \bar{y}_0$	= 0.88			

Table of Multiplying Factors

n	1	2	3	4	5	6	7	8	9	10
$f\delta,n$		0.30	0.35	0.37	0.38	0.39	0.40	0.40	0.40	0.41
1 / n	1.00	0.50	0.33	0.25	0.20	0.17	0.14	0.12	0.11	0.10
1 / (n - 1)		1.00	0.50	0.33	0.25	0.20	0.17	0.14	0.12	0.11
$2/\sqrt{n}$	2.00	1.41	1.15	1.00	0.89	0.82	0.76	0.71	0.67	0.63
$1.79/\sqrt{n}$	1.79	1.26	1.03	0.89	0.80	0.73	0.68	0.63	0.60	0.57

Two variable EVOP program calculation worksheet

Figure 5.66 Work sheet cycle No. 10.

Table 5.45 2^2 BEVOP information board.

INFORMATION BOARD AFTER PHASE NO. II

Last Cycle Completed : n = 10

	RESPONSE No. 1	RESPONSE No. 2	RESPONSE No. 3
REQUIREMENT :	MAXIMIZE RESPONSE		
PROCESS AVERAGES :	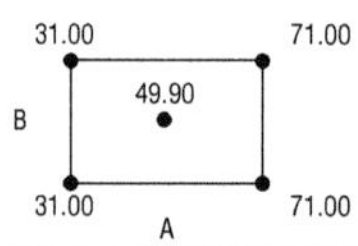		
2.S.E. LIMITS FOR AVERAGES	0.41		
EFFECTS WITH 2.S.E — PHASE MEAN	50.78		
FACTOR A	40.00 ± 0.41		
FACTOR B	0.00 ± 0.41		
INTERACTION AB	0.00 ± 0.41		
CHANGE IN MEAN	0.88 ± 0.37		
ESTIMATED STANDARD DEVIATION S	0.65		
PRIOR ESTIMATE OF STANDARD DEV.	1.00		

NOTE:

1. The Center of Experiment-Reference Conditions (The Current Best-Known Process Conditions)
2. The Change in Mean provides a measure of whether the conditions being run the current phase yield an average result better or worse than the reference conditions.
3. The Phase Mean is the Mean Response Over All Conditions Being Run in the present Phase.
4*. The change in Mean is a measure of the over-all Curvature of the Response Surface. It will be positive near a MINIMUM and negative near a MAXIMUM.
5. Sequencing of the Runs-Randomization, if runs can be randomized without too much difficulty, than this should be done.

*A Diagonally oriented saddle represented by equal coefficients of opposite sign could occure and yield a zero change in mean effect, but such a phenomenon would rarely be met in practice.

Solution

The *phase mean* yield was increased significantly from 7.0 to 50.0 at the end of phase III. This value was achieved at the last center point: $A_0 = 5.0$; $B_0 = 0.0$ (Figures 5.67 and 5.68).

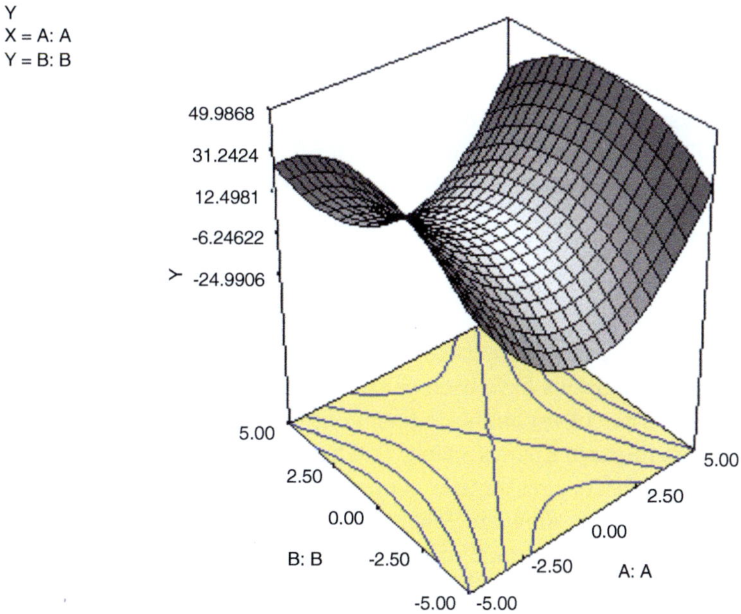

Figure 5.67 Response surface.

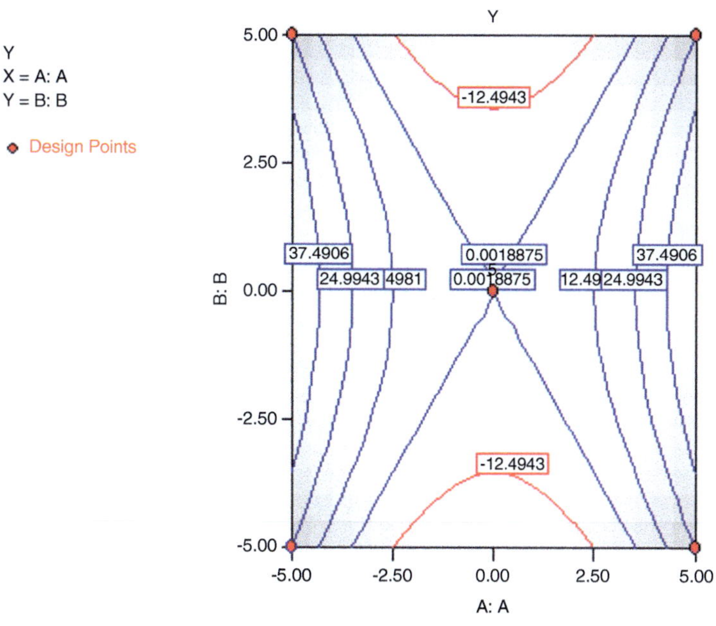

Figure 5.68 Contour diagram.

5.7.2 2^3 BEVOP Simulation

5.7.2.1 Simulation No. 4

In development of a certain batch process, three process factors were being studied. These were Temperature (A), Pressure (B), and Time (C) at which the reaction was conducted. The current best known process conditions are: $A_0 = 20$ [°C], $B_0 = 20$ [bar], $C_0 = 20$ [min], and corresponding product yield is around $y = 45$ [%]. Theoretically, yield range is: $0.0 < y < 100$ [%]. Start simulation with factor's intervals: $\Delta a = 5.0$; $\Delta b = 5.0$; $\Delta c = 5.0$. The current best process condition, factor's interval, and process noise measured by standard deviation can be set on different value to test EVOP efficiency. Experimental limits for temperature are $0.0 < A < 100$ [°C], pressure $0.0 < A < 100$ [bar], and time $0.0 < A < 100$ [min]. The appropriate mathematical function was used to calculate product response numbers plus noise (STDEV multiplied by Random Normal Deviates) to mask response and simulate real yield plant data. The objective of this study is to *maximize* product yield using EVOP.

Phase I: The settings (Center Point, Factor's Interval, and Process Noise) and yield for different process conditions have been calculated using Microsoft Excel simulation in Table 5.46. The corresponding calculations for ten cycles, phase I, have been shown in work sheets in Figures 5.69–5.71. The information board that would appear at the end of cycle 10, after transfer of the various calculated quantities from the appropriate worksheet, is shown in Table 5.47.

This simulation started with small factor's change (interval) and the noise (STDEV).

Table 5.46 2^3 BEVOP simulation.

PROJECT : S-4
PHASE : I
DATE :

BEVOP SIMULATION

I. CENTER OF EXPERIMENT : Ao = 20.00, Bo = 20.00, Co = 20.00 [Gen Random]

II. FACTOR'S INTERVAL OF VARIATION : Δa = 5.00, Δb = 5.00, Δc = 5.00

III. PROCESS NOISE : STDV = 3.0

EVOP WORK SHEET

TRIALS ORDER		Block	FACTORS			RESPONSE									
			A	B	C	n = 1	n = 2	n = 3	n = 4	n = 5	n = 6	n = 7	n = 8	n = 9	n = 10
0		Block I	20.00	20.00	20.00	47.20	39.73	46.33	43.56	47.88	45.03	54.67	45.40	53.70	40.70
1		Block I	15.00	15.00	15.00	26.82	26.87	25.31	28.29	22.85	25.59	28.81	32.02	28.74	24.04
2		Block I	25.00	25.00	15.00	48.37	53.22	47.42	49.82	54.80	53.06	50.88	52.22	53.09	53.42
3		Block I	25.00	15.00	25.00	48.53	50.28	51.23	53.11	48.20	54.28	49.98	52.35	54.15	49.94
4		Block I	15.00	25.00	25.00	52.37	47.43	51.78	49.57	48.86	50.35	49.32	48.94	52.63	55.98
Ø		Block II	20.00	20.00	20.00	50.55	44.06	46.73	43.38	45.90	42.87	47.05	50.01	50.42	47.56
5		Block II	15.00	15.00	25.00	37.70	40.77	45.90	40.68	35.50	37.57	38.00	42.38	40.64	37.59
6		Block II	25.00	25.00	25.00	60.95	56.31	58.36	61.39	60.92	60.24	63.14	61.63	67.69	67.85
7		Block II	25.00	15.00	15.00	41.73	40.57	37.48	37.96	39.87	40.66	34.71	38.26	35.56	37.94
8		Block II	15.00	25.00	15.00	37.92	41.89	39.32	37.70	41.43	36.27	32.88	40.89	36.23	40.24

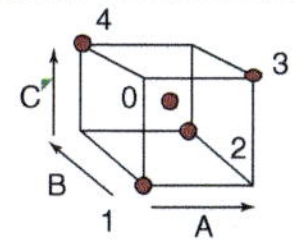

2³ Factorial in two blocks with reference run in each block

BLOCK I

CYCLE n: 1
RESPONSE :

Project S-4
Phase I
Date

Calculation of Averages						Calculation of Standard Deviation - Block I	
Operation condition	(0)	(1)	(2)	(3)	(4)		
(i) Previous sum for block I						Previous sum **s** (all blocks)	=
(ii) Previous average for block I							
(iii) New observations for block I	47.20	26.82	48.37	48.53	52.37	New **s** = range × ***fδ,n***	=
(iv) Differences (ii) less (iii)						Range	=
(v) New sums for block I	47.20	26.82	48.37	48.53	52.37	New sum **s** (all blocks)	=
(vi) New averages for block I	47.20	26.82	48.37	48.53	52.37		

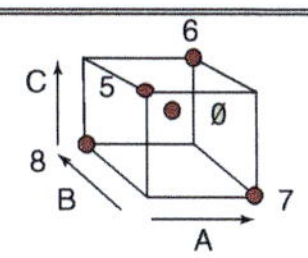

2³ Factorial in two blocks with reference run in each block

BLOCK II

CYCLE n: 1
RESPONSE :

Calculation of Averages						Calculation of Standard Deviation - Block II	
Operation condition	(Ø)	(5)	(6)	(7)	(8)	Prior estimate of σ	= 3.0
(i) Previous sum for block II						Previous sum **s** (all blocks)	=
(ii) Previous average for block II							
(iii) New observations for block II	50.55	37.70	60.95	41.73	37.92	New **s** = range × ***fδ,n***	=
(iv) Differences (ii) less (iii)						Range	=
(v) New sums for block II	50.55	37.70	60.95	41.73	37.92	New sum **s** (all blocks)	=
(vi) New averages for block II	50.55	37.70	60.95	41.73	37.92	Average **s** = {1/[2(n - 1)]} × new sum s	=

Calculation of Effects Using Yates' Algorithm

	$\bar{y}$	(i)	(ii)	(iii)	Multiplier	Effect	
(1)	26.82	68.55	154.83	354.38	0.125	**44.30**	Factorial mean
(7)	41.73	86.29	199.55	44.77	0.25	**11.19**	A
(8)	37.92	86.23	25.36	44.82	0.25	**11.20**	B
(2)	48.37	113.31	19.41	-6.72	0.25	**-1.68**	AB
(5)	37.70	14.91	17.74	44.71	0.25	**11.18**	C
(3)	48.53	10.45	27.08	-5.95	0.25	**-1.49**	AC
(4)	52.37	10.83	-4.46	9.34	0.25	**2.33**	BC
(6)	60.95	8.58	-2.26	2.21	0.25	**0.55**	ABC + blocks

Calculation of 2 S.E. Limits

For new average **s** :	$\pm (2/\sqrt{n})s$	= **6.00**
For new effects :	$\pm (1.41/\sqrt{n})s$	= **4.23**
For change in mean :	$\pm (1.26/\sqrt{n})s$	= **3.78**

		$\bar{y}$	Sum	Multiplier			
I	Reference (0) Conditions (Ø)	47.50 50.55	97.75	0.500	Reference mean	=	**48.87**
II	Factorial Conditions	(1 : 8) $\sum \bar{Y}$	354.38	0.125	Factorial mean	=	**44.30**
III	All conditions	(1 : 10) $\sum \bar{Y}$	452.13	0.100	Phase mean	=	**45.21**
IV	Change in mean	III - I			Change in mean	=	**-3.66**

Table of Multiplying Factors

n	1	2	3	4	5	6	7	8	9	10
fδ,n		0.30	0.35	0.37	0.38	0.39	0.40	0.40	0.40	0.41
1 / n	1.00	0.50	0.33	0.25	0.20	0.17	0.14	0.12	0.11	0.10
1 / [2(n - 1)]		0.50	0.25	0.17	0.12	0.10	0.08	0.07	0.06	0.06
$2/\sqrt{n}$	2.00	1.41	1.15	1.00	0.89	0.82	0.76	0.71	0.67	0.63
$1.41/\sqrt{n}$	1.41	1.00	0.82	0.71	0.63	0.58	1.86	1.99	2.10	2.24
$1.26/\sqrt{n}$	1.26	0.89	0.73	0.63	0.57	0.52	0.68	0.63	0.60	0.56

Three variable EVOP program calculation worksheet.

Figure 5.69 Work sheet cycle No. 1.

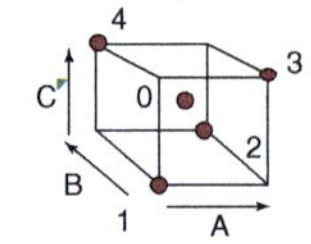

2^3 Factorial in two blocks with reference run in each block

BLOCK I

CYCLE n: 3
RESPONSE :

Project S-4
Phase I
Date

Calculation of Averages						Calculation of Standard Deviation - Block I	
Operation condition	(0)	(1)	(2)	(3)	(4)		
(i) Previous sum for block I	86.93	53.69	101.59	98.81	99.80	Previous sum **s** (all blocks)	= 6.84
(ii) Previous average for block I	43.46	26.84	50.79	49.41	49.90		
(iii) New observations for block I	46.33	25.31	47.42	51.23	51.78	New **s** = range × $f\delta,n$	= 2.19
(iv) Differences (ii) less (iii)	-2.87	1.53	3.38	-1.82	-1.88	Range	= 6.25
(v) New sums for block I	133.26	79.00	149.01	150.04	151.58	New sum **s** (all blocks)	= 9.02
(vi) New averages for block I	44.42	26.33	49.67	50.01	50.53		

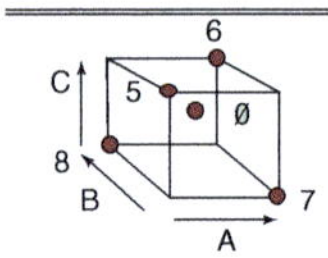

2^3 Factorial in two blocks with reference run in each block

BLOCK II

CYCLE n: 3
RESPONSE :

Calculation of Averages						Calculation of Standard Deviation - Block II	
Operation condition	(Ø)	(5)	(6)	(7)	(8)	Prior estimate of σ	= 3.0
(i) Previous sum for block II	94.61	78.47	117.26	82.30	79.81	Previous sum **s** (all blocks)	= 9.02
(ii) Previous average for block II	47.30	39.23	58.63	41.15	39.90		
(iii) New observations for block II	46.73	45.90	58.36	37.48	39.32	New **s** = range × $f\delta,n$	= 3.62
(iv) Differences (ii) less (iii)	0.57	-6.66	0.26	3.67	0.59	Range	= 10.34
(v) New sums for block II	141.34	124.37	175.62	119.77	119.13	New sum **s** (all blocks)	= 12.64
(vi) New averages for block II	47.11	41.46	58.54	39.92	39.71	Average **s** = {1/[2(n - 1)]} × new sum s	= 3.16

Calculation of Effects Using Yates' Algorithm								Calculation of 2 S.E. Limits	
	$\bar{y}$	(i)	(ii)	(iii)	Multiplier	Effect			
(1)	26.33	66.26	155.64	356.17	0.125	**44.52**	Factorial mean		
(7)	39.92	89.38	200.54	40.12	0.25	**10.03**	A	For new average **s** : $\pm(2/\sqrt{n})$s	= **3.65**
(8)	39.71	91.47	23.55	40.72	0.25	**10.18**	B		
(2)	49.67	109.07	16.57	-4.18	0.25	**-1.04**	AB	For new effects : $\pm(1.41/\sqrt{n})$s	= **2.57**
(5)	41.46	13.59	23.12	44.90	0.25	**11.23**	C		
(3)	50.01	9.96	17.60	-6.98	0.25	**-1.74**	AC	For change in mean : $\pm(1.26/\sqrt{n})$s	= **2.30**
(4)	50.53	8.56	-3.63	-5.52	0.25	**-1.38**	BC		
(6)	58.54	8.01	-0.55	3.09	0.25	**0.77**	ABC + blocks		

		$\bar{y}$	Sum	Multiplier			
I	Reference (0) Conditions (Ø)	44.42 47.11	91.53	0.500	Reference mean	=	**45.77**
II	Factorial Conditions	(1 : 8) $\sum\bar{Y}$	356.17	0.125	Factorial mean	=	**44.52**
III	All conditions	(1 : 10) $\sum\bar{Y}$	447.71	0.100	Phase mean	=	**44.77**
IV	Change in mean	III - I			Change in mean	=	**-1.00**

Table of Multiplying Factors

n	1	2	3	4	5	6	7	8	9	10
$f\delta,n$		0.30	0.35	0.37	0.38	0.39	0.40	0.40	0.40	0.41
1 / n	1.00	0.50	0.33	0.25	0.20	0.17	0.14	0.12	0.11	0.10
1 / [2(n - 1)]		0.50	0.25	0.17	0.12	0.10	0.08	0.07	0.06	0.06
$2/\sqrt{n}$	2.00	1.41	1.15	1.00	0.89	0.82	0.76	0.71	0.67	0.63
$1.41/\sqrt{n}$	1.41	1.00	0.82	0.71	0.63	0.58	1.86	1.99	2.10	2.24
$1.26/\sqrt{n}$	1.26	0.89	0.73	0.63	0.57	0.52	0.68	0.63	0.60	0.56

Three variable EVOP program calculation worksheet.

Figure 5.70 Work sheet cycle No. 3.

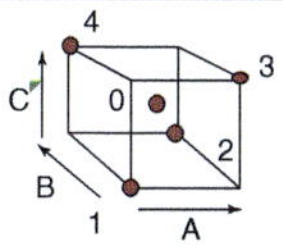

2^3 Factorial in two blocks with reference run in each block

BLOCK I
CYCLE n: 10
RESPONSE :

Project S-4
Phase I
Date

Calculation of Averages						Calculation of Standard Deviation - Block I	
Operation condition	(0)	(1)	(2)	(3)	(4)		
(i) Previous sum for block I	423.50	245.30	462.87	462.12	451.25	Previous sum **s** (all blocks)	= 47.26
(ii) Previous average for block I	47.06	27.26	51.43	51.35	50.14		
(iii) New observations for block I	40.70	24.04	53.42	49.94	55.98	New **s** = range × $f\delta,n$	= 5.00
(iv) Differences (ii) less (iii)	6.36	3.22	-1.99	1.41	-5.84	Range	= 12.20
(v) New sums for block I	464.20	269.34	516.29	512.05	507.23	New sum **s** (all blocks)	= 52.27
(vi) New averages for block I	46.42	26.93	51.63	51.21	50.72		

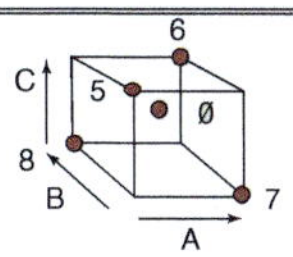

2^3 Factorial in two blocks with reference run in each block

BLOCK II
CYCLE n: 10
RESPONSE :

Calculation of Averages						Calculation of Standard Deviation - Block II	
Operation condition	(Ø)	(5)	(6)	(7)	(8)	Prior estimate of **σ**	= 3.0
(i) Previous sum for block II	420.98	359.13	550.63	346.78	344.51	Previous sum **s** (all blocks)	= 52.27
(ii) Previous average for block II	46.78	39.90	61.18	38.53	38.28		
(iii) New observations for block II	47.56	37.59	67.85	37.94	40.24	New **s** = range × $f\delta,n$	= 3.68
(iv) Differences (ii) less (iii)	-0.78	2.32	-6.67	0.59	-1.96	Range	= 8.99
(v) New sums for block II	468.53	396.71	618.48	384.73	384.76	New sum **s** (all blocks)	= 55.95
(vi) New averages for block II	46.85	39.67	61.85	38.47	38.48	Average **s** = {1/[2(n - 1)]} × new sum s	= 3.11

Calculation of Effects Using Yates' Algorithm

	$\bar{y}$	(i)	(ii)	(iii)	Multiplier	Effect	
(1)	26.93	65.41	155.51	358.96	0.125	**44.87**	Factorial mean
(7)	38.47	90.10	203.45	47.35	0.25	**11.84**	A
(8)	38.48	90.88	24.69	46.39	0.25	**11.60**	B
(2)	51.63	112.57	22.66	1.21	0.25	**0.30**	AB
(5)	39.67	11.54	24.70	47.94	0.25	**11.98**	C
(3)	51.21	13.15	21.69	-2.03	0.25	**-0.51**	AC
(4)	50.72	11.53	1.61	-3.00	0.25	**-0.75**	BC
(6)	61.85	11.13	-0.41	-2.02	0.25	**-0.51**	ABC + blocks

Calculation of 2 S.E. Limits

For new average **s** :	$\pm (2/\sqrt{n})s$	= **1.97**
For new effects :	$\pm (1.41/\sqrt{n})s$	= **1.39**
For change in mean :	$\pm (1.26/\sqrt{n})s$	= **1.24**

		$\bar{y}$	**Sum**	**Multiplier**			
I	Reference (0) Conditions (Ø)	46.42 46.85	93.27	0.500	Reference mean	=	**46.64**
II	Factorial Conditions	(1 : 8) $\sum\bar{Y}$	358.96	0.125	Factorial mean	=	**44.87**
III	All conditions	(1 : 10) $\sum\bar{Y}$	452.23	0.100	Phase mean	=	**45.22**
IV	Change in mean	III - I			Change in mean	=	**-1.41**

Table of Multiplying Factors

n	1	2	3	4	5	6	7	8	9	10
$f\delta,n$		0.30	0.35	0.37	0.38	0.39	0.40	0.40	0.40	0.41
1 / n	1.00	0.50	0.33	0.25	0.20	0.17	0.14	0.12	0.11	0.10
1 / [2(n - 1)]		0.50	0.25	0.17	0.12	0.10	0.08	0.07	0.06	0.06
$2/\sqrt{n}$	2.00	1.41	1.15	1.00	0.89	0.82	0.76	0.71	0.67	0.63
$1.41/\sqrt{n}$	1.41	1.00	0.82	0.71	0.63	0.58	1.86	1.99	2.10	2.24
$1.26/\sqrt{n}$	1.26	0.89	0.73	0.63	0.57	0.52	0.68	0.63	0.60	0.56

Three variable EVOP program calculation worksheet.

Figure 5.71 Work sheet cycle No. 10.

Table 5.47 2^3 BEVOP information board.

2^3 INFORMATION BOARD AFTER PHASE NO. I Last Cycle Completed : n = 10			
	RESPONSE No. 1	RESPONSE No. 2	RESPONSE No. 3
REQUIREMENT :	MAXIMIZE RESPONSE		
PROCESS AVERAGES :	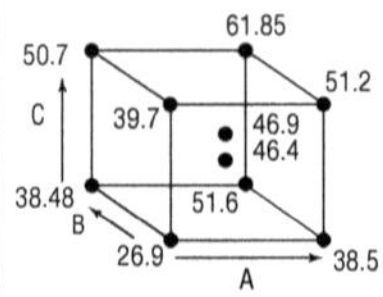		
2.S.E. LIMITS FOR AVERAGES	1.97		
EFFECTS WITH 2.S.E PHASE MEAN FACTOR A FACTOR B FACTOR C INTERACTION AB INTERACTION AC INTERACTION BC INTERACTION ABC CHANGE IN MEAN	 45.22 11.84 ± 1.39 11.60 ± 1.39 11.98 ± 1.39 0.30 ± 1.39 -0.51 ± 1.39 -0.75 ± 1.39 -0.51 ± 1.39 -1.41 ± 1.24		
ESTIMATED STANDARD DEVIATION S	3.11		
PRIOR ESTIMATE OF STANDARD DEV.	3.00		

NOTE:

1. The Center of Experiment-Reference Conditions (The Current Best-Known Process Conditions)
2. The Change in Mean provides a measure of whether the conditions being run the current phase yield an average result better or worse than the reference conditions.
3. The Phase Mean is the Mean Response Over All Conditions Being Run in the present Phase.
4*. The change in Mean is a measure of the over-all Curvature of the Response Surface. It will be positive near a MINIMUM and negative near a MAXIMUM.
5. Sequencing of the Runs-Randomization, if runs can be randomized without too much difficulty, than this should be done.

*A Diagonally oriented saddle represented by equal coefficients of opposite sign could occure and yield a zero change in mean effect, but such a phenomenon would rarely be met in practice.

Conclusions

- All three factors are highly significant with positive sign. It means higher factor's value will increase yield.
- The *change in mean* is also significant with negative sign. Maybe some slight curvature in yield response surface.;
- The estimated standard deviation is very close to the prior estimate of the standard deviation.

Recommendation for Phase II

- It is recommended to move the center of the experiment ($A_0 = 20$; $B_0 = 20$; $C_0 = 20$) with increase in both factors to the new position ($A_0 = 30$; $B_0 = 30$; $C_0 = 30$).
- To keep the same factor's variation 5.0.
- To keep the same noise level STDEV = 3.0 (Table 5.48, Figures 5.72–5.74, Table 5.49).

Table 5.48 2^3 BEVOP simulation.

PROJECT : S-4
PHASE : II
DATE :

BEVOP SIMULATION

I. CENTER OF EXPERIMENT : Ao = 30.00 Bo = 30.00 Co = 30.00 [Gen Random]

II. FACTOR'S INTERVAL OF VARIATION : Δa = 5.00 Δb = 5.00 Δc = 5.00

III. PROCESS NOISE : STDV = 3.0

EVOP WORK SHEET

TRIALS ORDER		Block	FACTORS			RESPONSE									
			A	B	C	n = 1	n = 2	n = 3	n = 4	n = 5	n = 6	n = 7	n = 8	n = 9	n = 10
0		Block I	30.00	30.00	30.00	77.20	69.73	76.33	73.56	77.88	75.03	84.67	75.40	83.70	70.70
1		Block I	25.00	25.00	25.00	62.82	62.87	61.31	64.29	58.85	61.59	64.81	68.02	64.74	60.04
2		Block I	35.00	35.00	25.00	76.37	81.22	75.42	77.82	82.80	81.06	78.88	80.22	81.09	81.42
3		Block I	35.00	25.00	35.00	76.53	78.28	79.23	81.11	76.20	82.28	77.98	80.35	82.15	77.94
4		Block I	25.00	35.00	35.00	80.37	75.43	79.78	77.57	76.86	78.35	77.32	76.94	80.63	83.98
Ø		Block II	30.00	30.00	30.00	80.55	74.06	76.73	73.38	75.90	72.87	77.05	80.01	80.42	77.56
5		Block II	25.00	25.00	35.00	69.70	72.77	77.90	72.68	67.50	69.57	70.00	74.38	72.64	69.59
6		Block II	35.00	35.00	35.00	84.95	80.31	82.36	85.39	84.92	84.24	87.14	85.63	91.69	91.85
7		Block II	35.00	25.00	25.00	73.73	72.57	69.48	69.96	71.87	72.66	66.71	70.26	67.56	69.94
8		Block II	25.00	35.00	25.00	69.92	73.89	71.32	69.70	73.43	68.27	64.88	72.89	68.23	72.24

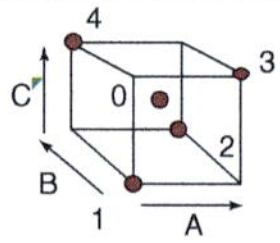

2^3 Factorial in two blocks with reference run in each block

BLOCK I

CYCLE n: 1
RESPONSE : Yield

Project S-4
Phase II
Date

Calculation of Averages					
Operation condition	(0)	(1)	(2)	(3)	(4)
(i) Previous sum for block I					
(ii) Previous average for block I					
(iii) New observations for block I	77.20	62.82	76.37	76.53	80.37
(iv) Differences (ii) less (iii)					
(v) New sums for block I	77.20	62.82	76.37	76.53	80.37
(vi) New averages for block I	77.20	62.82	76.37	76.53	80.37

Calculation of Standard Deviation - Block I	
Previous sum **s** (all blocks)	=
New **s** = range × $f\delta,n$	=
Range	=
New sum **s** (all blocks)	=

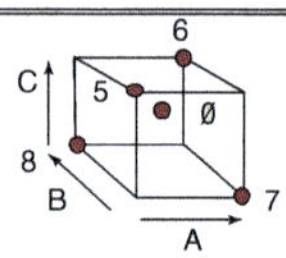

2^3 Factorial in two blocks with reference run in each block

BLOCK II

CYCLE n: 1
RESPONSE : Yield

Calculation of Averages					
Operation condition	(Ø)	(5)	(6)	(7)	(8)
(i) Previous sum for block II					
(ii) Previous average for block II					
(iii) New observations for block II	80.55	69.70	84.95	73.73	69.92
(iv) Differences (ii) less (iii)					
(v) New sums for block II	80.55	69.70	84.95	73.73	69.92
(vi) New averages for block II	80.55	69.70	84.95	73.73	69.92

Calculation of Standard Deviation - Block II	
Prior estimate of σ	= 3.0
Previous sum **s** (all blocks)	=
New **s** = range × $f\delta,n$	=
Range	=
New sum **s** (all blocks)	=
Average **s** = {1/[2(n - 1)]} × new sum s	=

Calculation of Effects Using Yates' Algorithm

	$\bar{y}$	(i)	(ii)	(iii)	Multiplier	Effect	
(1)	62.82	136.55	282.83	594.38	0.125	**74.30**	Factorial mean
(7)	73.73	146.29	311.55	28.77	0.25	**7.19**	A
(8)	69.92	146.23	17.36	28.82	0.25	**7.20**	B
(2)	76.37	165.31	11.41	-6.72	0.25	**-1.68**	AB
(5)	69.70	10.91	9.74	28.71	0.25	**7.18**	C
(3)	76.53	6.45	19.08	-5.95	0.25	**-1.49**	AC
(4)	80.37	6.83	-4.46	9.34	0.25	**2.33**	BC
(6)	84.95	4.58	-2.26	2.21	0.25	**0.55**	ABC + blocks

Calculation of 2 S.E. Limits

For new average **s** :	$\pm(2/\sqrt{n})s$	= **6.00**
For new effects :	$\pm(1.41/\sqrt{n})s$	= **4.23**
For change in mean :	$\pm(1.26/\sqrt{n})s$	= **3.78**

		$\bar{y}$	Sum	Multiplier
I	Reference (0)	77.20	157.75	0.500
	Conditions (Ø)	80.55		
II	Factorial Conditions	(1 : 8) $\sum\bar{Y}$	594.38	0.125
III	All conditions	(1 : 10) $\sum\bar{Y}$	752.13	0.100
IV	Change in mean	III - I		

Reference mean	=	**78.87**
Factorial mean	=	**74.30**
Phase mean	=	**75.21**
Change in mean	=	**-3.66**

Table of Multiplying Factors

n	1	2	3	4	5	6	7	8	9	10
$f\delta,n$		0.30	0.35	0.37	0.38	0.39	0.40	0.40	0.40	0.41
1 / n	1.00	0.50	0.33	0.25	0.20	0.17	0.14	0.12	0.11	0.10
1 / [2(n - 1)]		0.50	0.25	0.17	0.12	0.10	0.08	0.07	0.06	0.06
$2/\sqrt{n}$	2.00	1.41	1.15	1.00	0.89	0.82	0.76	0.71	0.67	0.63
$1.41/\sqrt{n}$	1.41	1.00	0.82	0.71	0.63	0.58	1.86	1.99	2.10	2.24
$1.26/\sqrt{n}$	1.26	0.89	0.73	0.63	0.57	0.52	0.68	0.63	0.60	0.56

Three variable EVOP program calculation worksheet.

Figure 5.72 Work sheet cycle No. 1.

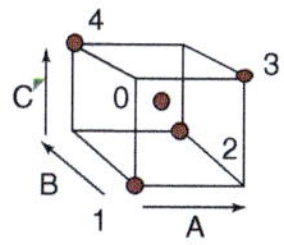

2^3 Factorial in two blocks with reference run in each block

BLOCK I

CYCLE n: 3
RESPONSE : Yield

Project S-4
Phase II
Date

Calculation of Averages						Calculation of Standard Deviation - Block I	
Operation condition	(0)	(1)	(2)	(3)	(4)		
(i) Previous sum for block I	146.93	125.69	157.59	154.81	155.80	Previous sum **s** (all blocks)	= 6.84
(ii) Previous average for block I	73.46	62.84	78.79	77.41	77.90		
(iii) New observations for block I	76.33	61.31	75.42	79.23	79.78	New **s** = range × $f\delta,n$	= 2.19
(iv) Differences (ii) less (iii)	-2.87	1.53	3.38	-1.82	-1.88	Range	= 6.25
(v) New sums for block I	223.26	187.00	233.01	234.04	235.58	New sum **s** (all blocks)	= 9.02
(vi) New averages for block I	74.42	62.33	77.67	78.01	78.53		

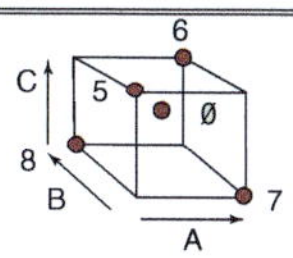

2^3 Factorial in two blocks with reference run in each block

BLOCK II

CYCLE n: 3
RESPONSE : Yield

Calculation of Averages						Calculation of Standard Deviation - Block II	
Operation condition	(Ø)	(5)	(6)	(7)	(8)	Prior estimate of σ	= 3.0
(i) Previous sum for block II	154.61	142.47	165.26	146.30	143.81	Previous sum **s** (all blocks)	= 9.02
(ii) Previous average for block II	77.30	71.23	82.63	73.15	71.90		
(iii) New observations for block II	76.73	77.90	82.36	69.48	71.32	New **s** = range × $f\delta,n$	= 3.62
(iv) Differences (ii) less (iii)	0.57	-6.66	0.26	3.67	0.59	Range	= 10.34
(v) New sums for block II	231.34	220.37	247.62	215.77	215.13	New sum **s** (all blocks)	= 12.64
(vi) New averages for block II	77.11	73.46	82.54	71.92	71.71	Average **s** = {1/[2(n - 1)]} × new sum s	= 3.16

Calculation of Effects Using Yates' Algorithm

	$\bar{y}$	(i)	(ii)	(iii)	Multiplier	Effect	
(1)	62.33	134.26	283.64	596.17	0.125	**74.52**	Factorial mean
(7)	71.92	149.38	312.54	24.12	0.25	**6.03**	A
(8)	71.71	151.47	15.55	24.72	0.25	**6.18**	B
(2)	77.67	161.07	8.57	-4.18	0.25	**-1.04**	AB
(5)	73.46	9.59	15.12	28.90	0.25	**7.23**	C
(3)	78.01	5.96	9.60	-6.98	0.25	**-1.74**	AC
(4)	78.53	4.56	-3.63	-5.52	0.25	**-1.38**	BC
(6)	82.54	4.01	-0.55	3.09	0.25	**0.77**	ABC + blocks

Calculation of 2 S.E. Limits

For new average **s** :	± $(2/\sqrt{n})$s	= **3.65**
For new effects :	± $(1.41/\sqrt{n})$s	= **2.57**
For change in mean :	± $(1.26/\sqrt{n})$s	= **2.30**

		$\bar{y}$	**Sum**	**Multiplier**			
I	Reference (0) Conditions (Ø)	74.42 77.11	151.53	0.500	Reference mean	=	**75.77**
II	Factorial Conditions	(1 : 8) $\sum\dot{Y}$	596.17	0.125	Factorial mean	=	**74.52**
III	All conditions	(1 : 10) $\sum\dot{Y}$	747.71	0.100	Phase mean	=	**74.77**
IV	Change in mean	III - I			Change in mean	=	**-1.00**

Table of Multiplying Factors

n	1	2	3	4	5	6	7	8	9	10
$f\delta,n$		0.30	0.35	0.37	0.38	0.39	0.40	0.40	0.40	0.41
1 / n	1.00	0.50	0.33	0.25	0.20	0.17	0.14	0.12	0.11	0.10
1 / [2(n - 1)]		0.50	0.25	0.17	0.12	0.10	0.08	0.07	0.06	0.06
$2/\sqrt{n}$	2.00	1.41	1.15	1.00	0.89	0.82	0.76	0.71	0.67	0.63
$1.41/\sqrt{n}$	1.41	1.00	0.82	0.71	0.63	0.58	1.86	1.99	2.10	2.24
$1.26/\sqrt{n}$	1.26	0.89	0.73	0.63	0.57	0.52	0.68	0.63	0.60	0.56

Three variable EVOP program calculation worksheet.

Figure 5.73 Work sheet cycle No. 3.

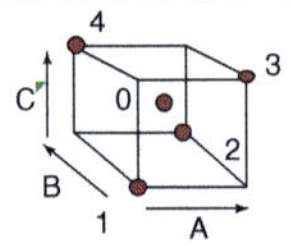

2^3 Factorial in two blocks with reference run in each block

BLOCK I

CYCLE n: 10
RESPONSE : Yield

Project S-4
Phase II
Date

Calculation of Averages						Calculation of Standard Deviation - Block I	
Operation condition	(0)	(1)	(2)	(3)	(4)		
(i) Previous sum for block I	693.50	569.30	714.87	714.12	703.25	Previous sum **s** (all blocks)	= 47.26
(ii) Previous average for block I	77.06	63.26	79.43	79.35	78.14		
(iii) New observations for block I	70.70	60.04	81.42	77.94	83.98	New **s** = range × ***fδ,n***	= 5.00
(iv) Differences (ii) less (iii)	6.36	3.22	-1.99	1.41	-5.84	Range	= 12.20
(v) New sums for block I	764.20	629.34	796.29	792.05	787.23	New sum **s** (all blocks)	= 52.27
(vi) New averages for block I	76.42	62.93	79.63	79.21	78.72		

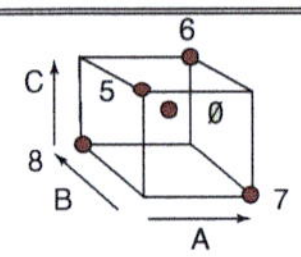

2^3 Factorial in two blocks with reference run in each block

BLOCK II

CYCLE n: 10
RESPONSE : Yield

Calculation of Averages						Calculation of Standard Deviation - Block II	
Operation condition	(Ø)	(5)	(6)	(7)	(8)	Prior estimate of **σ**	= 3.0
(i) Previous sum for block II	690.98	647.13	766.63	634.78	632.51	Previous sum **s** (all blocks)	= 52.27
(ii) Previous average for block II	76.78	71.90	85.18	70.53	70.28		
(iii) New observations for block II	77.56	69.59	91.85	69.94	72.24	New **s** = range × ***fδ,n***	= 3.68
(iv) Differences (ii) less (iii)	-0.78	2.32	-6.67	0.59	-1.96	Range	= 8.99
(v) New sums for block II	768.53	716.71	858.48	704.73	704.76	New sum **s** (all blocks)	= 55.95
(vi) New averages for block II	76.85	71.67	85.85	70.47	70.48	Average **s** = {1/[2(n - 1)]} × new sum s	= 3.11

Calculation of Effects Using Yates' Algorithm

	$\bar{y}$	(i)	(ii)	(iii)	Multiplier	Effect	
(1)	62.93	133.41	283.51	598.96	0.125	**74.87**	Factorial mean
(7)	70.47	150.10	315.45	31.35	0.25	**7.84**	A
(8)	70.48	150.88	16.69	30.39	0.25	**7.60**	B
(2)	79.63	164.57	14.66	1.21	0.25	**0.30**	AB
(5)	71.67	7.54	16.70	31.94	0.25	**7.98**	C
(3)	79.21	9.15	13.69	-2.03	0.25	**-0.51**	AC
(4)	78.72	7.53	1.61	-3.00	0.25	**-0.75**	BC
(6)	85.85	7.13	-0.41	-2.02	0.25	**-0.51**	ABC + blocks

Calculation of 2 S.E. Limits

For new average **s** :	± $(2/\sqrt{n})$s	= **1.97**
For new effects :	± $(1.41/\sqrt{n})$s	= **1.39**
For change in mean :	± $(1.26/\sqrt{n})$s	= **1.24**

		$\bar{y}$	Sum	Multiplier			
I	Reference (0) Conditions (Ø)	76.42 76.85	153.27	0.500	Reference mean	=	**76.64**
II	Factorial Conditions	(1 : 8) $\sum \dot{Y}$	598.96	0.125	Factorial mean	=	**74.87**
III	All conditions	(1 : 10) $\sum \dot{Y}$	752.23	0.100	Phase mean	=	**75.22**
IV	Change in mean	III - I			Change in mean	=	**-1.41**

Table of Multiplying Factors

n	1	2	3	4	5	6	7	8	9	10
fδ,n		0.30	0.35	0.37	0.38	0.39	0.40	0.40	0.40	0.41
1 / n	1.00	0.50	0.33	0.25	0.20	0.17	0.14	0.12	0.11	0.10
1 / [2(n - 1)]		0.50	0.25	0.17	0.12	0.10	0.08	0.07	0.06	0.06
$2/\sqrt{n}$	2.00	1.41	1.15	1.00	0.89	0.82	0.76	0.71	0.67	0.63
$1.41/\sqrt{n}$	1.41	1.00	0.82	0.71	0.63	0.58	1.86	1.99	2.10	2.24
$1.26/\sqrt{n}$	1.26	0.89	0.73	0.63	0.57	0.52	0.68	0.63	0.60	0.56

Three variable EVOP program calculation worksheet.

Figure 5.74 Work sheet cycle No. 10.

Table 5.49 2^3 BEVOP information board.

2^3 INFORMATION BOARD AFTER PHASE NO. II Last Cycle Completed : n = 10			
	RESPONSE No. 1	RESPONSE No. 2	RESPONSE No. 3
REQUIREMENT :	MAXIMIZE RESPONSE		
PROCESS AVERAGES :	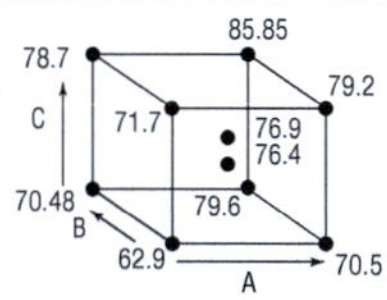		
2.S.E. LIMITS FOR AVERAGES	1.97		
EFFECTS WITH 2.S.E — PHASE MEAN FACTOR A FACTOR B FACTOR C INTERACTION AB INTERACTION AC INTERACTION BC INTERACTION ABC CHANGE IN MEAN	75.22 7.84 ± 1.39 7.60 ± 1.39 7.98 ± 1.39 0.30 ± 1.39 -0.51 ± 1.39 -0.75 ± 1.39 -0.51 ± 1.39 -1.41 ± 1.24		
ESTIMATED STANDARD DEVIATION S	3.11		
PRIOR ESTIMATE OF STANDARD DEV.	3.00		

NOTE:

1. The Center of Experiment-Reference Conditions (The Current Best-Known Process Conditions)
2. The Change in Mean provides a measure of whether the conditions being run the current phase yield an average result better or worse than the reference conditions.
3. The Phase Mean is the Mean Response Over All Conditions Being Run in the present Phase.
4*. The change in Mean is a measure of the over-all Curvature of the Response Surface. It will be positive near a MINIMUM and negative near a MAXIMUM.
5. Sequencing of the Runs-Randomization, if runs can be randomized without too much difficulty, than this should be done.

*A Diagonally oriented saddle represented by equal coefficients of opposite sign could occure and yield a zero change in mean effect, but such a phenomenon would rarely be met in practice.

Conclusions

- All three factors are highly significant with positive sign. It means higher factor's value will increase yield. We are moving toward optimum (maximum yield).
- The *phase mean* is increased significantly.
- The *change in mean* is also significant with negative sign. There is a slight curvature in yield response surface.
- The estimated standard deviation is very close to the prior estimate of the standard deviation.

Recommendation for Phase III

- It is recommended to move the center of the experiment ($A_0 = 30$; $B_0 = 30$; $C_0 = 30$) with increase in both factors to the a new position ($A_0 = 40$; $B_0 = 40$; $C_0 = 40$).
- To keep the same factor's variation 5.0.
- To keep the same noise level STDEV = 3.0 (Table 5.50, Figures 5.75–5.77, Table 5.51).

Table 5.50 2^3 BEVOP simulation.

PROJECT : S-4
PHASE : III
DATE :

BEVOP SIMULATION

I. CENTER OF EXPERIMENT : Ao = 40.00, Bo = 40.00, Co = 40.00 [Gen Random]

II. FACTOR'S INTERVAL OF VARIATION : Δ a = 5.00, Δ b = 5.00, Δ c = 5.00

III. PROCESS NOISE : STDV = 3.0

EVOP WORK SHEET

TRIALS ORDER		Block	FACTORS			RESPONSE									
			A	B	C	n = 1	n = 2	n = 3	n = 4	n = 5	n = 6	n = 7	n = 8	n = 9	n = 10
0		Block I	40.00	40.00	40.00	95.20	87.73	94.33	91.56	95.88	93.03	102.67	93.40	101.70	88.70
1		Block I	35.00	35.00	35.00	86.82	86.87	85.31	88.29	82.85	85.59	88.81	92.02	88.74	84.04
2		Block I	45.00	45.00	35.00	92.37	97.22	91.42	93.82	98.80	97.06	94.88	96.22	97.09	97.42
3		Block I	45.00	35.00	45.00	92.53	94.28	95.23	97.11	92.20	98.28	93.98	96.35	98.15	93.94
4		Block I	35.00	45.00	45.00	96.37	91.43	95.78	93.57	92.86	94.35	93.32	92.94	96.63	99.98
Ø		Block II	40.00	40.00	40.00	98.55	92.06	94.73	91.38	93.90	90.87	95.05	98.01	98.42	95.56
5		Block II	35.00	35.00	45.00	89.70	92.77	97.90	92.68	87.50	89.57	90.00	94.38	92.64	89.59
6		Block II	45.00	45.00	45.00	96.95	92.31	94.36	97.39	96.92	96.24	99.14	97.63	103.69	103.85
7		Block II	45.00	35.00	35.00	93.73	92.57	89.48	89.96	91.87	92.66	86.71	90.26	87.56	89.94
8		Block II	35.00	45.00	35.00	89.92	93.89	91.32	89.70	93.43	88.27	84.88	92.89	88.23	92.24

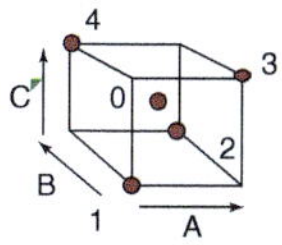

2^3 Factorial in two blocks with reference run in each block

BLOCK I

CYCLE n: 1
RESPONSE : Yield

Project S-4
Phase III
Date

Calculation of Averages						Calculation of Standard Deviation - Block I	
Operation condition	(0)	(1)	(2)	(3)	(4)		
(i) Previous sum for block I						Previous sum **s** (all blocks)	=
(ii) Previous average for block I							
(iii) New observations for block I	95.20	86.82	92.37	92.53	96.37	New **s** = range × ***fδ,n***	=
(iv) Differences (ii) less (iii)						Range	=
(v) New sums for block I	95.20	86.82	92.37	92.53	96.37	New sum **s** (all blocks)	=
(vi) New averages for block I	95.20	86.82	92.37	92.53	96.37		

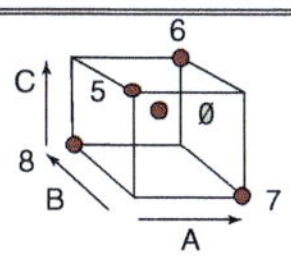

2^3 Factorial in two blocks with reference run in each block

BLOCK II

CYCLE n: 1
RESPONSE : Yield

Calculation of Averages						Calculation of Standard Deviation - Block II	
Operation condition	(Ø)	(5)	(6)	(7)	(8)	Prior estimate of σ	= 3.0
(i) Previous sum for block II						Previous sum **s** (all blocks)	=
(ii) Previous average for block II							
(iii) New observations for block II	98.55	89.70	96.95	93.73	89.92	New **s** = range × ***fδ,n***	=
(iv) Differences (ii) less (iii)						Range	=
(v) New sums for block II	98.55	89.70	96.95	93.73	89.92	New sum **s** (all blocks)	=
(vi) New averages for block II	98.55	89.70	96.95	93.73	89.92	Average **s** = {1/[2(n - 1)]} × new sum s	=

Calculation of Effects Using Yates' Algorithm

	$\bar{y}$	(i)	(ii)	(iii)	Multiplier	Effect	
(1)	86.82	180.55	362.83	738.38	0.125	**92.30**	Factorial mean
(7)	93.73	182.29	375.55	12.77	0.25	**3.19**	A
(8)	89.92	182.23	9.36	12.82	0.25	**3.20**	B
(2)	92.37	193.31	3.41	-6.72	0.25	**-1.68**	AB
(5)	89.70	6.91	1.74	12.71	0.25	**3.18**	C
(3)	92.53	2.45	11.08	-5.95	0.25	**-1.49**	AC
(4)	96.37	2.83	-4.46	9.34	0.25	**2.33**	BC
(6)	96.95	0.58	-2.26	2.21	0.25	**0.55**	ABC + blocks

Calculation of 2 S.E. Limits

For new average **s** :	$\pm (2/\sqrt{n})s$	= **6.00**
For new effects :	$\pm (1.41/\sqrt{n})s$	= **4.23**
For change in mean :	$\pm (1.26/\sqrt{n})s$	= **3.78**

		$\bar{y}$	Sum	Multiplier			
I	Reference (0)	95.20	193.75	0.500	Reference mean	=	**96.87**
	Conditions (Ø)	98.55					
II	Factorial Conditions	(1 : 8) $\sum \bar{Y}$	738.38	0.125	Factorial mean	=	**92.30**
III	All conditions	(1 : 10) $\sum \bar{Y}$	932.13	0.100	Phase mean	=	**93.21**
IV	Change in mean	III - I			Change in mean	=	**-3.66**

Table of Multiplying Factors

n	1	2	3	4	5	6	7	8	9	10
fδ,n		0.30	0.35	0.37	0.38	0.39	0.40	0.40	0.40	0.41
$1/n$	1.00	0.50	0.33	0.25	0.20	0.17	0.14	0.12	0.11	0.10
$1/[2(n-1)]$		0.50	0.25	0.17	0.12	0.10	0.08	0.07	0.06	0.06
$2/\sqrt{n}$	2.00	1.41	1.15	1.00	0.89	0.82	0.76	0.71	0.67	0.63
$1.41/\sqrt{n}$	1.41	1.00	0.82	0.71	0.63	0.58	1.86	1.99	2.10	2.24
$1.26/\sqrt{n}$	1.26	0.89	0.73	0.63	0.57	0.52	0.68	0.63	0.60	0.56

Three variable EVOP program calculation worksheet.

Figure 5.75 Work sheet cycle No. 1.

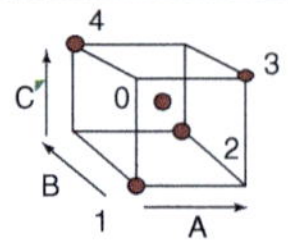

2^3 Factorial in two blocks with reference run in each block

BLOCK I

CYCLE n: 3
RESPONSE : Yield

Project S-4
Phase III
Date

Calculation of Averages						Calculation of Standard Deviation - Block I	
Operation condition	(0)	(1)	(2)	(3)	(4)		
(i) Previous sum for block I	182.93	173.69	189.59	186.81	187.80	Previous sum **s** (all blocks)	= 6.84
(ii) Previous average for block I	91.46	86.84	94.79	93.41	93.90		
(iii) New observations for block I	94.33	85.31	91.42	95.23	95.78	New **s** = range × $f\delta,n$	= 2.19
(iv) Differences (ii) less (iii)	-2.87	1.53	3.38	-1.82	-1.88	Range	= 6.25
(v) New sums for block I	277.26	259.00	281.01	282.04	283.58	New sum **s** (all blocks)	= 9.02
(vi) New averages for block I	92.42	86.33	93.67	94.01	94.53		

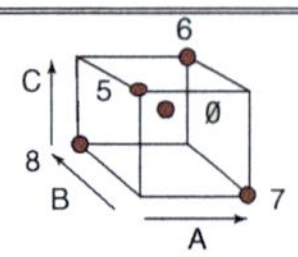

2^3 Factorial in two blocks with reference run in each block

BLOCK II

CYCLE n: 3
RESPONSE : Yield

Calculation of Averages						Calculation of Standard Deviation - Block II	
Operation condition	(Ø)	(5)	(6)	(7)	(8)	Prior estimate of σ	= 3.0
(i) Previous sum for block II	190.61	182.47	189.26	186.30	183.81	Previous sum **s** (all blocks)	= 9.02
(ii) Previous average for block II	95.30	91.23	94.63	93.15	91.90		
(iii) New observations for block II	94.73	97.90	94.36	89.48	91.32	New **s** = range × $f\delta,n$	= 3.62
(iv) Differences (ii) less (iii)	0.57	-6.66	0.26	3.67	0.59	Range	= 10.34
(v) New sums for block II	285.34	280.37	283.62	275.77	275.13	New sum **s** (all blocks)	= 12.64
(vi) New averages for block II	95.11	93.46	94.54	91.92	91.71	Average **s** = {1/[2(n - 1)]} × new sum s	= 3.16

Calculation of Effects Using Yates' Algorithm

	$\bar{y}$	(i)	(ii)	(iii)	Multiplier	Effect	
(1)	86.33	178.26	363.64	740.17	0.125	**92.52**	Factorial mean
(7)	91.92	185.38	376.54	8.12	0.25	**2.03**	A
(8)	91.71	187.47	7.55	8.72	0.25	**2.18**	B
(2)	93.67	189.07	0.57	-4.18	0.25	**-1.04**	AB
(5)	93.46	5.59	7.12	12.90	0.25	**3.23**	C
(3)	94.01	1.96	1.60	-6.98	0.25	**-1.74**	AC
(4)	94.53	0.56	-3.63	-5.52	0.25	**-1.38**	BC
(6)	94.54	0.01	-0.55	3.09	0.25	**0.77**	ABC + blocks

Calculation of 2 S.E. Limits

For new average **s** :	$\pm (2/\sqrt{n})s$	= **3.65**
For new effects :	$\pm (1.41/\sqrt{n})s$	= **2.57**
For change in mean :	$\pm (1.26/\sqrt{n})s$	= **2.30**

		$\bar{y}$	Sum	Multiplier			
I	Reference (0)	92.42	187.53	0.500	Reference mean	=	**93.77**
	Conditions (Ø)	95.11					
II	Factorial Conditions	(1 : 8) $\sum\acute{Y}$	740.17	0.125	Factorial mean	=	**92.52**
III	All conditions	(1 : 10) $\sum\acute{Y}$	927.71	0.100	Phase mean	=	**92.77**
IV	Change in mean	III - I			Change in mean	=	**-1.00**

Table of Multiplying Factors

n	1	2	3	4	5	6	7	8	9	10
$f\delta,n$		0.30	0.35	0.37	0.38	0.39	0.40	0.40	0.40	0.41
1 / n	1.00	0.50	0.33	0.25	0.20	0.17	0.14	0.12	0.11	0.10
1 / [2(n - 1)]		0.50	0.25	0.17	0.12	0.10	0.08	0.07	0.06	0.06
$2/\sqrt{n}$	2.00	1.41	1.15	1.00	0.89	0.82	0.76	0.71	0.67	0.63
$1.41/\sqrt{n}$	1.41	1.00	0.82	0.71	0.63	0.58	1.86	1.99	2.10	2.24
$1.26/\sqrt{n}$	1.26	0.89	0.73	0.63	0.57	0.52	0.68	0.63	0.60	0.56

Three variable EVOP program calculation worksheet.

Figure 5.76 Work sheet cycle No. 3.

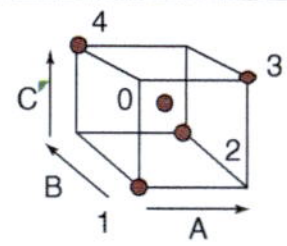

2^3 Factorial in two blocks with reference run in each block

BLOCK I

CYCLE n: 10
RESPONSE : Yield

Project S-4
Phase III
Date

Calculation of Averages						Calculation of Standard Deviation - Block I	
Operation condition	(0)	(1)	(2)	(3)	(4)		
(i) Previous sum for block I	855.50	785.30	858.87	858.12	847.25	Previous sum **s** (all blocks)	= 47.26
(ii) Previous average for block I	95.06	87.26	95.43	95.35	94.14		
(iii) New observations for block I	88.70	84.04	97.42	93.94	99.98	New **s** = range × ***fδ,n***	= 5.00
(iv) Differences (ii) less (iii)	6.36	3.22	-1.99	1.41	-5.84	Range	= 12.20
(v) New sums for block I	944.20	869.34	956.29	952.05	947.23	New sum **s** (all blocks)	= 52.27
(vi) New averages for block I	94.42	86.93	95.63	95.21	94.72		

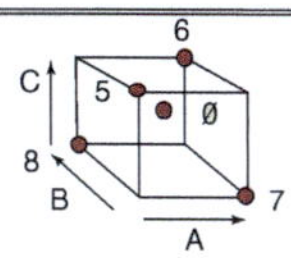

2^3 Factorial in two blocks with reference run in each block

BLOCK II

CYCLE n: 10
RESPONSE : Yield

Calculation of Averages						Calculation of Standard Deviation - Block II	
Operation condition	(Ø)	(5)	(6)	(7)	(8)	Prior estimate of **σ**	= 3.0
(i) Previous sum for block II	852.98	827.13	874.63	814.78	812.51	Previous sum **s** (all blocks)	= 52.27
(ii) Previous average for block II	94.78	91.90	97.18	90.53	90.28		
(iii) New observations for block II	95.56	89.59	103.85	89.94	92.24	New **s** = range × ***fδ,n***	= 3.68
(iv) Differences (ii) less (iii)	-0.78	2.32	-6.67	0.59	-1.96	Range	= 8.99
(v) New sums for block II	948.53	916.71	978.48	904.73	904.76	New sum **s** (all blocks)	= 55.95
(vi) New averages for block II	94.85	91.67	97.85	90.47	90.48	Average **s** = {1/[2(n - 1)]} × new sum s	= 3.11

Calculation of Effects Using Yates' Algorithm

	$\bar{y}$	(i)	(ii)	(iii)	Multiplier	Effect	
(1)	86.93	177.41	363.51	742.96	0.125	**92.87**	Factorial mean
(7)	90.47	186.10	379.45	15.35	0.25	**3.84**	A
(8)	90.48	186.88	8.69	14.39	0.25	**3.60**	B
(2)	95.63	192.57	6.66	1.21	0.25	**0.30**	AB
(5)	91.67	3.54	8.70	15.94	0.25	**3.98**	C
(3)	95.21	5.15	5.69	-2.03	0.25	**-0.51**	AC
(4)	94.72	3.53	1.61	-3.00	0.25	**-0.75**	BC
(6)	97.85	3.13	-0.41	-2.02	0.25	**-0.51**	ABC + blocks

Calculation of 2 S.E. Limits

For new average **s** :	$\pm (2/\sqrt{n})s$	= **1.97**
For new effects :	$\pm (1.41/\sqrt{n})s$	= **1.39**
For change in mean :	$\pm (1.26/\sqrt{n})s$	= **1.24**

		$\bar{y}$	Sum	Multiplier			
I	Reference (0) Conditions (Ø)	94.42 94.85	189.27	0.500	Reference mean	=	**94.64**
II	Factorial Conditions	$(1:8)\sum \bar{Y}$	742.96	0.125	Factorial mean	=	**92.87**
III	All conditions	$(1:10)\sum \bar{Y}$	932.23	0.100	Phase mean	=	**93.22**
IV	Change in mean	III - I			Change in mean	=	**-1.41**

Table of Multiplying Factors

n	1	2	3	4	5	6	7	8	9	10
fδ,n		0.30	0.35	0.37	0.38	0.39	0.40	0.40	0.40	0.41
1 / n	1.00	0.50	0.33	0.25	0.20	0.17	0.14	0.12	0.11	0.10
1 / [2(n - 1)]		0.50	0.25	0.17	0.12	0.10	0.08	0.07	0.06	0.06
$2/\sqrt{n}$	2.00	1.41	1.15	1.00	0.89	0.82	0.76	0.71	0.67	0.63
$1.41/\sqrt{n}$	1.41	1.00	0.82	0.71	0.63	0.58	1.86	1.99	2.10	2.24
$1.26/\sqrt{n}$	1.26	0.89	0.73	0.63	0.57	0.52	0.68	0.63	0.60	0.56

Three variable EVOP program calculation worksheet.

Figure 5.77 Work sheet cycle No. 10.

Table 5.51 2^3 BEVOP information board.

2^3 INFORMATION BOARD AFTER PHASE NO. III Last Cycle Completed : n = 10			
	RESPONSE No. 1	RESPONSE No. 2	RESPONSE No. 3
REQUIREMENT :	MAXIMIZE RESPONSE		
PROCESS AVERAGES :	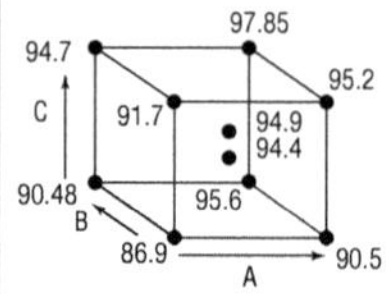		
2.S.E. LIMITS FOR AVERAGES	1.97		
EFFECTS WITH 2.S.E PHASE MEAN FACTOR A FACTOR B FACTOR C INTERACTION AB INTERACTION AC INTERACTION BC INTERACTION ABC CHANGE IN MEAN	 93.22 3.84 ± 1.39 3.60 ± 1.39 3.98 ± 1.39 0.30 ± 1.39 -0.51 ± 1.39 -0.75 ± 1.39 -0.51 ± 1.39 -1.41 ± 1.24		
ESTIMATED STANDARD DEVIATION S	3.11		
PRIOR ESTIMATE OF STANDARD DEV.	3.00		

NOTE:

1. The Center of Experiment-Reference Conditions (The Current Best-Known Process Conditions)
2. The Change in Mean provides a measure of whether the conditions being run the current phase yield an average result better or worse than the reference conditions.
3. The Phase Mean is the Mean Response Over All Conditions Being Run in the present Phase.
4*. The change in Mean is a measure of the over-all Curvature of the Response Surface. It will be positive near a MINIMUM and negative near a MAXIMUM.
5. Sequencing of the Runs-Randomization, if runs can be randomized without too much difficulty, than this should be done.

*A Diagonally oriented saddle represented by equal coefficients of opposite sign could occure and yield a zero change in mean effect, but such a phenomenon would rarely be met in practice.

Conclusions

- All three factors are highly significant with positive sign. It means higher factor's value will increase yield. We are moving toward optimum (maximum yield).
- The *phase mean* is increased significantly.
- The *change in mean* is significant with negative sign, and negative sign means a maximum yield.
- The estimated standard deviation is very close to the prior estimate of the standard deviation.

Recommendation for Phase IV

- It is recommended to move the center of the experiment ($A_0 = 40$; $B_0 = 40$; $C_0 = 40$) with increase in both factors to a new position ($A_0 = 50$; $B_0 = 50$; $C_0 = 50$).
- To keep the same factor's variation 5.0.
- To keep the same noise level STDEV = 3.0 (Table 5.52, Figures 5.78–5.80, Table 5.53).

Table 5.52 2^3 BEVOP simulation.

PROJECT : S-4
PHASE : IV
DATE :

BEVOP SIMULATION

I. CENTER OF EXPERIMENT : Ao = 50.00, Bo = 50.00, Co = 50.00 [Gen Random]

II. FACTOR'S INTERVAL OF VARIATION : Δa = 5.00, Δb = 5.00, Δc = 5.00

III. PROCESS NOISE : STDV = 3.0

EVOP WORK SHEET

TRIALS ORDER		Block	FACTORS			RESPONSE									
			A	B	C	n = 1	n = 2	n = 3	n = 4	n = 5	n = 6	n = 7	n = 8	n = 9	n = 10
0		Block I	50.00	50.00	50.00	101.20	93.73	100.33	97.56	101.88	99.03	108.67	99.40	107.70	94.70
1		Block I	45.00	45.00	45.00	98.82	98.87	97.31	100.29	94.85	97.59	100.81	104.02	100.74	96.04
2		Block I	55.00	55.00	45.00	96.37	101.22	95.42	97.82	102.80	101.06	98.88	100.22	101.09	101.42
3		Block I	55.00	45.00	55.00	96.53	98.28	99.23	101.11	96.20	102.28	97.98	100.35	102.15	97.94
4		Block I	45.00	55.00	55.00	100.37	95.43	99.78	97.57	96.86	98.35	97.32	96.94	100.63	103.98
Ø		Block II	50.00	50.00	50.00	104.55	98.06	100.73	97.38	99.90	96.87	101.05	104.01	104.42	101.56
5		Block II	45.00	45.00	55.00	97.70	100.77	105.90	100.68	95.50	97.57	98.00	102.38	100.64	97.59
6		Block II	55.00	55.00	55.00	96.95	92.31	94.36	97.39	96.92	96.24	99.14	97.63	103.69	103.85
7		Block II	55.00	45.00	45.00	101.73	100.57	97.48	97.96	99.87	100.66	94.71	98.26	95.56	97.94
8		Block II	45.00	55.00	45.00	97.92	101.89	99.32	97.70	101.43	96.27	92.88	100.89	96.23	100.24

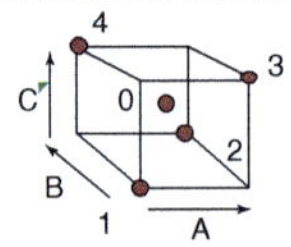

2³ Factorial in two blocks with reference run in each block

BLOCK I

CYCLE n: 1
RESPONSE : Yield

Project S-4
Phase IV
Date

Calculation of Averages					
Operation condition	(0)	(1)	(2)	(3)	(4)
(i) Previous sum for block I					
(ii) Previous average for block I					
(iii) New observations for block I	101.20	98.82	96.37	96.53	100.37
(iv) Differences (ii) less (iii)					
(v) New sums for block I	101.20	98.82	96.37	96.53	100.37
(vi) New averages for block I	101.20	98.82	96.37	96.53	100.37

Calculation of Standard Deviation - Block I	
Previous sum **s** (all blocks)	=
New **s** = range × ***fδ,n***	=
Range	=
New sum **s** (all blocks)	=

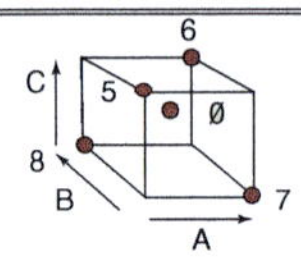

2³ Factorial in two blocks with reference run in each block

BLOCK II

CYCLE n: 1
RESPONSE : Yield

Calculation of Averages					
Operation condition	(Ø)	(5)	(6)	(7)	(8)
(i) Previous sum for block II					
(ii) Previous average for block II					
(iii) New observations for block II	104.55	97.70	96.95	101.73	97.92
(iv) Differences (ii) less (iii)					
(v) New sums for block II	104.55	97.70	96.95	101.73	97.92
(vi) New averages for block II	104.55	97.70	96.95	101.73	97.92

Calculation of Standard Deviation - Block II	
Prior estimate of **σ**	= 3.0
Previous sum **s** (all blocks)	=
New **s** = range × ***fδ,n***	=
Range	=
New sum **s** (all blocks)	=
Average **s** = {1/[2(n - 1)]} × new sum s	=

Calculation of Effects Using Yates' Algorithm

	$\bar{y}$	(i)	(ii)	(iii)	Multiplier	Effect	
(1)	98.82	200.55	394.83	786.38	0.125	**98.30**	Factorial mean
(7)	101.73	194.29	391.55	-3.23	0.25	**-0.81**	A
(8)	97.92	194.23	1.36	-3.18	0.25	**-0.80**	B
(2)	96.37	197.31	-4.59	-6.72	0.25	**-1.68**	AB
(5)	97.70	2.91	-6.26	-3.29	0.25	**-0.82**	C
(3)	96.53	-1.55	3.08	-5.95	0.25	**-1.49**	AC
(4)	100.37	-1.17	-4.46	9.34	0.25	**2.33**	BC
(6)	96.95	-3.42	-2.26	2.21	0.25	**0.55**	ABC + blocks

Calculation of 2 S.E. Limits

For new average **s** :	$\pm(2/\sqrt{n})s$	= **6.00**
For new effects :	$\pm(1.41/\sqrt{n})s$	= **4.23**
For change in mean :	$\pm(1.26/\sqrt{n})s$	= **3.78**

		$\bar{y}$	Sum	Multiplier
I	Reference (0)	101.20	205.75	0.500
	Conditions (Ø)	104.55		
II	Factorial Conditions	(1 : 8) $\sum \bar{Y}$	786.38	0.125
III	All conditions	(1 : 10) $\sum \bar{Y}$	992.13	0.100
IV	Change in mean	III - I		

Reference mean	=	**102.87**
Factorial mean	=	**98.30**
Phase mean	=	**99.21**
Change in mean	=	**-3.66**

Table of Multiplying Factors

n	1	2	3	4	5	6	7	8	9	10
fδ,n		0.30	0.35	0.37	0.38	0.39	0.40	0.40	0.40	0.41
1 / n	1.00	0.50	0.33	0.25	0.20	0.17	0.14	0.12	0.11	0.10
1 / [2(n - 1)]		0.50	0.25	0.17	0.12	0.10	0.08	0.07	0.06	0.06
$2/\sqrt{n}$	2.00	1.41	1.15	1.00	0.89	0.82	0.76	0.71	0.67	0.63
$1.41/\sqrt{n}$	1.41	1.00	0.82	0.71	0.63	0.58	1.86	1.99	2.10	2.24
$1.26/\sqrt{n}$	1.26	0.89	0.73	0.63	0.57	0.52	0.68	0.63	0.60	0.56

Three variable EVOP program calculation worksheet.

Figure 5.78 Work sheet cycle No. 1.

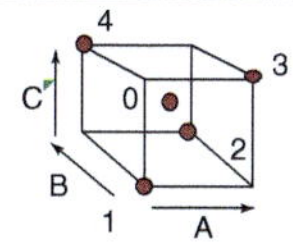

2^3 Factorial in two blocks with reference run in each block

BLOCK I

CYCLE n: 3
RESPONSE : Yield

Project S-4
Phase IV
Date

Calculation of Averages					
Operation condition	(0)	(1)	(2)	(3)	(4)
(i) Previous sum for block I	194.93	197.69	197.59	194.81	195.80
(ii) Previous average for block I	97.46	98.84	98.79	97.41	97.90
(iii) New observations for block I	100.33	97.31	95.42	99.23	99.78
(iv) Differences (ii) less (iii)	-2.87	1.53	3.38	-1.82	-1.88
(v) New sums for block I	295.26	295.00	293.01	294.04	295.58
(vi) New averages for block I	98.42	98.33	97.67	98.01	98.53

Calculation of Standard Deviation - Block I	
Previous sum **s** (all blocks)	= 6.84
New **s** = range × ***f δ,n***	= 2.19
Range	= 6.25
New sum **s** (all blocks)	= 9.02

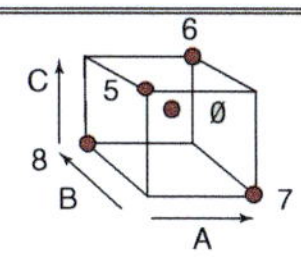

2^3 Factorial in two blocks with reference run in each block

BLOCK II

CYCLE n: 3
RESPONSE : Yield

Calculation of Averages					
Operation condition	(Ø)	(5)	(6)	(7)	(8)
(i) Previous sum for block II	202.61	198.47	189.26	202.30	199.81
(ii) Previous average for block II	101.30	99.23	94.63	101.15	99.90
(iii) New observations for block II	100.73	105.90	94.36	97.48	99.32
(iv) Differences (ii) less (iii)	0.57	-6.66	0.26	3.67	0.59
(v) New sums for block II	303.34	304.37	283.62	299.77	299.13
(vi) New averages for block II	101.11	101.46	94.54	99.92	99.71

Calculation of Standard Deviation - Block II	
Prior estimate of σ	= 3.0
Previous sum **s** (all blocks)	= 9.02
New **s** = range × ***f δ,n***	= 3.62
Range	= 10.34
New sum **s** (all blocks)	= 12.64
Average **s** = {1/[2(n - 1)]} × new sum s	= 3.16

Calculation of Effects Using Yates' Algorithm

	$\bar{y}$	(i)	(ii)	(iii)	Multiplier	Effect	
(1)	98.33	198.26	395.64	788.17	0.125	**98.52**	Factorial mean
(7)	99.92	197.38	392.54	-7.88	0.25	**-1.97**	A
(8)	99.71	199.47	-0.45	-7.28	0.25	**-1.82**	B
(2)	97.67	193.07	-7.43	-4.18	0.25	**-1.04**	AB
(5)	101.46	1.59	-0.88	-3.10	0.25	**-0.77**	C
(3)	98.01	-2.04	-6.40	-6.98	0.25	**-1.74**	AC
(4)	98.53	-3.44	-3.63	-5.52	0.25	**-1.38**	BC
(6)	94.54	-3.99	-0.55	3.09	0.25	**0.77**	ABC + blocks

Calculation of 2 S.E. Limits

For new average **s** :	$\pm (2/\sqrt{n})s$	= **3.65**
For new effects :	$\pm (1.41/\sqrt{n})s$	= **2.57**
For change in mean :	$\pm (1.26/\sqrt{n})s$	= **2.30**

		$\bar{y}$	**Sum**	**Multiplier**
I	Reference (0) Conditions (Ø)	98.42 101.11	199.53	0.500
II	Factorial Conditions	(1 : 8) $\sum \acute{Y}$	788.17	0.125
III	All conditions	(1 : 10) $\sum \acute{Y}$	987.71	0.100
IV	Change in mean	III - I		

Reference mean	=	**99.77**
Factorial mean	=	**98.52**
Phase mean	=	**98.77**
Change in mean	=	**-1.00**

Table of Multiplying Factors

n	1	2	3	4	5	6	7	8	9	10
f δ,n		0.30	0.35	0.37	0.38	0.39	0.40	0.40	0.40	0.41
1 / n	1.00	0.50	0.33	0.25	0.20	0.17	0.14	0.12	0.11	0.10
1 / [2(n - 1)]		0.50	0.25	0.17	0.12	0.10	0.08	0.07	0.06	0.06
$2/\sqrt{n}$	2.00	1.41	1.15	1.00	0.89	0.82	0.76	0.71	0.67	0.63
$1.41/\sqrt{n}$	1.41	1.00	0.82	0.71	0.63	0.58	1.86	1.99	2.10	2.24
$1.26/\sqrt{n}$	1.26	0.89	0.73	0.63	0.57	0.52	0.68	0.63	0.60	0.56

Three variable EVOP program calculation worksheet.

Figure 5.79 Work sheet cycle No. 3.

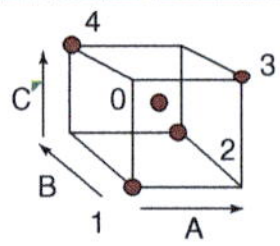

2^3 Factorial in two blocks with reference run in each block

BLOCK I

CYCLE n: 10
RESPONSE : Yield

Project S-4
Phase IV
Date

Calculation of Averages						Calculation of Standard Deviation - Block I	
Operation condition	(0)	(1)	(2)	(3)	(4)		
(i) Previous sum for block I	909.50	893.30	894.87	894.12	883.25	Previous sum **s** (all blocks)	= 47.26
(ii) Previous average for block I	101.06	99.26	99.43	99.35	98.14		
(iii) New observations for block I	94.70	96.04	101.42	97.94	103.98	New **s** = range × ***fδ,n***	= 5.00
(iv) Differences (ii) less (iii)	6.36	3.22	-1.99	1.41	-5.84	Range	= 12.20
(v) New sums for block I	1004.20	989.34	996.29	992.05	987.23	New sum **s** (all blocks)	= 52.27
(vi) New averages for block I	100.42	98.93	99.63	99.21	98.72		

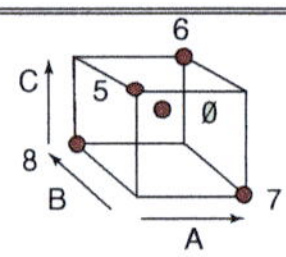

2^3 Factorial in two blocks with reference run in each block

BLOCK II

CYCLE n: 10
RESPONSE : Yield

Calculation of Averages						Calculation of Standard Deviation - Block II	
Operation condition	(Ø)	(5)	(6)	(7)	(8)	Prior estimate of **σ**	= 3.0
(i) Previous sum for block II	906.98	899.13	874.63	886.78	884.51	Previous sum **s** (all blocks)	= 52.27
(ii) Previous average for block II	100.78	99.90	97.18	98.53	98.28		
(iii) New observations for block II	101.56	97.59	103.85	97.94	100.24	New **s** = range × ***fδ,n***	= 3.68
(iv) Differences (ii) less (iii)	-0.78	2.32	-6.67	0.59	-1.96	Range	= 8.99
(v) New sums for block II	1008.53	996.71	978.48	984.73	984.76	New sum **s** (all blocks)	= 55.95
(vi) New averages for block II	100.85	99.67	97.85	98.47	98.48	Average **s** = {1/[2(n - 1)]} × new sum s	= 3.11

Calculation of Effects Using Yates' Algorithm								Calculation of 2 S.E. Limits		
	$\bar{y}$	(i)	(ii)	(iii)	Multiplier	Effect				
(1)	98.93	197.41	395.51	790.96	0.125	**98.87**	Factorial mean			
(7)	98.47	198.10	395.45	-0.65	0.25	**-0.16**	A	For new average **s** :	$\pm(2/\sqrt{n})s$	= **1.97**
(8)	98.48	198.88	0.69	-1.61	0.25	**-0.40**	B			
(2)	99.63	196.57	-1.34	1.21	0.25	**0.30**	AB	For new effects :	$\pm(1.41/\sqrt{n})s$	= **1.39**
(5)	99.67	-0.46	0.70	-0.06	0.25	**-0.02**	C			
(3)	99.21	1.15	-2.31	-2.03	0.25	**-0.51**	AC	For change in mean :	$\pm(1.26/\sqrt{n})s$	= **1.24**
(4)	98.72	-0.47	1.61	-3.00	0.25	**-0.75**	BC			
(6)	97.85	-0.87	-0.41	-2.02	0.25	**-0.51**	ABC + blocks			

		$\bar{y}$	Sum	Multiplier			
I	Reference (0) Conditions (Ø)	100.42 100.85	201.27	0.500	Reference mean	=	**100.64**
II	Factorial Conditions	$(1:8)\sum\bar{Y}$	790.96	0.125	Factorial mean	=	**98.87**
III	All conditions	$(1:10)\sum\bar{Y}$	992.23	0.100	Phase mean	=	**99.22**
IV	Change in mean	III - I			Change in mean	=	**-1.41**

Table of Multiplying Factors

n	1	2	3	4	5	6	7	8	9	10
fδ,n		0.30	0.35	0.37	0.38	0.39	0.40	0.40	0.40	0.41
1 / n	1.00	0.50	0.33	0.25	0.20	0.17	0.14	0.12	0.11	0.10
1 / [2(n - 1)]		0.50	0.25	0.17	0.12	0.10	0.08	0.07	0.06	0.06
$2/\sqrt{n}$	2.00	1.41	1.15	1.00	0.89	0.82	0.76	0.71	0.67	0.63
$1.41/\sqrt{n}$	1.41	1.00	0.82	0.71	0.63	0.58	1.86	1.99	2.10	2.24
$1.26/\sqrt{n}$	1.26	0.89	0.73	0.63	0.57	0.52	0.68	0.63	0.60	0.56

Three variable EVOP program calculation worksheet.

Figure 5.80 Work sheet cycle No. 10.

Table 5.53 2^3 BEVOP information board.

2^3 INFORMATION BOARD AFTER PHASE NO. IV

Last Cycle Completed : n = 10

	RESPONSE No. 1	RESPONSE No. 2	RESPONSE No. 3
REQUIREMENT :	MAXIMIZE RESPONSE		
PROCESS AVERAGES :	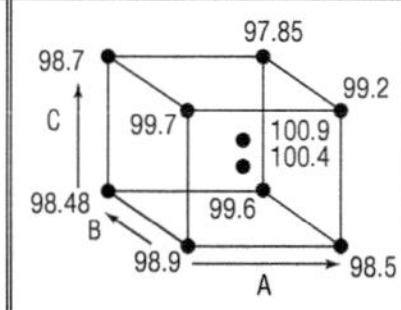		
2.S.E. LIMITS FOR AVERAGES	1.97		
EFFECTS WITH 2.S.E — PHASE MEAN	99.22		
FACTOR A	-0.16 ± 1.39		
FACTOR B	-0.40 ± 1.39		
FACTOR C	-0.02 ± 1.39		
INTERACTION AB	0.30 ± 1.39		
INTERACTION AC	-0.51 ± 1.39		
INTERACTION BC	-0.75 ± 1.39		
INTERACTION ABC	-0.51 ± 1.39		
CHANGE IN MEAN	-1.41 ± 1.24		
ESTIMATED STANDARD DEVIATION S	3.11		
PRIOR ESTIMATE OF STANDARD DEV.	3.00		

NOTE:

1. The Center of Experiment-Reference Conditions (The Current Best-Known Process Conditions)
2. The Change in Mean provides a measure of whether the conditions being run the current phase yield an average result better or worse than the reference conditions.
3. The Phase Mean is the Mean Response Over All Conditions Being Run in the present Phase.
4*. The change in Mean is a measure of the over-all Curvature of the Response Surface. It will be positive near a MINIMUM and negative near a MAXIMUM.
5. Sequencing of the Runs-Randomization, if runs can be randomized without too much difficulty, than this should be done.

*A Diagonally oriented saddle represented by equal coefficients of opposite sign could occure and yield a zero change in mean effect, but such a phenomenon would rarely be met in practice.

Conclusions

- All three factors are not significant, EVOP is sitting on the top of the response surface. We have reached optimum, maximum yield 100 [%].
- The *phase mean* reached maximum.
- The *change in mean* is significant with negative sign, and negative sign means a maximum yield.
- The estimated standard deviation is very close to the prior estimate of the standard deviation.

Solution

The response was increased significantly from 45 to 100 [%] at the end of the phase IV. This value was achieved at the last center point: $A_0 = 50$; $B_0 = 50$; $C_0 = 50$ (Figures 5.81 and 5.82).

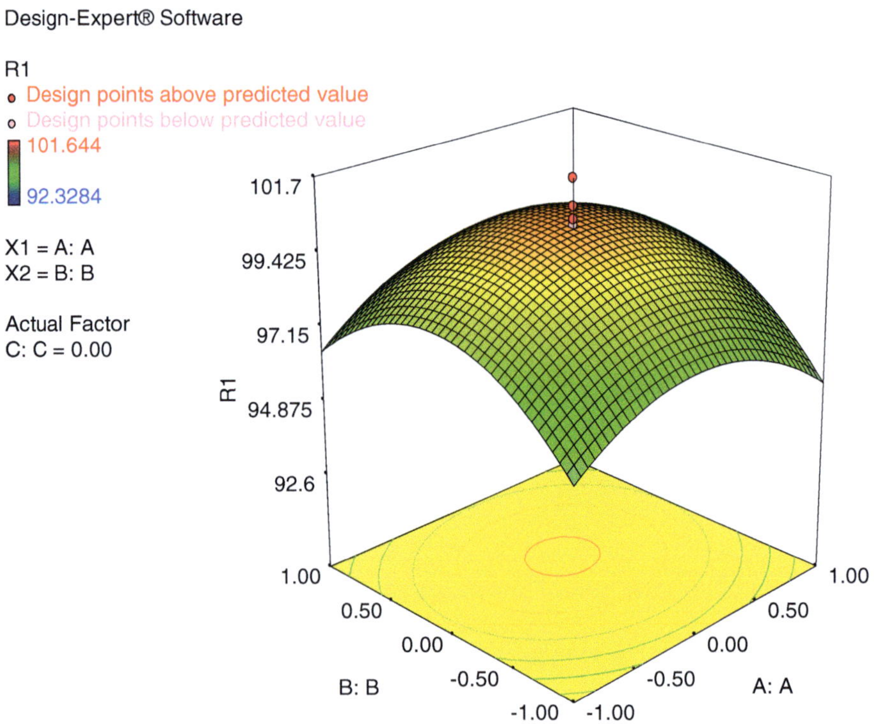

Figure 5.81 Response surface.

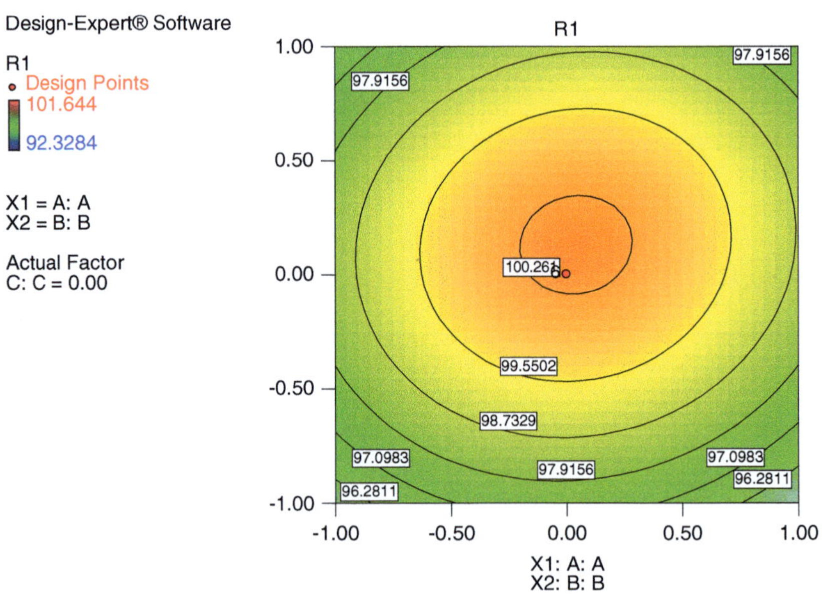

Figure 5.82 Contour diagram.

5.8 Rotating Square Evolutionary Operation – ROVOP

5.8.1 2^2 ROVOP

ROVOP attempts to eliminate uncertainty because of the size of variant used. As with EVOP, experimentation is restricted to two or three factors so that the experimental design remains simple; designs with more than three factors can be handled but get very complicated.

The procedure consists of starting with very small variants in a body center 2^n factorial design, then for each succeeding cycle on the same center enlarging the variant by $\sqrt{2}$ and rotating the design through an angle of 45°, as shown in Figure 5.83. This permits effects to converge more quickly on some predetermined level of significance and provides information on the quadratic effects.

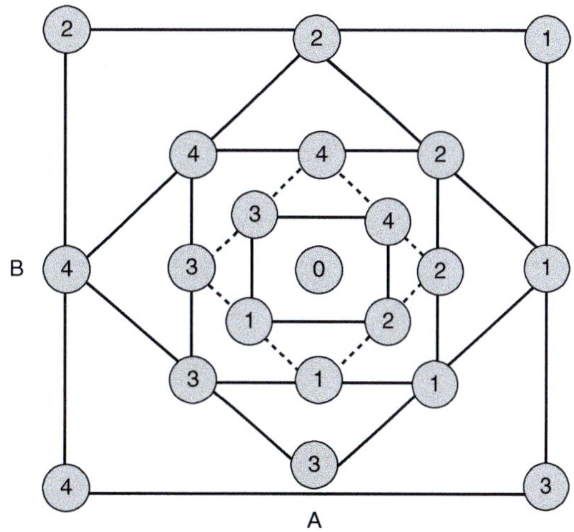

Figure 5.83 Five successive 2^2 ROVOP patterns.

Experimentation continues with ROVOP until a response effect, shown to be valid at the predetermined level of significance, is developed; at this point the pattern is moved, conservatively, in the direction of optimization and ROVOP continued until no further movement is indicated. When the operator moves to new phase (new cycles conditions), at least two points of the new phase should lie on or within the bounds of experimental space already explored, Figures 5.84 and 5.85. Also, the interval between the high and low levels of the factor to be changed should be reduced by at least a factor of $\sqrt{2}$; *levels of factors shown to be not significant need not be reduced.* Occasionally, and ideally, the response pattern follows a hill contour in which the peak represents the optimum conditions. With ROVOP the peak is approached gradually and eventually is straddled; now the quadratic terms of the response equation take control and movement *toward the center of the design is indicated.* Continuation of ROVOP beyond this point will provide adjustment for long-term changes only, such as changes in raw materials and may or may not be worthwhile.

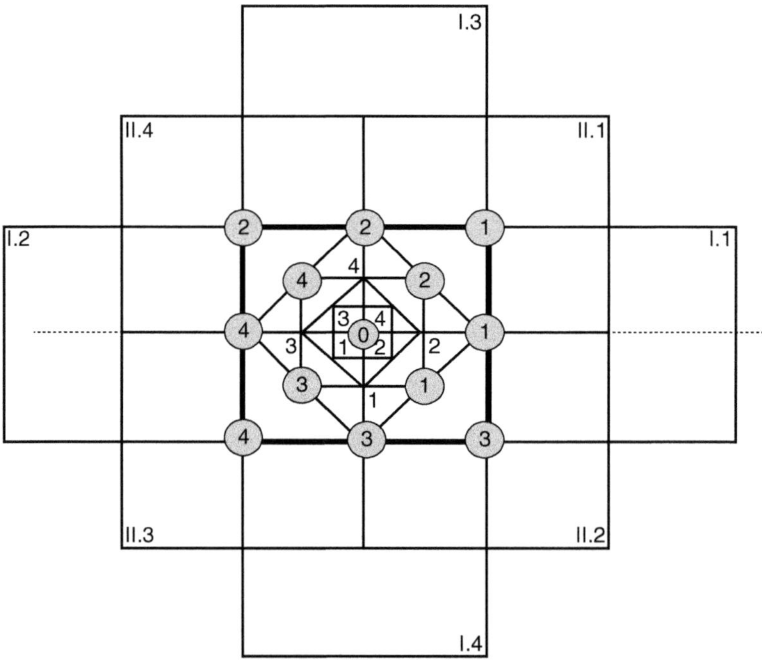

Figure 5.84 Moving center of the ROVOP without reduction of interval.

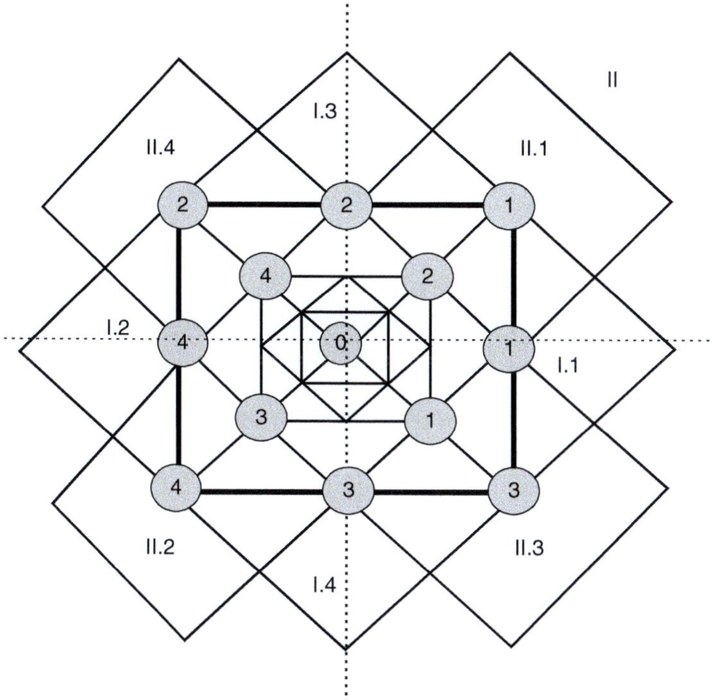

Figure 5.85 Moving center of the ROVOP with reduction of interval.

More commonly, the optimum is a compromise between several rising and falling surfaces. Although it may be of doubtful validity, it would be advantageous to define a “desirability” function expressing the several responses as a single number. This function would be expected to have the ideal single-peaked contour with a definable optimum. Without such a function, the optimum must be defined as the most desirable combination of responses, including cost and quality Appendix E [3, 42–44].

5.8.2 Method of Analysis

Because of the complex pattern of ROVOP, multiple regression analysis is necessary and access to a computer is essential. The response equation will include the linear and quadratic terms of the factors. The response surface is interpreted most easily by means of a response matrix consisting of calculated responses for predetermined combinations of the control factors. Response surfaces may be drawn through points of equal response (Contour Graph).

When a process is stable in time, each new cycle can be added to the old with increasing precision in the estimation of the response equation. However, when a process is not stable in time as a result of changing raw materials, seasonal weather conditions, catalyst depletion, scale build-up, etc. this greater precision will not be achieved and the new data may quite thoroughly confuse the analysis of previous data.

One solution on this problem is to analyze the residuals on a control chart. If an assignable cause of variation, in particular trends or runs above or below apparent by the usual tests, this problem must be settled before ROVOP continues. Alternatively, the results of old cycles may be deleted so that only the latest operating conditions are reflected in the response equation [11].

The advantages of ROVOP are:

1. Starting levels for ROVOP are not of great importance and can be very close to the existing operating conditions.
2. The rotating pattern provides quadratic terms immediately and this is very useful if the process already is operating close to its optimum.
3. The pattern provides good exploration of the experimental space.
4. The expanding pattern of experimentation rapidly provides a response surface.

ROVOP has the following disadvantages:

1. Analysis of data is not easy and requires a computer.
2. Only quantitative factors may be included.
3. For more than three factors, the design becomes very complicated.

As far as the author is aware, no publications have been made in which these ROVOP modifications of BEVOP have been used and this was already indicated in 1974 by Lowe [33]. Readers who are interested in the details of these EVOP modifications are referred to the following references [1, 11, 33]. To the author's best knowledge, ROVOP had never been used in industry and the best way to test this idea is to run simulation, which was already verified by BEVOP method.

5.8.3 2^2 ROVOP Simulation

5.8.3.1 Simulation No. 5

The mathematical formula from simulation No. 1, Chapter 5.6, was used to demonstrate efficiency of the ROVOP method. Simulation started with ROVOP until significant factor effect was detected and after ROVOP was switched to BEVOP until optimum was achieved.

Phase I: The settings (Center Point, Factor's Interval, and Process Noise) and yield for different process conditions have been calculated using Microsoft Excel simulation in Table 5.54. With this ROVOP simulation, one more parameter in settings was introduced: N – number of replicates-repetitions of ROVOP cycles. The corresponding calculations for ten replicates in cycle 1, phase I, have been shown in work sheet in Figure 5.86. The information board that would appear at the end of the replica No. 10, after transfer of the various calculated quantities from the appropriate worksheet, is shown in Table 5.55.

This simulation started with small factor's change (interval) and the large noise (STDEV).

Table 5.54 2^2 ROVOP simulation.

PROJECT : S-5
PHASE : I
CYCLE : 1

ROVOP SIMULATION

S-5

I. CENTER OF EXPERIMENT : Ao = 20.00; Bo = 20.00

Gen Random

II. FACTOR'S INTERVAL OF VARIATION : Δa = 0.50; Δb = 0.50

III. PROCESS (PROPERTY-RESPONSE) NOISE : STDV = 10.00

EVOP WORK SHEET

TRIALS ORDER			FACTOR		RESPONSE									
			A	B	n = 1	n = 2	n = 3	n = 4	n = 5	n = 6	n = 7	n = 8	n = 9	n = 10
0		Block 1	**20.00**	**20.00**	**65.38**	**51.68**	**54.83**	**71.17**	**53.07**	**66.56**	**49.44**	**69.62**	**55.37**	**59.95**
1		Block 1	19.50	19.50	**73.23**	50.85	74.15	57.89	57.69	64.86	75.56	56.23	62.80	**65.42**
2		Block 1	20.50	19.50	**49.97**	50.86	65.44	71.63	61.81	64.64	76.81	73.69	72.45	**58.02**
3		Block 1	19.50	20.50	**73.18**	86.71	66.57	64.63	67.34	68.77	78.24	87.89	89.10	**54.21**
4		Block 1	20.50	20.50	**68.66**	73.30	79.08	42.40	49.54	56.50	69.68	82.00	61.01	**74.61**

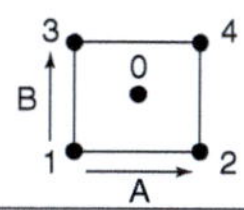

2^2 Factorial with added reference condition

Replica N :
10
Response:YIELD

Project S-5
Phase I
Cycle: 1

Calculation of Averages						Calculation of Standard Deviation	
Operation condition	(0)	(1)	(2)	(3)	(4)	Prior estimate of σ	= 10.00
(i) Previous cycle sum	118.17	125.46	144.48	169.72	133.26	Previous sum **s**	= 78.17
(ii) Previous cycle average	59.09	62.73	72.24	84.86	66.63	Previous average **s**	= 9.77
(iii) New observations	59.95	65.42	58.02	54.21	74.61	New **s** = range × ***fδ,n***	= 15.84
(iv) Differences (ii) less (iii)	-0.86	-2.69	14.22	30.65	-7.98	Range	= 38.63
(v) New sums	119.03	128.15	130.26	139.07	141.25	New sum **s**	= 94.01
(vi) New averages : $\bar{y}i$	59.52	64.08	65.13	69.54	70.62	New average **s** = new sum **s** / n - 1	= 10.45

Calculation of Effects			Calculation of 2 S.E. Limits
Phase mean	= *1/5 ($\bar{y}o$ + $\bar{y}1$ + $\bar{y}2$ + $\bar{y}3$ + $\bar{y}4$)*	= 65.78	
(A) effect	= *½ ($\bar{y}2$ + $\bar{y}4$ - $\bar{y}1$ - $\bar{y}3$)*	= 1.07	For new average **s** : $\pm(2/\sqrt{n})s$ = 6.58
(B) effect	= *½ ($\bar{y}3$ + $\bar{y}4$ - $\bar{y}1$ - $\bar{y}2$)*	= 5.48	For new effects : $\pm(2/\sqrt{n})s$ = 6.58
(AB) effect	= *½ ($\bar{y}1$ + $\bar{y}4$ - $\bar{y}3$ - $\bar{y}2$)*	= 0.02	For change in mean : $\pm(1.79/\sqrt{n})s$ = 5.95
Change in mean effect	= phase mean - *$\bar{y}o$*	= 6.26	

Table of Multiplying Factors

n	1	2	3	4	5	6	7	8	9	10
fδ,n		0.30	0.35	0.37	0.38	0.39	0.40	0.40	0.40	0.41
1 / n	1.00	0.50	0.33	0.25	0.20	0.17	0.14	0.12	0.11	0.10
1 / (n - 1)		1.00	0.50	0.33	0.25	0.20	0.17	0.14	0.12	0.11
$2/\sqrt{n}$	2.00	1.41	1.15	1.00	0.89	0.82	0.76	0.71	0.67	0.63
$1.79/\sqrt{n}$	1.79	1.26	1.03	0.89	0.80	0.73	0.68	0.63	0.60	0.57

Two variable EVOP program calculation worksheet

Figure 5.86 Work sheet replica No. 10.

Table 5.55 2^2 ROVOP information board.

INFORMATION BOARD AFTER PHASE NO. I

Last Replica Completed: 10 Last Cycle Completed: 1

	RESPONSE No. 1	RESPONSE No. 2	RESPONSE No. 3
REQUIREMENT :	MAXIMIZE RESPONSE		
PROCESS AVERAGES :	69.54 (top left), 70.62 (top right), 59.52 (centre), 64.08 (bottom left), 65.13 (bottom right); B (vertical axis), A (horizontal axis)		
2.S.E. LIMITS FOR AVERAGES	6.58		
EFFECTS WITH 2.S.E — PHASE MEAN	65.78		
FACTOR A	1.07 ± 6.58		
FACTOR B	5.48 ± 6.58		
INTERACTION AB	0.02 ± 6.58		
CHANGE IN MEAN	6.26 ± 5.95		
ESTIMATED STANDARD DEVIATION S	10.45		
PRIOR ESTIMATE OF STANDARD DEV.	10.00		

NOTE:

1. The Center of Experiment-Reference Conditions (The Current Best-Known Process Conditions)
2. The Change in Mean provides a measure of whether the conditions being run the current phase yield an average result better or worse than the reference conditions.
3. The Phase Mean is the Mean Response Over All Conditions Being Run in the present Phase.
4*. The change in Mean is a measure of the over-all Curvature of the Response Surface. It will be positive near a MINIMUM and negative near a MAXIMUM.
5. Sequencing of the Runs-Randomization, if runs can be randomized without too much difficulty, than this should be done.

*A Diagonally oriented saddle represented by equal coefficients of opposite sign could occure and yield a zero change in mean effect, but such a phenomenon would rarely be met in practice.

Conclusions

- There is no significant factor's effect due to high noise after 10 replicates.
- The *change in mean* effect is significant and is a measure of the overall curvature of the response surface. It will be positive near a MINIMUM and negative near a MAXIMUM.
- The estimated standard deviation is very close to the prior estimate of the standard deviation.

Recommendation for Phase I, Cycle 2

- Following the ROVOP rules, 2^2 Factorial Design was rotated through an angle of 45^0 and factor's interval of variation multiplied by $\sqrt{2}$.
- To enlarge factor's variation from 0.5 to 0.71 (Table 5.56, Figure 5.87, Table 5.57).

Table 5.56 2^2 ROVOP simulation.

PROJECT : S-5
PHASE : I
CYCLE : 2

ROVOP SIMULATION

S-5

I. CENTER OF EXPERIMENT : Ao = 20.00; Bo = 20.00 [Gen Random]

II. FACTOR'S INTERVAL OF VARIATION : Δa = 0.71; Δb = 0.71

III. PROCESS (PROPERTY-RESPONSE) NOISE : STDV = 10.00

EVOP WORK SHEET

TRIALS ORDER			FACTOR		RESPONSE									
			A	B	n = 1	n = 2	n = 3	n = 4	n = 5	n = 6	n = 7	n = 8	n = 9	n = 10
0		Block 1	**20.00**	**20.00**	**65.38**	**51.68**	**54.83**	**71.17**	**53.07**	**66.56**	**49.44**	**69.62**	**55.37**	**59.95**
1		Block 1	19.29	19.29	**72.72**	50.34	73.64	57.38	57.18	64.35	75.05	55.72	62.29	**64.91**
2		Block 1	20.71	19.29	**49.96**	50.35	65.43	71.62	61.80	64.63	76.79	73.68	72.44	**58.01**
3		Block 1	19.29	20.71	**73.17**	86.70	66.56	64.62	67.33	68.76	78.23	87.88	89.09	**54.20**
4		Block 1	20.71	20.71	**69.15**	73.80	79.57	42.89	50.04	57.00	70.18	82.50	61.51	**75.11**

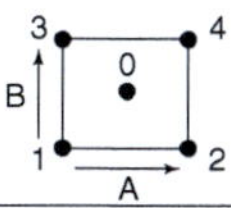

2^2 *Factorial with added reference condition*

Replica N :
10
Response:YIELD

Project S-5
Phase I
Cycle 2

Calculation of Averages					
Operation condition	(0)	(1)	(2)	(3)	(4)
(i) Previous cycle sum	118.17	124.43	144.46	169.70	134.25
(ii) Previous cycle average	59.09	62.22	72.23	84.85	67.13
(iii) New observations	59.95	64.91	58.01	54.20	75.11
(iv) Differences (ii) less (iii)	-0.86	-2.69	14.22	30.65	-7.98
(v) New sums	119.03	127.12	130.24	139.05	142.23
(vi) New averages : $\ddot{y}i$	59.52	63.56	65.12	69.53	71.12

Calculation of Standard Deviation	
Prior estimate of σ	= 10.00
Previous sum **s**	= 78.17
Previous average **s**	= 9.77
New **s** = range × $f\delta,n$	= 15.84
Range	= 38.63
New sum **s**	= 94.01
New average **s** = $\frac{\text{new sum } s}{n-1}$	= 10.45

Calculation of Effects		
Phase mean	= *1/5 ($\ddot{y}o$ + $\ddot{y}1$ + $\ddot{y}2$ + $\ddot{y}3$ + $\ddot{y}4$)*	= 65.77
(A) effect	= *½ ($\ddot{y}2$ + $\ddot{y}4$ - $\ddot{y}1$ - $\ddot{y}3$)*	= 1.58
(B) effect	= *½ ($\ddot{y}3$ + $\ddot{y}4$ - $\ddot{y}1$ - $\ddot{y}2$)*	= 5.98
(AB) effect	= *½ ($\ddot{y}1$ + $\ddot{y}4$ - $\ddot{y}3$ - $\ddot{y}2$)*	= 0.02
Change in mean effect	= phase mean - $\ddot{y}o$	= 6.25

Calculation of 2 S.E. Limits		
For new average **s** :	$\pm (2/\sqrt{n})s$	= 6.58
For new effects :	$\pm (2/\sqrt{n})s$	= 6.58
For change in mean :	$\pm (1.79/\sqrt{n})s$	= 5.95

Table of Multiplying Factors

n	1	2	3	4	5	6	7	8	9	10
$f\delta,n$		0.30	0.35	0.37	0.38	0.39	0.40	0.40	0.40	0.41
1 / n	1.00	0.50	0.33	0.25	0.20	0.17	0.14	0.12	0.11	0.10
1 / (n - 1)		1.00	0.50	0.33	0.25	0.20	0.17	0.14	0.12	0.11
$2/\sqrt{n}$	2.00	1.41	1.15	1.00	0.89	0.82	0.76	0.71	0.67	0.63
$1.79/\sqrt{n}$	1.79	1.26	1.03	0.89	0.80	0.73	0.68	0.63	0.60	0.57

Two variable EVOP program calculation worksheet

Figure 5.87 Work sheet replica No. 10.

Table 5.57 2^2 ROVOP information board.

INFORMATION BOARD AFTER PHASE NO. I

Last Replica Completed: 10 Last Cycle Completed: 2

		RESPONSE No. 1	RESPONSE No. 2	RESPONSE No. 3
REQUIREMENT :		MAXIMIZE RESPONSE		
PROCESS AVERAGES :		69.53 71.12 59.52 B 63.56 A 65.12		
2.S.E. LIMITS FOR AVERAGES		6.58		
EFFECTS WITH 2.S.E	PHASE MEAN	65.77		
	FACTOR A	1.58 ± 6.58		
	FACTOR B	5.98 ± 6.58		
	INTERACTION AB	0.02 ± 6.58		
	CHANGE IN MEAN	6.25 ± 5.95		
ESTIMATED STANDARD DEVIATION S		10.45		
PRIOR ESTIMATE OF STANDARD DEV.		10.00		

NOTE:

1. The Center of Experiment-Reference Conditions (The Current Best-Known Process Conditions)
2. The Change in Mean provides a measure of whether the conditions being run the current phase yield an average result better or worse than the reference conditions.
3. The Phase Mean is the Mean Response Over All Conditions Being Run in the present Phase.
4*. The change in Mean is a measure of the over-all Curvature of the Response Surface. It will be positive near a MINIMUM and negative near a MAXIMUM.
5. Sequencing of the Runs-Randomization, if runs can be randomized without too much difficulty, than this should be done.

*A Diagonally oriented saddle represented by equal coefficients of opposite sign could occure and yield a zero change in mean effect, but such a phenomenon would rarely be met in practice.

Conclusions

- There is no significant factor's effect due to high noise after ten replicas.
- The *change in mean* effect is significant and is a measure of the overall curvature of the response surface. It will be positive near a MINIMUM and negative near a MAXIMUM.
- The estimated standard deviation is very close to the prior estimate of the standard deviation.

Recommendation for Phase I, Cycle 3

- Following the ROVOP rules, 2^2 Factorial Design was rotated through an angle of 45° and factor's interval of variation multiplied by $\sqrt{2}$.
- To enlarge factor's variation from 0.71 to 1.0 (Table 5.58, Figure 5.88, Table 5.59).

Table 5.58 2^2 ROVOP simulation.

PROJECT : S-5
PHASE : I
CYCLE : 3

ROVOP SIMULATION

S-5

I. CENTER OF EXPERIMENT : Ao = 20.00 ; Bo = 20.00 — Gen Random

II. FACTOR'S INTERVAL OF VARIATION : Δa = 1.00 ; Δb = 1.00

III. PROCESS (PROPERTY-RESPONSE) NOISE : STDV = 10.00

EVOP WORK SHEET

TRIALS ORDER			FACTOR		RESPONSE									
			A	B	n = 1	n = 2	n = 3	n = 4	n = 5	n = 6	n = 7	n = 8	n = 9	n = 10
0		Block 1	**20.00**	**20.00**	**65.38**	**51.68**	**54.83**	**71.17**	**53.07**	**66.56**	**49.44**	**69.62**	**55.37**	**59.95**
1		Block 1	19.00	19.00	**72.00**	49.62	72.92	56.66	56.46	63.63	74.33	55.00	61.57	**64.19**
2		Block 1	21.00	19.00	**49.94**	50.33	65.41	71.60	61.78	64.61	76.78	73.66	72.42	**57.99**
3		Block 1	19.00	21.00	**73.15**	86.68	66.54	64.60	67.31	68.74	78.21	87.86	89.07	**54.18**
4		Block 1	21.00	21.00	**69.83**	74.47	80.25	43.57	50.71	57.67	70.85	83.17	62.18	**75.78**

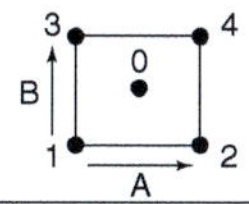

2^2 Factorial with added reference condition

Replica N : *10*
Response:YIELD

Project S-5
Phase I
Cycle 3

Calculation of Averages						Calculation of Standard Deviation	
Operation condition	(0)	(1)	(2)	(3)	(4)	Prior estimate of σ	= 10.00
(i) Previous cycle sum	118.17	123.00	144.42	169.66	135.60	Previous sum **s**	= 78.17
(ii) Previous cycle average	59.09	61.50	72.21	84.83	67.80	Previous average **s**	= 9.77
(iii) New observations	59.95	64.19	57.99	54.18	75.78	New **s** = range × $f\delta,n$	= 15.84
(iv) Differences (ii) less (iii)	-0.86	-2.69	14.22	30.65	-7.98	Range	= 38.63
(v) New sums	119.03	125.69	130.20	139.01	143.59	New sum **s**	= 94.01
(vi) New averages : $\bar{y}i$	59.52	62.85	65.10	69.51	71.79	New average **s** = new sum **s** / n - 1	= 10.45

Calculation of Effects			Calculation of 2 S.E. Limits	
Phase mean	$= 1/5\ (\bar{y}o + \bar{y}1 + \bar{y}2 + \bar{y}3 + \bar{y}4)$	= 65.75		
(A) effect	$= ½\ (\bar{y}2 + \bar{y}4 - \bar{y}1 - \bar{y}3)$	= 2.27	For new average **s** : $\pm (2/\sqrt{n})s$	= 6.58
(B) effect	$= ½\ (\bar{y}3 + \bar{y}4 - \bar{y}1 - \bar{y}2)$	= 6.68	For new effects : $\pm (2/\sqrt{n})s$	= 6.58
(AB) effect	$= ½\ (\bar{y}1 + \bar{y}4 - \bar{y}3 - \bar{y}2)$	= 0.02	For change in mean : $\pm (1.79/\sqrt{n})s$	= 5.95
Change in mean effect	= phase mean - $\bar{y}o$	= 6.24		

Table of Multiplying Factors

n	1	2	3	4	5	6	7	8	9	10
$f\delta,n$		0.30	0.35	0.37	0.38	0.39	0.40	0.40	0.40	0.41
1 / n	1.00	0.50	0.33	0.25	0.20	0.17	0.14	0.12	0.11	0.10
1 / (n - 1)		1.00	0.50	0.33	0.25	0.20	0.17	0.14	0.12	0.11
$2/\sqrt{n}$	2.00	1.41	1.15	1.00	0.89	0.82	0.76	0.71	0.67	0.63
$1.79/\sqrt{n}$	1.79	1.26	1.03	0.89	0.80	0.73	0.68	0.63	0.60	0.57

Two variable EVOP program calculation worksheet

Figure 5.88 Work sheet replica No. 10.

Table 5.59 2^2 ROVOP information board.

INFORMATION BOARD AFTER PHASE NO. I Last Replica Completed: 10 Last Cycle Completed: 3			
	RESPONSE No. 1	RESPONSE No. 2	RESPONSE No. 3
REQUIREMENT :	MAXIMIZE RESPONSE		
PROCESS AVERAGES :	69.51 71.79 59.52 B 62.85 65.10 A		
2.S.E. LIMITS FOR AVERAGES	6.58		
EFFECTS WITH 2.S.E — PHASE MEAN	65.75		
FACTOR A	2.27 ± 6.58		
FACTOR B	6.68 ± 6.58		
INTERACTION AB	0.02 ± 6.58		
CHANGE IN MEAN	6.24 ± 5.95		
ESTIMATED STANDARD DEVIATION S	10.45		
PRIOR ESTIMATE OF STANDARD DEV.	10.00		

NOTE:

1. The Center of Experiment-Reference Conditions (The Current Best-Known Process Conditions)
2. The Change in Mean provides a measure of whether the conditions being run the current phase yield an average result better or worse than the reference conditions.
3. The Phase Mean is the Mean Response Over All Conditions Being Run in the present Phase.
4*. The change in Mean is a measure of the over-all Curvature of the Response Surface. It will be positive near a MINIMUM and negative near a MAXIMUM.
5. Sequencing of the Runs-Randomization, if runs can be randomized without too much difficulty, than this should be done.

*A Diagonally oriented saddle represented by equal coefficients of opposite sign could occure and yield a zero change in mean effect, but such a phenomenon would rarely be met in practice.

Conclusions

- There is one factor B greater than limits but very close. The second factor is not significant.
- The *change in mean* effect is significant and is a measure of the overall curvature of the response surface. It will be positive near a MINIMUM and negative near a MAXIMUM.
- The estimated standard deviation is very close to the prior estimate of the standard deviation.

Recommendation for Phase I, Cycle 4

- Following the ROVOP rules, 2^2 Factorial Design was rotated through an angle of 45^0 and factor's interval of variation multiplied by $\sqrt{2}$.
- To enlarge factor's variation from 1.00 to 1.41 (Table 5.60, Figure 5.89, Table 5.61).

Table 5.60 2^2 ROVOP simulation.

PROJECT : S-5
PHASE : I
CYCLE : 4

ROVOP SIMULATION

S-5

I. CENTER OF EXPERIMENT : Ao = 20.00 Bo = 20.00 [Gen Random]

II. FACTOR'S INTERVAL OF VARIATION : Δa = 1.41 Δb = 1.41

III. PROCESS (PROPERTY-RESPONSE) NOISE : STDV = 10.00

EVOP WORK SHEET

TRIALS ORDER			FACTOR		RESPONSE									
			A	B	n = 1	n = 2	n = 3	n = 4	n = 5	n = 6	n = 7	n = 8	n = 9	n = 10
0		Block 1	**20.00**	**20.00**	**65.38**	**51.68**	**54.83**	**71.17**	**53.07**	**66.56**	**49.44**	**69.62**	**55.37**	**59.95**
1		Block 1	18.59	18.59	**70.98**	48.60	71.90	55.64	55.44	62.61	73.31	53.98	60.55	**63.17**
2		Block 1	21.41	18.59	**49.90**	50.29	65.37	71.56	61.74	64.57	76.74	73.62	72.38	**57.95**
3		Block 1	18.59	21.41	**73.11**	86.64	66.50	64.56	67.27	68.70	78.17	87.82	89.03	**54.14**
4		Block 1	21.41	21.41	**70.78**	75.42	81.19	44.51	51.66	58.62	71.80	84.12	63.13	**76.73**

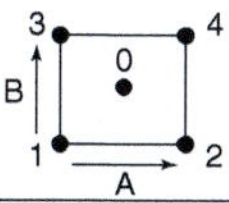

2^2 *Factorial with added reference condition*

Replica N :
10
Response:YIELD

Project S-5
Phase I
Cycle 4

Calculation of Averages						Calculation of Standard Deviation	
Operation condition	(0)	(1)	(2)	(3)	(4)	Prior estimate of σ	= 10.00
(i) Previous cycle sum	118.17	120.95	144.34	169.59	137.49	Previous sum **s**	= 78.17
(ii) Previous cycle average	59.09	60.48	72.17	84.79	68.75	Previous average **s**	= 9.77
(iii) New observations	59.95	63.17	57.95	54.14	76.73	New **s** = range × ***fδ,n***	= 15.84
(iv) Differences (ii) less (iii)	-0.86	-2.69	14.22	30.65	-7.98	Range	= 38.63
(v) New sums	119.03	123.65	130.12	138.93	145.47	New sum **s**	= 94.01
(vi) New averages : *ӱi*	59.52	61.82	65.06	69.47	72.74	New average **s** = new sum **s** / n - 1	= 10.45

Calculation of Effects			Calculation of 2 S.E. Limits	
Phase mean	= *1/5 (ӱo + ӱ1 + ӱ2 + ӱ3 + ӱ4)*	= 65.72		
(A) ——— effect	= *½ (ӱ2 + ӱ4 - ӱ1 - ӱ3)*	= 3.26	For new average **s** : ± $(2/\sqrt{n})$s	= 6.58
(B) ——— effect	= *½ (ӱ3 + ӱ4 - ӱ1 - ӱ2)*	= 7.66	For new effects : ± $(2/\sqrt{n})$s	= 6.58
(AB) ——— effect	= *½ (ӱ1 + ӱ4 - ӱ3 - ӱ2)*	= 0.02	For change in mean : ± $(1.79/\sqrt{n})$s	= 5.95
Change in mean effect	= phase mean - *ӱo*	= 6.20		

Table of Multiplying Factors

n	1	2	3	4	5	6	7	8	9	10
fδ,n		0.30	0.35	0.37	0.38	0.39	0.40	0.40	0.40	0.41
1 / n	1.00	0.50	0.33	0.25	0.20	0.17	0.14	0.12	0.11	0.10
1 / (n - 1)		1.00	0.50	0.33	0.25	0.20	0.17	0.14	0.12	0.11
$2/\sqrt{n}$	2.00	1.41	1.15	1.00	0.89	0.82	0.76	0.71	0.67	0.63
$1.79/\sqrt{n}$	1.79	1.26	1.03	0.89	0.80	0.73	0.68	0.63	0.60	0.57

Two variable EVOP program calculation worksheet

Figure 5.89 Work sheet replica No. 10.

Table 5.61 2^2 ROVOP information board.

INFORMATION BOARD AFTER PHASE NO. I

Last Replica Completed: 10 **Last Cycle Completed: 4**

		RESPONSE No. 1	RESPONSE No. 2	RESPONSE No. 3
REQUIREMENT :		MAXIMIZE RESPONSE		
PROCESS AVERAGES :		69.47 72.74 B 59.52 61.82 A 65.06		
2.S.E. LIMITS FOR AVERAGES		6.58		
EFFECTS WITH 2.S.E	PHASE MEAN FACTOR A FACTOR B INTERACTION AB CHANGE IN MEAN	65.72 3.26 ± 6.58 7.66 ± 6.58 0.02 ± 6.58 6.20 ± 5.95		
ESTIMATED STANDARD DEVIATION S		10.45		
PRIOR ESTIMATE OF STANDARD DEV.		10.00		

NOTE:

1. The Center of Experiment-Reference Conditions (The Current Best-Known Process Conditions)
2. The Change in Mean provides a measure of whether the conditions being run the current phase yield an average result better or worse than the reference conditions.
3. The Phase Mean is the Mean Response Over All Conditions Being Run in the present Phase.
4*. The change in Mean is a measure of the over-all Curvature of the Response Surface. It will be positive near a MINIMUM and negative near a MAXIMUM.
5. Sequencing of the Runs-Randomization, if runs can be randomized without too much difficulty, than this should be done.

*A Diagonally oriented saddle represented by equal coefficients of opposite sign could occure and yield a zero change in mean effect, but such a phenomenon would rarely be met in practice.

Conclusions

- Only factor B is significant. The second factor is still not significant.
- The *change in mean* effect is significant and is a measure of the overall curvature of the response surface. It will be positive near a MINIMUM and negative near a MAXIMUM.
- The estimated standard deviation is very close to the prior estimate of the standard deviation.

Recommendation for Phase I, Cycle 5

- Following the ROVOP rules, 2^2 Factorial Design was rotated through an angle of 45° and factor's interval of variation multiplied by $\sqrt{2}$.
- To enlarge factor's variation from 1.41 to 2.0 (Table 5.62, Figure 5.90, Table 5.63).

Table 5.62 2^2 ROVOP simulation.

PROJECT : S-5
PHASE : I
CYCLE : 5

ROVOP SIMULATION

S-5

I. CENTER OF EXPERIMENT : Ao = 20.00; Bo = 20.00 [Gen Random]

II. FACTOR'S INTERVAL OF VARIATION : Δa = 2.00; Δb = 2.00

III. PROCESS (PROPERTY-RESPONSE) NOISE : STDV = 10.00

EVOP WORK SHEET

TRIALS ORDER		FACTOR A	FACTOR B	RESPONSE n = 1	n = 2	n = 3	n = 4	n = 5	n = 6	n = 7	n = 8	n = 9	n = 10
0	Block 1	20.00	20.00	65.38	51.68	54.83	71.17	53.07	66.56	49.44	69.62	55.37	59.95
1	Block 1	18.00	18.00	69.48	47.10	70.40	54.14	53.94	61.11	71.81	52.48	59.05	61.67
2	Block 1	22.00	18.00	49.82	50.21	65.29	71.48	61.66	64.49	76.66	73.54	72.30	57.87
3	Block 1	18.00	22.00	73.03	86.56	66.42	64.48	67.19	68.62	78.09	87.74	88.95	54.06
4	Block 1	22.00	22.00	72.11	76.75	82.53	45.85	52.99	59.95	73.13	85.45	64.46	78.06

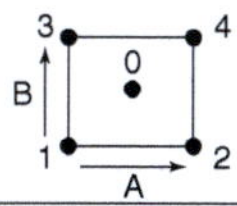

2^2 Factorial with added reference condition

Replica N :
10
Response:YIELD

Project S-5
Phase I
Cycle 5

Calculation of Averages					
Operation condition	(0)	(1)	(2)	(3)	(4)
(i) Previous cycle sum	118.17	117.96	144.18	169.42	140.16
(ii) Previous cycle average	59.09	58.98	72.09	84.71	70.08
(iii) New observations	59.95	61.67	57.87	54.06	78.06
(iv) Differences (ii) less (iii)	-0.86	-2.69	14.22	30.65	-7.98
(v) New sums	119.03	120.65	129.96	138.77	148.15
(vi) New averages : $\bar{y}i$	59.52	60.33	64.98	69.39	74.07

Calculation of Standard Deviation	
Prior estimate of σ	= 10.00
Previous sum **s**	= 78.17
Previous average **s**	= 9.77
New **s** = range × $f\delta,n$	= 15.84
Range	= 38.63
New sum **s**	= 94.01
New average **s** = new sum **s** / (n - 1)	= 10.45

Calculation of Effects		
Phase mean	= 1/5 ($\bar{y}o + \bar{y}1 + \bar{y}2 + \bar{y}3 + \bar{y}4$)	= 65.66
(A) effect	= ½ ($\bar{y}2 + \bar{y}4 - \bar{y}1 - \bar{y}3$)	= 4.67
(B) effect	= ½ ($\bar{y}3 + \bar{y}4 - \bar{y}1 - \bar{y}2$)	= 9.08
(AB) effect	= ½ ($\bar{y}1 + \bar{y}4 - \bar{y}3 - \bar{y}2$)	= 0.02
Change in mean effect	= phase mean - $\bar{y}o$	= 6.14

Calculation of 2 S.E. Limits		
For new average **s** :	$\pm(2/\sqrt{n})s$	= 6.58
For new effects :	$\pm(2/\sqrt{n})s$	= 6.58
For change in mean :	$\pm(1.79/\sqrt{n})s$	= 5.95

Table of Multiplying Factors

n	1	2	3	4	5	6	7	8	9	10
$f\delta,n$		0.30	0.35	0.37	0.38	0.39	0.40	0.40	0.40	0.41
1 / n	1.00	0.50	0.33	0.25	0.20	0.17	0.14	0.12	0.11	0.10
1 / (n - 1)		1.00	0.50	0.33	0.25	0.20	0.17	0.14	0.12	0.11
$2/\sqrt{n}$	2.00	1.41	1.15	1.00	0.89	0.82	0.76	0.71	0.67	0.63
$1.79/\sqrt{n}$	1.79	1.26	1.03	0.89	0.80	0.73	0.68	0.63	0.60	0.57

Two variable EVOP program calculation worksheet

Figure 5.90 Work sheet replica No. 10.

Table 5.63 2^2 ROVOP information board.

INFORMATION BOARD AFTER PHASE NO. I

Last Replica Completed: 10 **Last Cycle Completed: 5**

	RESPONSE No. 1	RESPONSE No. 2	RESPONSE No. 3
REQUIREMENT :	MAXIMIZE RESPONSE		
PROCESS AVERAGES :	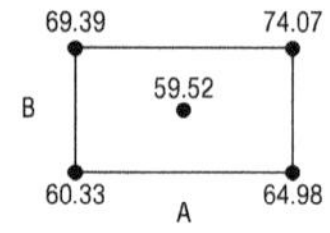		
2.S.E. LIMITS FOR AVERAGES	6.58		
EFFECTS WITH 2.S.E — PHASE MEAN	65.66		
FACTOR A	4.67 ± 6.58		
FACTOR B	9.08 ± 6.58		
INTERACTION AB	0.02 ± 6.58		
CHANGE IN MEAN	6.14 ± 5.95		
ESTIMATED STANDARD DEVIATION S	10.45		
PRIOR ESTIMATE OF STANDARD DEV.	10.00		

NOTE:

1. The Center of Experiment-Reference Conditions (The Current Best-Known Process Conditions)
2. The Change in Mean provides a measure of whether the conditions being run the current phase yield an average result better or worse than the reference conditions.
3. The Phase Mean is the Mean Response Over All Conditions Being Run in the present Phase.
4*. The change in Mean is a measure of the over-all Curvature of the Response Surface. It will be positive near a MINIMUM and negative near a MAXIMUM.
5. Sequencing of the Runs-Randomization, if runs can be randomized without too much difficulty, than this should be done.

*A Diagonally oriented saddle represented by equal coefficients of opposite sign could occure and yield a zero change in mean effect, but such a phenomenon would rarely be met in practice.

Conclusions

- Only factor B is significant. The second factor is still not significant. One more cycle is needed to make sure.
- The *change in mean* effect is significant and is a measure of the overall curvature of the response surface. It will be positive near a MINIMUM and negative near a MAXIMUM.
- The estimated standard deviation is very close to the prior estimate of the standard deviation.

Recommendation for Phase I, Cycle 6

- Following the ROVOP rules, 2^2 Factorial Design was rotated through an angle of 45° and factor's interval of variation multiplied by $\sqrt{2}$.
- To enlarge factor's variation from 2.0 to 2.82 (Table 5.64, Figure 5.91, Table 5.65).

Table 5.64 2^2 ROVOP simulation.

PROJECT : S-5
PHASE : I
CYCLE : 6

ROVOP SIMULATION

S-5

Gen Random

I. CENTER OF EXPERIMENT : Ao = 20.00; Bo = 20.00

II. FACTOR'S INTERVAL OF VARIATION : Δa = 2.82; Δb = 2.82

III. PROCESS (PROPERTY-RESPONSE) NOISE : STDV = 10.00

EVOP WORK SHEET

TRIALS ORDER			FACTOR		RESPONSE									
			A	B	n = 1	n = 2	n = 3	n = 4	n = 5	n = 6	n = 7	n = 8	n = 9	n = 10
0		Block 1	**20.00**	**20.00**	**65.38**	**51.68**	**54.83**	**71.17**	**53.07**	**66.56**	**49.44**	**69.62**	**55.37**	**59.95**
1		Block 1	17.18	17.18	**67.36**	44.98	68.28	52.02	51.81	58.99	69.69	50.36	56.92	**59.55**
2		Block 1	22.82	17.18	**49.66**	50.05	65.13	71.33	61.50	64.33	76.50	73.38	72.14	**57.71**
3		Block 1	17.18	22.82	**72.88**	86.40	66.27	64.32	67.03	68.47	77.93	87.58	88.80	**53.90**
4		Block 1	22.82	22.82	**73.92**	78.56	84.33	47.66	54.80	61.76	74.94	87.26	66.27	**79.87**

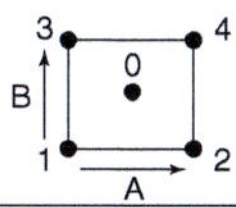

2^2 Factorial with added reference condition

Replica N :
10
Response:YIELD

Project S-5
Phase I
Cycle 6

Calculation of Averages						Calculation of Standard Deviation	
Operation condition	(0)	(1)	(2)	(3)	(4)	Prior estimate of σ	= 10.00
(i) Previous cycle sum	118.17	113.71	143.86	169.11	143.78	Previous sum **s**	= 78.17
(ii) Previous cycle average	59.09	56.85	71.93	84.55	71.89	Previous average **s**	= 9.77
(iii) New observations	59.95	59.55	57.71	53.90	79.87	New **s** = range × $f\delta,n$	= 15.84
(iv) Differences (ii) less (iii)	-0.86	-2.69	14.22	30.65	-7.98	Range	= 38.63
(v) New sums	119.03	116.40	129.65	138.45	151.77	New sum **s**	= 94.01
(vi) New averages : $\bar{y}i$	59.52	58.20	64.82	69.23	75.88	New average **s** = new sum **s** / (n - 1)	= 10.45

Calculation of Effects			Calculation of 2 S.E. Limits
Phase mean	= 1/5 ($\bar{y}o$ + $\bar{y}1$ + $\bar{y}2$ + $\bar{y}3$ + $\bar{y}4$)	= 65.53	
(A) ———— effect	= ½ ($\bar{y}2$ + $\bar{y}4$ - $\bar{y}1$ - $\bar{y}3$)	= 6.64	For new average **s** : $\pm (2/\sqrt{n})$s = 6.58
(B) ———— effect	= ½ ($\bar{y}3$ + $\bar{y}4$ - $\bar{y}1$ - $\bar{y}2$)	= 11.04	For new effects : $\pm (2/\sqrt{n})$s = 6.58
(AB) ———— effect	= ½ ($\bar{y}1$ + $\bar{y}4$ - $\bar{y}3$ - $\bar{y}2$)	= 0.02	For change in mean : $\pm (1.79/\sqrt{n})$s = 5.95
Change in mean effect	= phase mean - $\bar{y}o$	= 6.01	

Table of Multiplying Factors

n	1	2	3	4	5	6	7	8	9	10
$f\delta,n$		0.30	0.35	0.37	0.38	0.39	0.40	0.40	0.40	0.41
1 / n	1.00	0.50	0.33	0.25	0.20	0.17	0.14	0.12	0.11	0.10
1 / (n - 1)		1.00	0.50	0.33	0.25	0.20	0.17	0.14	0.12	0.11
$2/\sqrt{n}$	2.00	1.41	1.15	1.00	0.89	0.82	0.76	0.71	0.67	0.63
$1.79/\sqrt{n}$	1.79	1.26	1.03	0.89	0.80	0.73	0.68	0.63	0.60	0.57

Two variable EVOP program calculation worksheet

Figure 5.91 Work sheet replica No. 10.

Table 5.65 2^2 ROVOP information board.

INFORMATION BOARD AFTER PHASE NO. I

Last Replica Completed: 10 **Last Cycle Completed: 6**

	RESPONSE No. 1	RESPONSE No. 2	RESPONSE No. 3
REQUIREMENT :	MAXIMIZE RESPONSE		
PROCESS AVERAGES :	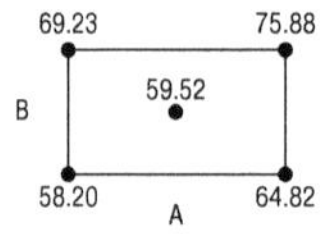		
2.S.E. LIMITS FOR AVERAGES	6.58		
EFFECTS WITH 2.S.E — PHASE MEAN	65.53		
FACTOR A	6.64 ± 6.58		
FACTOR B	11.04 ± 6.58		
INTERACTION AB	0.02 ± 6.58		
CHANGE IN MEAN	6.01 ± 5.95		
ESTIMATED STANDARD DEVIATION S	10.45		
PRIOR ESTIMATE OF STANDARD DEV.	10.00		

NOTE:

1. The Center of Experiment-Reference Conditions (The Current Best-Known Process Conditions)
2. The Change in Mean provides a measure of whether the conditions being run the current phase yield an average result better or worse than the reference conditions.
3. The Phase Mean is the Mean Response Over All Conditions Being Run in the present Phase.
4*. The change in Mean is a measure of the over-all Curvature of the Response Surface. It will be positive near a MINIMUM and negative near a MAXIMUM.
5. Sequencing of the Runs-Randomization, if runs can be randomized without too much difficulty, than this should be done.

*A Diagonally oriented saddle represented by equal coefficients of opposite sign could occure and yield a zero change in mean effect, but such a phenomenon would rarely be met in practice.

Conclusions

- Both factors are significant with positive sign. It is time to move center of the experiment in positive direction.
- The *change in mean* effect is still significant and is a measure of the overall curvature of the response surface. It will be positive near a MINIMUM and negative near a MAXIMUM.
- The estimated standard deviation is very close to the prior estimate of the standard deviation.

Recommendation for Phase II, Cycle 1

- Move the center of the experiment from $A_0 = 20$; $B_0 = 20$ to $A_0 = 22.82$; $B_0 = 22.82$ and start a phase II.
- Reduce factor's interval of variation from $\Delta a = 2.82$; $\Delta b = 2.82$ to $\Delta a = 2.00$; $\Delta b = 2.00$ following the ROVOP rules from Figure 5.85 (Table 5.66, Figure 5.92, Table 5.67).

Table 5.66 2^2 ROVOP simulation.

PROJECT : S-5
PHASE : II
CYCLE : 1

ROVOP SIMULATION

S-5

I. CENTER OF EXPERIMENT : Ao = 22.82 ; Bo = 22.82 | Gen Random

II. FACTOR'S INTERVAL OF VARIATION : Δa = 2.00 ; Δb = 2.00

III. PROCESS (PROPERTY-RESPONSE) NOISE : STDV = 10.00

EVOP WORK SHEET

TRIALS ORDER			FACTOR		RESPONSE									
			A	B	n = 1	n = 2	n = 3	n = 4	n = 5	n = 6	n = 7	n = 8	n = 9	n = 10
0		Block 1	**22.82**	**22.82**	**71.83**	**58.13**	**61.28**	**77.62**	**59.52**	**73.01**	**55.89**	**76.07**	**61.82**	**66.40**
1		Block 1	20.82	20.82	**76.38**	54.00	77.30	61.04	60.84	68.01	78.71	59.38	65.95	**68.57**
2		Block 1	24.82	20.82	**56.27**	56.66	71.74	77.93	68.11	70.94	83.10	79.99	78.75	**64.32**
3		Block 1	20.82	24.82	**79.48**	93.01	72.87	70.93	73.64	75.07	84.54	94.19	95.40	**60.51**
4		Block 1	24.82	24.82	**78.11**	82.75	88.52	51.84	58.99	65.95	79.13	91.45	70.46	**84.06**

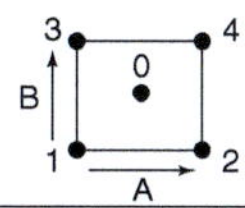

2^2 Factorial with added reference condition

Replica N :
10
Response:YIELD

Project S-5
Phase II
Cycle 1

Calculation of Averages						Calculation of Standard Deviation	
Operation condition	(0)	(1)	(2)	(3)	(4)	Prior estimate of σ	= 10.00
(i) Previous cycle sum	131.07	131.76	157.08	182.32	152.16	Previous sum **s**	= 78.17
(ii) Previous cycle average	65.54	65.88	78.54	91.16	76.08	Previous average **s**	= 9.77
(iii) New observations	66.40	68.57	64.32	60.51	84.06	New **s** = range × $f\delta,n$	= 15.84
(iv) Differences (ii) less (iii)	-0.86	-2.69	14.22	30.65	-7.98	Range	= 38.63
(v) New sums	131.93	134.46	142.86	151.67	160.14	New sum **s**	= 94.01
(vi) New averages : $\bar{y}i$	65.97	67.23	71.43	75.84	80.07	New average **s** = new sum **s** / (n - 1)	= 10.45

Calculation of Effects			Calculation of 2 S.E. Limits	
Phase mean	$= 1/5\,(\bar{y}o + \bar{y}1 + \bar{y}2 + \bar{y}3 + \bar{y}4)$	= 72.11		
(A) ———— effect	$= ½\,(\bar{y}2 + \bar{y}4 - \bar{y}1 - \bar{y}3)$	= 4.22	For new average **s** : $\pm(2/\sqrt{n})s$	= 6.58
(B) ———— effect	$= ½\,(\bar{y}3 + \bar{y}4 - \bar{y}1 - \bar{y}2)$	= 8.62	For new effects : $\pm(2/\sqrt{n})s$	= 6.58
(AB) ———— effect	$= ½\,(\bar{y}1 + \bar{y}4 - \bar{y}3 - \bar{y}2)$	= 0.02	For change in mean : $\pm(1.79/\sqrt{n})s$	= 5.95
Change in mean effect	= phase mean - $\bar{y}o$	= 6.14		

Table of Multiplying Factors

n	1	2	3	4	5	6	7	8	9	10
$f\delta,n$		0.30	0.35	0.37	0.38	0.39	0.40	0.40	0.40	0.41
1 / n	1.00	0.50	0.33	0.25	0.20	0.17	0.14	0.12	0.11	0.10
1 / (n - 1)		1.00	0.50	0.33	0.25	0.20	0.17	0.14	0.12	0.11
$2/\sqrt{n}$	2.00	1.41	1.15	1.00	0.89	0.82	0.76	0.71	0.67	0.63
$1.79/\sqrt{n}$	1.79	1.26	1.03	0.89	0.80	0.73	0.68	0.63	0.60	0.57

Two variable EVOP program calculation worksheet

Figure 5.92 Work sheet replica No. 10.

Table 5.67 2^2 ROVOP information board.

INFORMATION BOARD AFTER PHASE NO. II

Last Replica Completed: 10 Last Cycle Completed: 1

	RESPONSE No. 1	RESPONSE No. 2	RESPONSE No. 3
REQUIREMENT :	MAXIMIZE RESPONSE		
PROCESS AVERAGES :	75.84 80.07 65.97 B 67.23 71.43 A		
2.S.E. LIMITS FOR AVERAGES	6.58		
EFFECTS WITH 2.S.E — PHASE MEAN	72.11		
FACTOR A	4.22 ± 6.58		
FACTOR B	8.62 ± 6.58		
INTERACTION AB	0.02 ± 6.58		
CHANGE IN MEAN	6.14 ± 5.95		
ESTIMATED STANDARD DEVIATION S	10.45		
PRIOR ESTIMATE OF STANDARD DEV.	10.00		

NOTE:

1. The Center of Experiment-Reference Conditions (The Current Best-Known Process Conditions)
2. The Change in Mean provides a measure of whether the conditions being run the current phase yield an average result better or worse than the reference conditions.
3. The Phase Mean is the Mean Response Over All Conditions Being Run in the present Phase.
4*. The change in Mean is a measure of the over-all Curvature of the Response Surface. It will be positive near a MINIMUM and negative near a MAXIMUM.
5. Sequencing of the Runs-Randomization, if runs can be randomized without too much difficulty, than this should be done.

*A Diagonally oriented saddle represented by equal coefficients of opposite sign could occure and yield a zero change in mean effect, but such a phenomenon would rarely be met in practice.

Conclusions

- Only factor B is significant. The second factor is not significant.
- The *phase mean* is increased significantly.
- The *change in mean* effect is significant and is a measure of the overall curvature of the response surface. It will be positive near a MINIMUM and negative near a MAXIMUM.
- The estimated standard deviation is very close to the prior estimate of the standard deviation.

Recommendation for Phase II, Cycle 2

- Following the ROVOP rules, 2^2 Factorial Design was rotated through an angle of 45° and factor's interval of variation multiplied by $\sqrt{2}$.
- To enlarge factor's variation from 2.0 to 2.82 (Table 5.68, Figure 5.93, Table 5.69).

Table 5.68 2^2 ROVOP simulation.

PROJECT : S-5
PHASE : II
CYCLE : 2

ROVOP SIMULATION

S-5

I. CENTER OF EXPERIMENT : Ao = 22.82, Bo = 22.82 (Gen Random)

II. FACTOR'S INTERVAL OF VARIATION : Δa = 2.82, Δb = 2.82

III. PROCESS (PROPERTY-RESPONSE) NOISE : STDV = 10.00

EVOP WORK SHEET

TRIALS ORDER			FACTOR		RESPONSE									
			A	B	n = 1	n = 2	n = 3	n = 4	n = 5	n = 6	n = 7	n = 8	n = 9	n = 10
0		Block 1	22.82	22.82	71.83	58.13	61.28	77.62	59.52	73.01	55.89	76.07	61.82	66.40
1		Block 1	20.00	20.00	74.44	52.06	75.36	59.10	58.90	66.07	76.77	57.44	64.01	66.63
2		Block 1	25.64	20.00	56.11	56.50	71.58	77.78	67.95	70.78	82.95	79.83	78.59	64.16
3		Block 1	20.00	25.64	79.33	92.85	72.72	70.77	73.48	74.92	84.38	94.03	95.25	60.35
4		Block 1	25.64	25.64	79.73	84.38	90.15	53.47	60.62	67.58	80.76	93.08	72.09	85.69

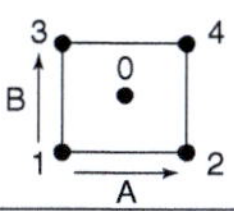

2^2 Factorial with added reference condition

Replica N :
10
Response:YIELD

Project S-5
Phase II
Cycle 2

Calculation of Averages					
Operation condition	(0)	(1)	(2)	(3)	(4)
(i) Previous cycle sum	131.07	127.88	156.76	182.01	155.41
(ii) Previous cycle average	65.54	63.94	78.38	91.00	77.71
(iii) New observations	66.40	66.63	64.16	60.35	85.69
(iv) Differences (ii) less (iii)	-0.86	-2.69	14.22	30.65	-7.98
(v) New sums	131.93	130.57	142.55	151.35	163.39
(vi) New averages : ȳi	65.97	65.29	71.27	75.68	81.70

Calculation of Standard Deviation	
Prior estimate of σ	= 10.00
Previous sum **s**	= 78.17
Previous average **s**	= 9.77
New **s** = range × **fδ,n**	= 15.84
Range	= 38.63
New sum **s**	= 94.01
New average **s** = new sum **s** / (n - 1)	= 10.45

Calculation of Effects

Phase mean = *1/5 (ȳo + ȳ1 + ȳ2 + ȳ3 + ȳ4)* = 71.98
(A) effect = *½ (ȳ2 + ȳ4 - ȳ1 - ȳ3)* = 6.00
(B) effect = *½ (ȳ3 + ȳ4 - ȳ1 - ȳ2)* = 10.41
(AB) effect = *½ (ȳ1 + ȳ4 - ȳ3 - ȳ2)* = 0.02
Change in mean effect = phase mean - ȳo = 6.01

Calculation of 2 S.E. Limits

For new average **s** : $\pm (2/\sqrt{n})s$ = 6.58
For new effects : $\pm (2/\sqrt{n})s$ = 6.58
For change in mean : $\pm (1.79/\sqrt{n})s$ = 5.95

Table of Multiplying Factors

n	1	2	3	4	5	6	7	8	9	10
fδ,n		0.30	0.35	0.37	0.38	0.39	0.40	0.40	0.40	0.41
1 / n	1.00	0.50	0.33	0.25	0.20	0.17	0.14	0.12	0.11	0.10
1 / (n - 1)		1.00	0.50	0.33	0.25	0.20	0.17	0.14	0.12	0.11
$2/\sqrt{n}$	2.00	1.41	1.15	1.00	0.89	0.82	0.76	0.71	0.67	0.63
$1.79/\sqrt{n}$	1.79	1.26	1.03	0.89	0.80	0.73	0.68	0.63	0.60	0.57

Two variable EVOP program calculation worksheet

Figure 5.93 Work sheet replica No. 10.

Table 5.69 2^2 ROVOP information board.

INFORMATION BOARD AFTER PHASE NO. II

Last Replica Completed: 10 Last Cycle Completed: 2

	RESPONSE No. 1	RESPONSE No. 2	RESPONSE No. 3
REQUIREMENT :	MAXIMIZE RESPONSE		
PROCESS AVERAGES :	75.68 (B high, A low); 81.70 (B high, A high); 65.97 (centre); 65.29 (B low, A low); 71.27 (B low, A high)		
2.S.E. LIMITS FOR AVERAGES	6.58		
EFFECTS WITH 2.S.E — PHASE MEAN	71.98		
FACTOR A	6.00 ± 6.58		
FACTOR B	10.41 ± 6.58		
INTERACTION AB	0.02 ± 6.58		
CHANGE IN MEAN	6.01 ± 5.95		
ESTIMATED STANDARD DEVIATION S	10.45		
PRIOR ESTIMATE OF STANDARD DEV.	10.00		

NOTE:

1. The Center of Experiment-Reference Conditions (The Current Best-Known Process Conditions)
2. The Change in Mean provides a measure of whether the conditions being run the current phase yield an average result better or worse than the reference conditions.
3. The Phase Mean is the Mean Response Over All Conditions Being Run in the present Phase.
4*. The change in Mean is a measure of the over-all Curvature of the Response Surface. It will be positive near a MINIMUM and negative near a MAXIMUM.
5. Sequencing of the Runs-Randomization, if runs can be randomized without too much difficulty, than this should be done.

*A Diagonally oriented saddle represented by equal coefficients of opposite sign could occure and yield a zero change in mean effect, but such a phenomenon would rarely be met in practice.

Conclusions

- Only factor B is significant. The second factor is still not significant. One more cycle is needed to make sure.
- The *change in mean* effect is significant and is a measure of the overall curvature of the response surface. It will be positive near a MINIMUM and negative near a MAXIMUM.
- The estimated standard deviation is very close to the prior estimate of the standard deviation.

Recommendation for Phase II, Cycle 3

- Following the ROVOP rules, 2^2 Factorial Design was rotated through an angle of 45° and factor's interval of variation multiplied by $\sqrt{2}$.
- To enlarge factor's variation from 2.82 to 4.0 (Table 5.70, Figure 5.94, Table 5.71).

Table 5.70 2^2 ROVOP simulation.

PROJECT : S-5
PHASE : II
CYCLE : 3

ROVOP SIMULATION

S-5

I. CENTER OF EXPERIMENT : Ao = 22.82; Bo = 22.82 [Gen Random]

II. FACTOR'S INTERVAL OF VARIATION : Δa = 4.00; Δb = 4.00

III. PROCESS (PROPERTY-RESPONSE) NOISE : STDV = 10.00

EVOP WORK SHEET

TRIALS ORDER			FACTOR		RESPONSE									
			A	B	n = 1	n = 2	n = 3	n = 4	n = 5	n = 6	n = 7	n = 8	n = 9	n = 10
0		Block 1	**22.82**	**22.82**	**71.83**	**58.13**	**61.28**	**77.62**	**59.52**	**73.01**	**55.89**	**76.07**	**61.82**	**66.40**
1		Block 1	18.82	18.82	**71.56**	49.18	72.47	56.21	56.01	63.18	73.89	54.56	61.12	**63.75**
2		Block 1	26.82	18.82	**55.79**	56.18	71.26	77.45	67.63	70.46	82.62	79.51	78.27	**63.84**
3		Block 1	18.82	26.82	**79.00**	92.53	72.39	70.45	73.16	74.59	84.06	93.71	94.92	**60.03**
4		Block 1	26.82	26.82	**81.98**	86.62	92.39	55.71	62.86	69.82	83.00	95.32	74.33	**87.93**

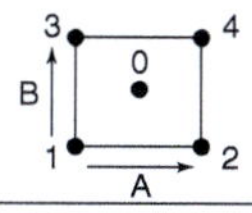

2^2 Factorial with added reference condition

Replica N :
10
Response:YIELD

Project: S-5
Phase: II
Cycle: 3

Calculation of Averages

Operation condition	(0)	(1)	(2)	(3)	(4)
(i) Previous cycle sum	131.07	122.10	156.12	181.36	159.90
(ii) Previous cycle average	65.54	61.05	78.06	90.68	79.95
(iii) New observations	66.40	63.75	63.84	60.03	87.93
(iv) Differences (ii) less (iii)	-0.86	-2.69	14.22	30.65	-7.98
(v) New sums	131.93	124.80	141.90	150.71	167.88
(vi) New averages : $\bar{y}_i$	65.97	62.40	70.95	75.36	83.94

Calculation of Standard Deviation

Prior estimate of σ	= 10.00
Previous sum **s**	= 78.17
Previous average **s**	= 9.77
New **s** = range × ***fδ,n***	= 15.84
Range	= 38.63
New sum **s**	= 94.01
New average **s** = new sum **s** / (n - 1)	= 10.45

Calculation of Effects

Phase mean	$= 1/5\,(\bar{y}_0 + \bar{y}_1 + \bar{y}_2 + \bar{y}_3 + \bar{y}_4)$	= 71.72
(A) effect	$= \tfrac{1}{2}\,(\bar{y}_2 + \bar{y}_4 - \bar{y}_1 - \bar{y}_3)$	= 8.57
(B) effect	$= \tfrac{1}{2}\,(\bar{y}_3 + \bar{y}_4 - \bar{y}_1 - \bar{y}_2)$	= 12.97
(AB) effect	$= \tfrac{1}{2}\,(\bar{y}_1 + \bar{y}_4 - \bar{y}_3 - \bar{y}_2)$	= 0.02
Change in mean effect	= phase mean - $\bar{y}_0$	= 5.76

Calculation of 2 S.E. Limits

For new average **s** :	$\pm (2/\sqrt{n})s$	= 6.58
For new effects :	$\pm (2/\sqrt{n})s$	= 6.58
For change in mean :	$\pm (1.79/\sqrt{n})s$	= 5.95

Table of Multiplying Factors

n	1	2	3	4	5	6	7	8	9	10
fδ,n		0.30	0.35	0.37	0.38	0.39	0.40	0.40	0.40	0.41
1 / n	1.00	0.50	0.33	0.25	0.20	0.17	0.14	0.12	0.11	0.10
1 / (n - 1)		1.00	0.50	0.33	0.25	0.20	0.17	0.14	0.12	0.11
$2/\sqrt{n}$	2.00	1.41	1.15	1.00	0.89	0.82	0.76	0.71	0.67	0.63
$1.79/\sqrt{n}$	1.79	1.26	1.03	0.89	0.80	0.73	0.68	0.63	0.60	0.57

Two variable EVOP program calculation worksheet

Figure 5.94 Work sheet replica No. 10.

Table 5.71 2^2 ROVOP information board.

INFORMATION BOARD AFTER PHASE NO. II Last Replica Completed: 10 Last Cycle Completed: 3			
	RESPONSE No. 1	RESPONSE No. 2	RESPONSE No. 3
REQUIREMENT :	MAXIMIZE RESPONSE		
PROCESS AVERAGES :	75.36 83.94 65.97 B 62.40 70.95 A		
2.S.E. LIMITS FOR AVERAGES	6.58		
EFFECTS WITH 2.S.E — PHASE MEAN FACTOR A FACTOR B INTERACTION AB CHANGE IN MEAN	71.72 8.57 ± 6.58 12.97 ± 6.58 0.02 ± 6.58 5.76 ± 5.95		
ESTIMATED STANDARD DEVIATION S	10.45		
PRIOR ESTIMATE OF STANDARD DEV.	10.00		

NOTE:

1. The Center of Experiment-Reference Conditions (The Current Best-Known Process Conditions)
2. The Change in Mean provides a measure of whether the conditions being run the current phase yield an average result better or worse than the reference conditions.
3. The Phase Mean is the Mean Response Over All Conditions Being Run in the present Phase.
4*. The change in Mean is a measure of the over-all Curvature of the Response Surface. It will be positive near a MINIMUM and negative near a MAXIMUM.
5. Sequencing of the Runs-Randomization, if runs can be randomized without too much difficulty, than this should be done.

*A Diagonally oriented saddle represented by equal coefficients of opposite sign could occure and yield a zero change in mean effect, but such a phenomenon would rarely be met in practice.

Conclusions

- Both factors are significant with positive sign. It is time to move the center of the experiment in positive direction.
- The *change in mean* effect is not significant.
- The estimated standard deviation is very close to the prior estimate of the standard deviation.

Recommendation for Phase III

- Move the center of the experiment from $A_0 = 20$; $B_0 = 20$ to $A_0 = 30$; $B_0 = 30$ and start a phase III.
- Keep the same factor's interval of variation.
- Switch from ROVOP to BEVOP (Table 5.72, Figure 5.95, Table 5.73).

Table 5.72 2^2 BEVOP simulation.

PROJECT : S-5
PHASE : III
CYCLE :

BEVOP SIMULATION

S-5

I. CENTER OF EXPERIMENT : Ao = 30.00; Bo = 30.00 [Gen Random]

II. FACTOR'S INTERVAL OF VARIATION : Δa = 4.00; Δb = 4.00

III. PROCESS (PROPERTY-RESPONSE) NOISE : STDV = 10.00

EVOP WORK SHEET

TRIALS ORDER			FACTOR		RESPONSE									
			A	B	n = 1	n = 2	n = 3	n = 4	n = 5	n = 6	n = 7	n = 8	n = 9	n = 10
0		Block 1	**30.00**	**30.00**	**85.38**	**71.68**	**74.83**	**91.17**	**73.07**	**86.56**	**69.44**	**89.62**	**75.37**	**79.95**
1		Block 1	26.00	26.00	**87.40**	65.02	88.32	72.06	71.86	79.03	89.73	70.40	76.97	**79.59**
2		Block 1	34.00	26.00	**69.34**	69.73	84.81	91.00	81.18	84.01	96.18	93.06	91.82	**77.39**
3		Block 1	26.00	34.00	**92.55**	106.08	85.94	84.00	86.71	88.14	97.61	107.26	108.47	**73.58**
4		Block 1	34.00	34.00	**93.23**	97.87	103.65	66.97	74.11	81.07	94.25	106.57	85.58	**99.18**

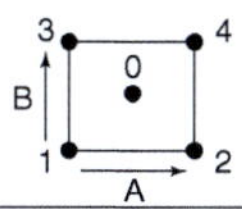

2^2 Factorial with added reference condition

Cycle N : *10*
Response:YIELD

Project S-5
Phase III

Calculation of Averages						Calculation of Standard Deviation	
Operation condition	(0)	(1)	(2)	(3)	(4)	Prior estimate of σ	= 10.00
(i) Previous cycle sum	158.17	153.80	183.22	208.46	182.40	Previous sum **s**	= 78.17
(ii) Previous cycle average	79.09	76.90	91.61	104.23	91.20	Previous average **s**	= 9.77
(iii) New observations	79.95	79.59	77.39	73.58	99.18	New **s** = range × ***fδ,n***	= 15.84
(iv) Differences (ii) less (iii)	-0.86	-2.69	14.22	30.65	-7.98	Range	= 38.63
(v) New sums	159.03	156.49	169.00	177.81	190.39	New sum **s**	= 94.01
(vi) New averages : *ÿi*	79.52	78.25	84.50	88.91	95.19	New average **s** = new sum **s** / (n - 1)	= 10.45

Calculation of Effects			Calculation of 2 S.E. Limits
Phase mean	= *1/5 (ÿo + ÿ1 + ÿ2 + ÿ3 + ÿ4)*	= 85.27	
(A) ———— effect	= *½ (ÿ2 + ÿ4 - ÿ1 - ÿ3)*	= 6.27	For new average **s** : ± (2/√n)s = 6.58
(B) ———— effect	= *½ (ÿ3 + ÿ4 - ÿ1 - ÿ2)*	= 10.68	For new effects : ± (2/√n)s = 6.58
(AB) ———— effect	= *½ (ÿ1 + ÿ4 - ÿ3 - ÿ2)*	= 0.02	For change in mean : ± (1.79/√n)s = 5.95
Change in mean effect	= phase mean - *ÿo*	= 5.76	

Table of Multiplying Factors

n	1	2	3	4	5	6	7	8	9	10
fδ,n		0.30	0.35	0.37	0.38	0.39	0.40	0.40	0.40	0.41
1 / n	1.00	0.50	0.33	0.25	0.20	0.17	0.14	0.12	0.11	0.10
1 / (n - 1)		1.00	0.50	0.33	0.25	0.20	0.17	0.14	0.12	0.11
2 / √n	2.00	1.41	1.15	1.00	0.89	0.82	0.76	0.71	0.67	0.63
1.79 / √n	1.79	1.26	1.03	0.89	0.80	0.73	0.68	0.63	0.60	0.57

Two variable EVOP program calculation worksheet

Figure 5.95 Work sheet cycle No. 10.

Table 5.73 2^2 BEVOP information board.

INFORMATION BOARD AFTER PHASE NO. III

Last Replica Completed: **Last Cycle Completed: 10**

		RESPONSE No. 1	RESPONSE No. 2	RESPONSE No. 3
REQUIREMENT :		MAXIMIZE RESPONSE		
PROCESS AVERAGES :		88.91 95.19 79.52 B 78.25 84.50 A		
2.S.E. LIMITS FOR AVERAGES		6.58		
EFFECTS WITH 2.S.E	PHASE MEAN FACTOR A FACTOR B INTERACTION AB CHANGE IN MEAN	85.27 6.27 ± 6.58 10.68 ± 6.58 0.02 ± 6.58 5.76 ± 5.95		
ESTIMATED STANDARD DEVIATION S		10.45		
PRIOR ESTIMATE OF STANDARD DEV.		10.00		

NOTE:

1. The Center of Experiment-Reference Conditions (The Current Best-Known Process Conditions)
2. The Change in Mean provides a measure of whether the conditions being run the current phase yield an average result better or worse than the reference conditions.
3. The Phase Mean is the Mean Response Over All Conditions Being Run in the present Phase.
4*. The change in Mean is a measure of the over-all Curvature of the Response Surface. It will be positive near a MINIMUM and negative near a MAXIMUM.
5. Sequencing of the Runs-Randomization, if runs can be randomized without too much difficulty, than this should be done.

*A Diagonally oriented saddle represented by equal coefficients of opposite sign could occure and yield a zero change in mean effect, but such a phenomenon would rarely be met in practice.

Conclusions

- Only factor *B* is significant. The second factor is not significant.
- The *phase mean* is increased significantly.
- The *change in mean* effect is not significant.
- The estimated standard deviation is very close to the prior estimate of the standard deviation.

Recommendation for Phase IV, Cycle 1

- Move the center of the experiment from $A_0 = 30$; $B_0 = 30$ to $A_0 = 40$; $B_0 = 40$ and start a phase IV.
- Keep the same factor's interval of variation (Table 5.74, Figure 5.96, Table 5.75).

Table 5.74 2^2 BEVOP simulation.

PROJECT : S-5
PHASE : IV
CYCLE :

BEVOP SIMULATION

S-5

I. CENTER OF EXPERIMENT : Ao = 40.00, Bo = 40.00 [Gen Random]

II. FACTOR'S INTERVAL OF VARIATION : Δa = 4.00, Δb = 4.00

III. PROCESS (PROPERTY-RESPONSE) NOISE : STDV = 10.00

EVOP WORK SHEET

TRIALS ORDER			FACTOR		RESPONSE									
			A	B	n = 1	n = 2	n = 3	n = 4	n = 5	n = 6	n = 7	n = 8	n = 9	n = 10
0		Block 1	40.00	40.00	97.38	83.68	86.83	103.17	85.07	98.56	81.44	101.62	87.37	91.95
1		Block 1	36.00	36.00	102.60	80.22	103.52	87.26	87.06	94.23	104.93	85.60	92.17	94.79
2		Block 1	44.00	36.00	81.34	81.73	96.31	103.00	93.18	96.01	108.18	105.06	103.82	89.39
3		Block 1	36.00	44.00	104.55	118.08	97.94	96.00	98.71	100.14	109.61	119.26	120.47	85.58
4		Block 1	44.00	44.00	102.03	106.67	112.45	75.77	82.91	89.87	103.05	115.37	94.38	107.98

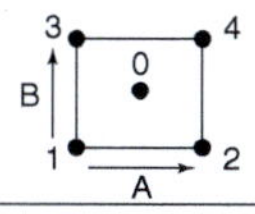

2^2 *Factorial with added reference condition*

Cycle N : *10*
Response:YIELD

Project S-5
Phase IV
Replica

Calculation of Averages

Operation condition	(0)	(1)	(2)	(3)	(4)
(i) Previous cycle sum	182.17	184.20	207.22	232.46	200.00
(ii) Previous cycle average	91.09	92.10	103.61	116.23	100.00
(iii) New observations	91.95	94.79	89.39	85.58	107.98
(iv) Differences (ii) less (iii)	-0.86	-2.69	14.22	30.65	-7.98
(v) New sums	183.03	186.89	193.00	201.81	207.99
(vi) New averages : $\bar{y}i$	91.52	93.45	96.50	100.91	103.99

Calculation of Standard Deviation

Prior estimate of σ	= 10.00
Previous sum **s**	= 78.17
Previous average **s**	= 9.77
New **s** = range × $f\delta,n$	= 15.84
Range	= 38.63
New sum **s**	= 94.01
New average **s** = $\frac{\text{new sum } s}{n-1}$	= 10.45

Calculation of Effects

Phase mean	$= 1/5\,(\bar{y}0 + \bar{y}1 + \bar{y}2 + \bar{y}3 + \bar{y}4)$	= 97.27
(A) effect	$= \frac{1}{2}(\bar{y}2 + \bar{y}4 - \bar{y}1 - \bar{y}3)$	= 3.07
(B) effect	$= \frac{1}{2}(\bar{y}3 + \bar{y}4 - \bar{y}1 - \bar{y}2)$	= 7.48
(AB) effect	$= \frac{1}{2}(\bar{y}1 + \bar{y}4 - \bar{y}3 - \bar{y}2)$	= 0.02
Change in mean effect	= phase mean - $\bar{y}0$	= 5.76

Calculation of 2 S.E. Limits

For new average **s** :	$\pm (2/\sqrt{n})s$	= 6.58
For new effects :	$\pm (2/\sqrt{n})s$	= 6.58
For change in mean :	$\pm (1.79/\sqrt{n})s$	= 5.95

Table of Multiplying Factors

n	1	2	3	4	5	6	7	8	9	10
$f\delta,n$		0.30	0.35	0.37	0.38	0.39	0.40	0.40	0.40	0.41
$1/n$	1.00	0.50	0.33	0.25	0.20	0.17	0.14	0.12	0.11	0.10
$1/(n-1)$		1.00	0.50	0.33	0.25	0.20	0.17	0.14	0.12	0.11
$2/\sqrt{n}$	2.00	1.41	1.15	1.00	0.89	0.82	0.76	0.71	0.67	0.63
$1.79/\sqrt{n}$	1.79	1.26	1.03	0.89	0.80	0.73	0.68	0.63	0.60	0.57

Two variable EVOP program calculation worksheet

Figure 5.96 Work sheet cycle No. 10.

Table 5.75 2^2 BEVOP information board.

INFORMATION BOARD AFTER PHASE NO. IV Last Replica Completed: Last Cycle Completed: 10				
		RESPONSE No. 1	RESPONSE No. 2	RESPONSE No. 3
REQUIREMENT :		MAXIMIZE RESPONSE		
PROCESS AVERAGES :		100.91 (top left), 103.99 (top right), 91.52 (center), 93.45 (bottom left), 96.50 (bottom right); B (vertical axis), A (horizontal axis)		
2.S.E. LIMITS FOR AVERAGES		6.58		
EFFECTS WITH 2.S.E	PHASE MEAN FACTOR A FACTOR B INTERACTION AB CHANGE IN MEAN	97.27 3.07 ± 6.58 7.48 ± 6.58 0.02 ± 6.58 5.76 ± 5.95		
ESTIMATED STANDARD DEVIATION S		10.45		
PRIOR ESTIMATE OF STANDARD DEV.		10.00		

NOTE:

1. The Center of Experiment-Reference Conditions (The Current Best-Known Process Conditions)
2. The Change in Mean provides a measure of whether the conditions being run the current phase yield an average result better or worse than the reference conditions.
3. The Phase Mean is the Mean Response Over All Conditions Being Run in the present Phase.
4*. The change in Mean is a measure of the over-all Curvature of the Response Surface. It will be positive near a MINIMUM and negative near a MAXIMUM.
5. Sequencing of the Runs-Randomization, if runs can be randomized without too much difficulty, than this should be done.

*A Diagonally oriented saddle represented by equal coefficients of opposite sign could occure and yield a zero change in mean effect, but such a phenomenon would rarely be met in practice.

Conclusions

- Only factor B is significant. The second factor is not significant.
- The *phase mean is* increased significantly.
- The *change in mean* effect is not significant.
- The estimated standard deviation is very close to the prior estimate of the standard deviation.

Recommendation for Phase V, Cycle 1

- The *phase mean* is increased significantly, since BEVOP has started. It means we are moving in the right direction.
- Move the center of the experiment from $A_0 = 40$; $B_0 = 40$ to $A_0 = 50$; $B_0 = 50$ and start a phase V (Table 5.76, Figure 5.97, Table 5.77).

Table 5.76 2^2 BEVOP simulation.

PROJECT : S-5
PHASE : V
CYCLE :

BEVOP SIMULATION

S-5

I. CENTER OF EXPERIMENT : Ao = 50.00
Bo = 50.00

Gen Random

II. FACTOR'S INTERVAL OF VARIATION : Δ a = 4.00
Δ b = 4.00

III. PROCESS (PROPERTY-RESPONSE) NOISE : STDV = 10.00

EVOP WORK SHEET

TRIALS ORDER			FACTOR		RESPONSE									
			A	B	n = 1	n = 2	n = 3	n = 4	n = 5	n = 6	n = 7	n = 8	n = 9	n = 10
0		Block 1	**50.00**	**50.00**	**101.38**	**87.68**	**90.83**	**107.17**	**89.07**	**102.56**	**85.44**	**105.62**	**91.37**	**95.95**
1		Block 1	46.00	46.00	**109.80**	87.42	110.72	94.46	94.26	101.43	112.13	92.80	99.37	**101.99**
2		Block 1	54.00	46.00	**85.34**	85.73	100.81	107.00	97.18	100.01	112.18	109.06	107.82	**93.39**
3		Block 1	46.00	54.00	**108.55**	122.08	101.94	100.00	102.71	104.14	113.61	123.26	124.47	**89.58**
4		Block 1	54.00	54.00	**102.83**	107.47	113.25	76.57	83.71	90.67	103.85	116.17	95.18	**108.78**

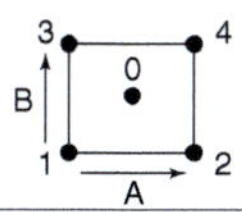

2^2 Factorial with added reference condition

Cycle N :
10
Response:YIELD

Project S-5
Phase V
Replica

Calculation of Averages						Calculation of Standard Deviation	
Operation condition	(0)	(1)	(2)	(3)	(4)	Prior estimate of **σ**	= 10.00
(i) Previous cycle sum	190.17	198.60	215.22	240.46	201.60	Previous sum **s**	= 78.17
(ii) Previous cycle average	95.09	99.30	107.61	120.23	100.80	Previous average **s**	= 9.77
(iii) New observations	95.95	101.99	93.39	89.58	108.78	New **s** = range × ***fδ,n***	= 15.84
(iv) Differences (ii) less (iii)	-0.86	-2.69	14.22	30.65	-7.98	Range	= 38.63
(v) New sums	191.03	201.29	201.00	209.81	209.59	New sum **s**	= 94.01
(vi) New averages : *ÿi*	95.52	100.65	100.50	104.91	104.79	New average **s** = new sum **s** / n - 1	= 10.45

Calculation of Effects			Calculation of 2 S.E. Limits
Phase mean	= *1/5 (ÿo + ÿ1 + ÿ2 + ÿ3 + ÿ4)*	= 101.27	
(A) ———— effect	= *½ (ÿ2 + ÿ4 - ÿ1 - ÿ3)*	= -0.13	For new average **s** : ± $(2/\sqrt{n})$s = 6.58
(B) ———— effect	= *½ (ÿ3 + ÿ4 - ÿ1 - ÿ2)*	= 4.28	For new effects : ± $(2/\sqrt{n})$s = 6.58
(AB) ———— effect	= *½ (ÿ1 + ÿ4 - ÿ3 - ÿ2)*	= 0.02	For change in mean : ± $(1.79/\sqrt{n})$s = 5.95
Change in mean effect	= phase mean - *ÿo*	= 5.76	

Table of Multiplying Factors

n	1	2	3	4	5	6	7	8	9	10
fδ,n		0.30	0.35	0.37	0.38	0.39	0.40	0.40	0.40	0.41
1 / n	1.00	0.50	0.33	0.25	0.20	0.17	0.14	0.12	0.11	0.10
1 / (n - 1)		1.00	0.50	0.33	0.25	0.20	0.17	0.14	0.12	0.11
$2/\sqrt{n}$	2.00	1.41	1.15	1.00	0.89	0.82	0.76	0.71	0.67	0.63
$1.79/\sqrt{n}$	1.79	1.26	1.03	0.89	0.80	0.73	0.68	0.63	0.60	0.57

Two variable EVOP program calculation worksheet

Figure 5.97 Work sheet cycle No. 10.

Table 5.77 2^2 BEVOP information board.

INFORMATION BOARD AFTER PHASE NO. V Last Replica Completed: Last Cycle Completed: 10			
	RESPONSE No. 1	RESPONSE No. 2	RESPONSE No. 3
REQUIREMENT :	MAXIMIZE RESPONSE		
PROCESS AVERAGES :	104.91 104.79 95.52 B 100.65 100.50 A		
2.S.E. LIMITS FOR AVERAGES	6.58		
EFFECTS WITH 2.S.E — PHASE MEAN FACTOR A FACTOR B INTERACTION AB CHANGE IN MEAN	101.27 -0.13 ± 6.58 4.28 ± 6.58 0.02 ± 6.58 5.76 ± 5.95		
ESTIMATED STANDARD DEVIATION S PRIOR ESTIMATE OF STANDARD DEV.	10.45 10.00		

NOTE:

1. The Center of Experiment-Reference Conditions (The Current Best-Known Process Conditions)
2. The Change in Mean provides a measure of whether the conditions being run the current phase yield an average result better or worse than the reference conditions.
3. The Phase Mean is the Mean Response Over All Conditions Being Run in the present Phase.
4*. The change in Mean is a measure of the over-all Curvature of the Response Surface. It will be positive near a MINIMUM and negative near a MAXIMUM.
5. Sequencing of the Runs-Randomization, if runs can be randomized without too much difficulty, than this should be done.

*A Diagonally oriented saddle represented by equal coefficients of opposite sign could occure and yield a zero change in mean effect, but such a phenomenon would rarely be met in practice.

Conclusions

- Both factors are not significant. It looks like we are at optimum.
- The *phase mean* reached optimum.
- The *change in mean* effect is not significant.
- The estimated standard deviation is very close to the prior estimate of the standard deviation.

Recommendation for Phase VI, Cycle 1

- The *phase mean* is increased significantly again, since BEVOP has started. It looks like optimum. To make sure, run phase VI.
- Move the center of the experiment far away from $A_0 = 50$; $B_0 = 50$ to $A_0 = 70$; $B_0 = 70$ and start a phase VI (Table 5.78, Figure 5.98, Table 5.79).

Table 5.78 2^2 BEVOP simulation.

PROJECT : S-5
PHASE : VI
CYCLE :

BEVOP SIMULATION

S-5

I. CENTER OF EXPERIMENT : Ao = 70.00; Bo = 70.00 [Gen Random]

II. FACTOR'S INTERVAL OF VARIATION : Δa = 4.00; Δb = 4.00

III. PROCESS (PROPERTY-RESPONSE) NOISE : STDV = 10.00

EVOP WORK SHEET

TRIALS ORDER			FACTOR		RESPONSE									
			A	B	n = 1	n = 2	n = 3	n = 4	n = 5	n = 6	n = 7	n = 8	n = 9	n = 10
0		Block 1	**70.00**	**70.00**	**85.38**	**71.68**	**74.83**	**91.17**	**73.07**	**86.56**	**69.44**	**89.62**	**75.37**	**79.95**
1		Block 1	66.00	66.00	**100.20**	77.82	101.12	84.86	84.66	91.83	102.53	83.20	89.77	**92.39**
2		Block 1	74.00	66.00	**69.34**	69.73	84.81	91.00	81.18	84.01	96.18	93.06	91.82	**77.39**
3		Block 1	66.00	74.00	**92.55**	106.08	85.94	84.00	86.71	88.14	97.61	107.26	108.47	**73.58**
4		Block 1	74.00	74.00	**80.43**	85.07	90.85	54.17	61.31	68.27	81.45	93.77	72.78	**86.38**

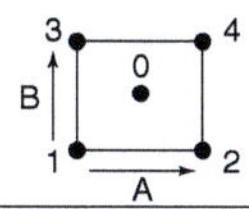

2^2 Factorial with added reference condition

Cycle N :
10
Response:YIELD

Project S-5
Phase VI
Replica

Calculation of Averages						Calculation of Standard Deviation	
Operation condition	(0)	(1)	(2)	(3)	(4)	Prior estimate of σ	= 10.00
(i) Previous cycle sum	158.17	179.40	183.22	208.46	156.80	Previous sum **s**	= 78.17
(ii) Previous cycle average	79.09	89.70	91.61	104.23	78.40	Previous average **s**	= 9.77
(iii) New observations	79.95	92.39	77.39	73.58	86.38	New **s** = range × ***fδ,n***	= 15.84
(iv) Differences (ii) less (iii)	-0.86	-2.69	14.22	30.65	-7.98	Range	= 38.63
(v) New sums	159.03	182.09	169.00	177.81	164.79	New sum **s**	= 94.01
(vi) New averages : $\bar{y}_i$	79.52	91.05	84.50	88.91	82.39	New average **s** = new sum **s** / (n - 1)	= 10.45

Calculation of Effects			Calculation of 2 S.E. Limits	
Phase mean	$= 1/5\,(\bar{y}_0 + \bar{y}_1 + \bar{y}_2 + \bar{y}_3 + \bar{y}_4)$	= 85.27		
(A) effect	$= ½\,(\bar{y}_2 + \bar{y}_4 - \bar{y}_1 - \bar{y}_3)$	= -6.53	For new average **s** : $\pm (2/\sqrt{n})s$	= 6.58
(B) effect	$= ½\,(\bar{y}_3 + \bar{y}_4 - \bar{y}_1 - \bar{y}_2)$	= -2.12	For new effects : $\pm (2/\sqrt{n})s$	= 6.58
(AB) effect	$= ½\,(\bar{y}_1 + \bar{y}_4 - \bar{y}_3 - \bar{y}_2)$	= 0.02	For change in mean : $\pm (1.79/\sqrt{n})s$	= 5.95
Change in mean effect	= phase mean - $\bar{y}_0$	= 5.76		

Table of Multiplying Factors

n	1	2	3	4	5	6	7	8	9	10
fδ,n		0.30	0.35	0.37	0.38	0.39	0.40	0.40	0.40	0.41
1 / n	1.00	0.50	0.33	0.25	0.20	0.17	0.14	0.12	0.11	0.10
1 / (n - 1)		1.00	0.50	0.33	0.25	0.20	0.17	0.14	0.12	0.11
$2/\sqrt{n}$	2.00	1.41	1.15	1.00	0.89	0.82	0.76	0.71	0.67	0.63
$1.79/\sqrt{n}$	1.79	1.26	1.03	0.89	0.80	0.73	0.68	0.63	0.60	0.57

Two variable EVOP program calculation worksheet

Figure 5.98 Work sheet cycle No. 10.

Table 5.79 2^2 BEVOP information board.

INFORMATION BOARD AFTER PHASE NO. VI

Last Replica Completed: **Last Cycle Completed: 10**

		RESPONSE No. 1	RESPONSE No. 2	RESPONSE No. 3
REQUIREMENT :		MAXIMIZE RESPONSE		
PROCESS AVERAGES :		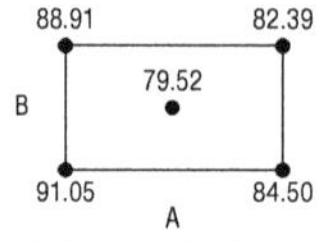		
2.S.E. LIMITS FOR AVERAGES		6.58		
EFFECTS WITH 2.S.E	PHASE MEAN	85.27		
	FACTOR A	-6.53 ± 6.58		
	FACTOR B	-2.12 ± 6.58		
	INTERACTION AB	0.02 ± 6.58		
	CHANGE IN MEAN	5.76 ± 5.95		
ESTIMATED STANDARD DEVIATION S		10.45		
PRIOR ESTIMATE OF STANDARD DEV.		10.00		

NOTE:

1. The Center of Experiment-Reference Conditions (The Current Best-Known Process Conditions)
2. The Change in Mean provides a measure of whether the conditions being run the current phase yield an average result better or worse than the reference conditions.
3. The Phase Mean is the Mean Response Over All Conditions Being Run in the present Phase.
4*. The change in Mean is a measure of the over-all Curvature of the Response Surface. It will be positive near a MINIMUM and negative near a MAXIMUM.
5. Sequencing of the Runs-Randomization, if runs can be randomized without too much difficulty, than this should be done.

*A Diagonally oriented saddle represented by equal coefficients of opposite sign could occure and yield a zero change in mean effect, but such a phenomenon would rarely be met in practice.

Conclusions

- Both factors are not significant though with negative sign. It seems we have moved away from optimum.
- The *phase mean* is reduced significantly.
- The *change in mean* effect is not significant.
- The estimated standard deviation is very close to the prior estimate of the standard deviation.

Recommendation

- Move the center of the experiment back to previous position (Figure 5.99).

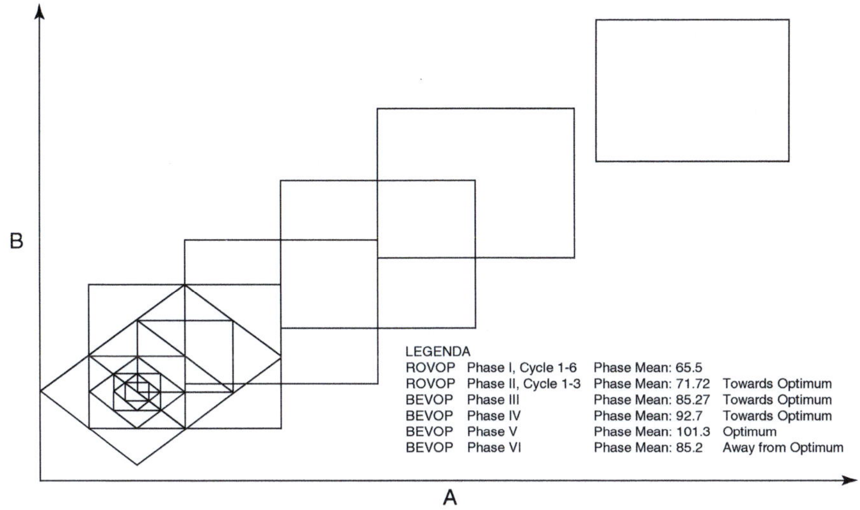

Figure 5.99 Graphical interpretation of ROVOP moving toward optimum.

Solution

It is already known from Simulation No.1 that optimum (maximum) yield is 100 [%] for factors settings $A_0 = 50$; $B_0 = 50$.

5.8.4 2^3 ROVOP Simulation

The procedure consists of starting with very small variants in a body center 2^3 factorial designs, then for each succeeding cycle in the same center enlarging the variant by $\sqrt{3}$ and rotating the design through an angle of 45°, as shown in Figure 5.100. This permits effects to converge more quickly on some predetermined level of significance and provides information on the quadratic effects.

Figure 5.100 Two successive 2^3 ROVOP patterns.

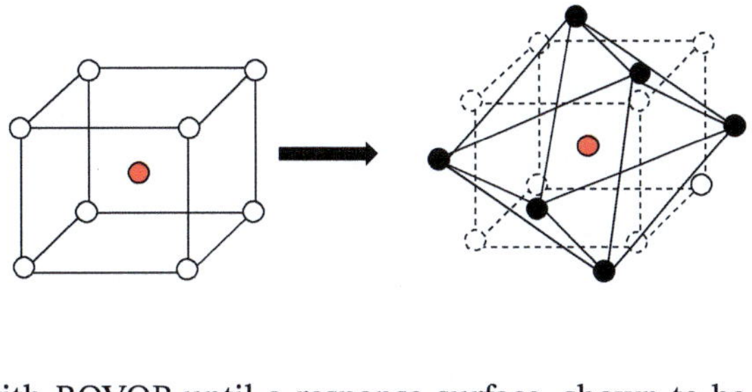

Experimentation continues with ROVOP until a response surface, shown to be valid at the predetermined level of significance, is developed; at this point the pattern is moved, conservatively, in the direction of optimization and ROVOP continued until no further movement is indicated. When the operator moves to new phase (new cycles conditions), at least two points of the new phase should lie on or within the bounds of experimental space already explored. Also, the interval between the high and low levels of the factor to be changed should be reduced by at least a factor of $\sqrt{3}$; *levels of factors shown to be not significant need not be reduced.* Occasionally, and ideally, the response pattern follows a hill contour where the peak represents the optimum conditions. With ROVOP the peak is approached gradually and eventually is straddled; now the quadratic terms of the response equation take control and movement *toward the center of the design is indicated.* Continuation of ROVOP beyond this point will provide adjustment for long-term changes only, such as changes in raw materials, which may or may not be worthwhile. Experimentation continues with switching ROVOP to BEVOP with larger factors change (interval) discovered from ROVOP.

5.8.4.1 Simulation No. 6: 2^3 ROVOP

The mathematical formula from Simulation No. 4 was used to demonstrate efficiency of the ROVOP method. This simulation started with ROVOP until significant factor effect was detected and after ROVOP was switched to BEVOP until optimum was achieved. Graphical movement was not presented because it is more complex and difficult to interpret in graphical form.

Phase I: The settings (Center Point, Factor's Interval, and Process Noise) and yield for different process conditions have been completed using Microsoft Excel simulation in Table 5.80. In this ROVOP simulation, one more parameter in settings was introduced: N – number of replicates-repetitions of ROVOP cycles. This time, minimum number of tables and figures will be shown below. The first phase with four cycles will be presented only with two tables: Simulation and Information Board.

This simulation started with small factor's change (interval) and the large noise (STDEV) (Tables 5.81–5.83).

Table 5.80 2^3 ROVOP simulation.

PROJECT : S-6
PHASE : I
CYCLE : 1

ROVOP SIMULATION

I. CENTER OF EXPERIMENT :	Ao = 20.00 Bo = 20.00 Co = 20.00	Gen Random
II. FACTOR'S INTERVAL OF VARIATION :	Δa = 5.00 Δb = 5.00 Δc = 5.00	
III. PROCESS NOISE :	STDV = 3.0	

EVOP WORK SHEET

TRIALS ORDER		Block	FACTORS			RESPONSE									
			A	B	C	n = 1	n = 2	n = 3	n = 4	n = 5	n = 6	n = 7	n = 8	n = 9	n = 10
0		Block I	20.00	20.00	20.00	45.74	44.80	44.46	46.28	45.09	47.30	42.08	36.29	54.38	48.31
1		Block I	19.50	19.50	19.50	32.77	48.25	38.43	57.88	45.16	41.71	57.66	49.62	41.10	40.37
2		Block I	20.50	20.50	19.50	48.83	52.24	45.95	47.33	40.68	30.18	46.39	49.41	43.54	46.50
3		Block I	20.50	19.50	20.50	51.45	51.46	46.92	55.57	51.66	54.68	54.32	55.55	49.25	40.47
4		Block I	19.50	20.50	20.50	50.35	50.01	52.50	48.14	47.45	39.62	45.32	48.13	49.03	37.23
Ø		Block II	20.00	20.00	20.00	46.54	49.37	41.30	43.91	43.81	32.31	41.84	54.45	40.92	41.11
5		Block II	19.50	19.50	20.50	43.42	39.25	41.33	37.98	46.24	47.64	45.63	38.37	50.33	46.00
6		Block II	20.50	20.50	20.50	54.11	47.34	48.19	44.36	42.87	45.01	49.43	54.93	43.36	43.27
7		Block II	20.50	19.50	19.50	38.90	44.06	41.21	40.57	40.27	53.11	47.83	47.13	40.64	35.22
8		Block II	19.50	20.50	19.50	47.56	46.74	47.68	42.60	36.55	43.74	39.48	47.67	56.78	44.48

Table 5.81 2^3 ROVOP information board.

2^3 INFORMATION BOARD AFTER PHASE NO. I

Last Replica Completed: 10 Last Cycle Completed: 1

	RESPONSE No. 1	RESPONSE No. 2	RESPONSE No. 3
REQUIREMENT :	MAXIMIZE RESPONSE		
PROCESS AVERAGES :	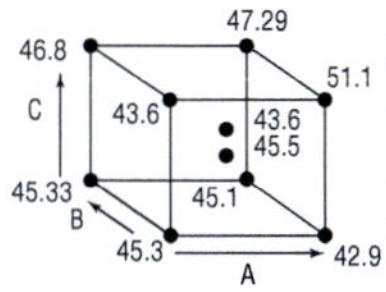		
2.S.E. LIMITS FOR AVERAGES	3.41		
EFFECTS WITH 2.S.E — PHASE MEAN	45.65		
FACTOR A	1.35 ± 2.40		
FACTOR B	0.39 ± 2.40		
FACTOR C	2.55 ± 2.40		
INTERACTION AB	-1.21 ± 2.40		
INTERACTION AC	2.66 ± 2.40		
INTERACTION BC	-0.73 ± 2.40		
INTERACTION ABC	-2.30 ± 2.40		
CHANGE IN MEAN	1.13 ± 2.15		
ESTIMATED STANDARD DEVIATION S	5.39		
PRIOR ESTIMATE OF STANDARD DEV.	5.00		

NOTE:

1. The Center of Experiment-Reference Conditions (The Current Best-Known Process Conditions)
2. The Change in Mean provides a measure of whether the conditions being run the current phase yield an average result better or worse than the reference conditions.
3. The Phase Mean is the Mean Response Over All Conditions Being Run in the present Phase.
4*. The change in Mean is a measure of the over-all Curvature of the Response Surface. It will be positive near a MINIMUM and negative near a MAXIMUM.
5. Sequencing of the Runs-Randomization, if runs can be randomized without too much difficulty, than this should be done.

*A Diagonally oriented saddle represented by equal coefficients of opposite sign could occure and yield a zero change in mean effect, but such a phenomenon would rarely be met in practice.

Table 5.82 2^3 ROVOP simulation.

PROJECT : S-6
PHASE : I
CYCLE : 4

ROVOP SIMULATION

I. CENTER OF EXPERIMENT : Ao = 20.00, Bo = 20.00, Co = 20.00 [Gen Random]

II. FACTOR'S INTERVAL OF VARIATION : Δ a = 2.59, Δ b = 2.59, Δ c = 2.59

III. PROCESS NOISE : STDV = 5.0

EVOP WORK SHEET

TRIALS ORDER		Block	FACTORS A	FACTORS B	FACTORS C	RESPONSE n = 1	n = 2	n = 3	n = 4	n = 5	n = 6	n = 7	n = 8	n = 9	n = 10
0		Block I	20.00	20.00	20.00	43.66	41.55	40.19	34.83	50.79	48.16	46.25	48.24	41.54	52.87
1		Block I	17.41	17.41	17.41	44.37	30.70	32.01	42.04	36.62	29.73	37.05	44.03	30.27	32.71
2		Block I	22 59	22.59	17.41	49.22	48.22	51.55	45.77	37.06	54.39	50.81	38.37	51.04	45.47
3		Block I	22.59	17.41	22.59	49.06	52.22	38.40	47.04	48.16	50.96	47.65	37.86	44.65	52.31
4		Block I	17.41	22.59	22.59	44.17	47.77	59.70	37.94	49.20	47.44	44.79	60.96	44.83	50.89
Ø		Block II	20.00	20.00	20.00	38.15	36.45	42.01	46.76	49.73	62.78	43.89	53.81	49.58	45.38
5		Block II	17.41	17.41	22.59	40.22	44.82	46.10	46.74	35.74	44.97	33.50	42.05	42.69	44.10
6		Block II	22.59	22.59	22.59	63.94	57.95	57.20	55.29	49.40	55.48	64.35	51.15	60.43	51.22
7		Block II	22.59	17.41	17.41	48.30	52.74	40.83	47.23	48.86	47.68	42.61	42.28	45.47	52.62
8		Block II	17.41	22.59	17.41	36.65	43.76	42.61	38.88	45.41	45.19	44.49	49.35	47.13	42.33

Table 5.83 2^3 ROVOP information board.

2^3 INFORMATION BOARD AFTER PHASE NO. I

Last Replica Completed: 10 Last Cycle Completed: 4

		RESPONSE No. 1	RESPONSE No. 2	RESPONSE No. 3
REQUIREMENT :		MAXIMIZE RESPONSE		
PROCESS AVERAGES :		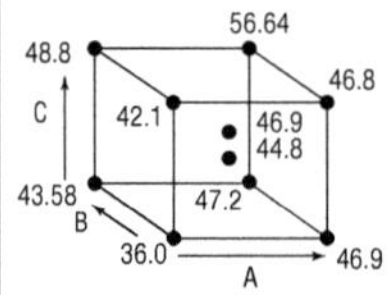		
2.S.E. LIMITS FOR AVERAGES		3.71		
EFFECTS WITH 2.S.E	PHASE MEAN FACTOR A FACTOR B FACTOR C INTERACTION AB INTERACTION AC INTERACTION BC INTERACTION ABC CHANGE IN MEAN	45.96 6.78 ± 2.62 6.11 ± 2.62 5.19 ± 2.62 -1.04 ± 2.62 -0.48 ± 2.62 2.13 ± 2.62 2.61 ± 2.62 0.13 ± 2.34		
ESTIMATED STANDARD DEVIATION S		5.87		
PRIOR ESTIMATE OF STANDARD DEV.		5.00		

NOTE:

1. The Center of Experiment-Reference Conditions (The Current Best-Known Process Conditions)
2. The Change in Mean provides a measure of whether the conditions being run the current phase yield an average result better or worse than the reference conditions.
3. The Phase Mean is the Mean Response Over All Conditions Being Run in the present Phase.
4*. The change in Mean is a measure of the over-all Curvature of the Response Surface. It will be positive near a MINIMUM and negative near a MAXIMUM.
5. Sequencing of the Runs-Randomization, if runs can be randomized without too much difficulty, than this should be done.

*A Diagonally oriented saddle represented by equal coefficients of opposite sign could occure and yield a zero change in mean effect, but such a phenomenon would rarely be met in practice.

Conclusions

- All three factors are significant with positive sign. It is time to move center of the experiment in positive direction.
- The *change in mean* effect is not significant.
- The estimated standard deviation is very close to the prior estimate of the standard deviation.

Recommendation for Phase II, Cycle 1

- Move the center of the experiment from $A_0 = 20$; $B_0 = 20$; $C_0 = 20$ to $A_0 = 22.59$; $B_0 = 22.59$; $C_0 = 22.59$, and start a phase II.
- Reduce factor's interval of variation from $\Delta a = 2.59$; $\Delta b = 2.59$; $\Delta c = 2.59$ to $\Delta a = 1.5$; $\Delta b = 1.5$; $\Delta c = 1.5$ following rules from Figure 5.88 (Tables 5.84 and 5.85).

Table 5.84 2^3 ROVOP simulation.

PROJECT : S-6
PHASE : II
CYCLE : 1

ROVOP SIMULATION

I. CENTER OF EXPERIMENT :

Ao = 22.59
Bo = 22.59
Co = 22.59

Gen Random

II. FACTOR'S INTERVAL OF VARIATION :

Δ a = 1.50
Δ b = 1.50
Δ c = 1.50

III. PROCESS NOISE :

STDV = 5.0

EVOP WORK SHEET

TRIALS ORDER		Block	FACTORS			RESPONSE									
			A	B	C	n = 1	n = 2	n = 3	n = 4	n = 5	n = 6	n = 7	n = 8	n = 9	n = 10
0		Block I	22.59	22.59	22.59	52.58	50.47	49.11	43.75	59.71	57.08	55.17	57.16	50.47	61.79
1		Block I	21.09	21.09	21.09	57.95	44.28	45.59	55.62	50.20	43.31	50.63	57.61	43.85	46.29
2		Block I	24.09	24.09	21.09	56.95	55.95	59.28	53.49	44.79	62.11	58.54	46.09	58.76	53.20
3		Block I	24.09	21.09	24.09	56.79	59.94	46.13	54.77	55.89	58.69	55.38	45.59	52.37	60.04
4		Block I	21.09	24.09	24.09	51.89	55.50	67.43	45.66	56.93	55.16	52.51	68.68	52.55	58.62
Ø		Block II	22.59	22.59	22.59	47.07	45.37	50.93	55.68	58.65	71.70	52.81	62.74	58.50	54.30
5		Block II	21.09	21.09	24.09	50.87	55.47	56.75	57.40	46.39	55.62	44.16	52.70	53.34	54.75
6		Block II	24.09	24.09	24.09	68.74	62.75	62.00	60.09	54.20	60.28	69.15	55.95	65.22	56.02
7		Block II	24.09	21.09	21.09	58.95	63.39	51.48	57.88	59.51	58.34	53.26	52.93	56.12	63.28
8		Block II	21.09	24.09	21.09	47.30	54.41	53.26	49.53	56.06	55.84	55.14	60.00	57.78	52.98

Table 5.85 2^3 ROVOP information board.

2^3 INFORMATION BOARD AFTER PHASE NO. II

Last Replica Completed: 10 **Last Cycle Completed: 1**

		RESPONSE No. 1	RESPONSE No. 2	RESPONSE No. 3
REQUIREMENT :		MAXIMIZE RESPONSE		
PROCESS AVERAGES :		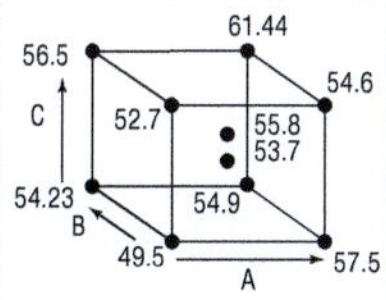		
2.S.E. LIMITS FOR AVERAGES		3.71		
EFFECTS WITH 2.S.E	PHASE MEAN	55.09		
	FACTOR A	3.86 ± 2.62		
	FACTOR B	3.18 ± 2.62		
	FACTOR C	2.26 ± 2.62		
	INTERACTION AB	-1.04 ± 2.62		
	INTERACTION AC	-0.48 ± 2.62		
	INTERACTION BC	2.13 ± 2.62		
	INTERACTION ABC	2.61 ± 2.62		
	CHANGE IN MEAN	0.34 ± 2.34		
ESTIMATED STANDARD DEVIATION S		5.87		
PRIOR ESTIMATE OF STANDARD DEV.		5.00		

NOTE:

1. The Center of Experiment-Reference Conditions (The Current Best-Known Process Conditions)
2. The Change in Mean provides a measure of whether the conditions being run the current phase yield an average result better or worse than the reference conditions.
3. The Phase Mean is the Mean Response Over All Conditions Being Run in the present Phase.
4*. The change in Mean is a measure of the over-all Curvature of the Response Surface. It will be positive near a MINIMUM and negative near a MAXIMUM.
5. Sequencing of the Runs-Randomization, if runs can be randomized without too much difficulty, than this should be done.

*A Diagonally oriented saddle represented by equal coefficients of opposite sign could occure and yield a zero change in mean effect, but such a phenomenon would rarely be met in practice.

Conclusions

- Both factors (*A* and *B*) are significant with positive sign. Move the center of the experiment in positive direction.
- The *change in mean* effect is not significant.
- The estimated standard deviation is very close to the prior estimate of the standard deviation.

Recommendation for Phase III, Cycle 1

- Move the center of the experiment from $A_0 = 22.59$; $B_0 = 22.59$; $C_0 = 22.59$ to $A_0 = 24.09$; $B_0 = 24.09$; $C_0 = 24.09$ and start a phase III.
- Enlarge factor's interval of variation from $\Delta a = 1.5$; $\Delta b = 1.5$; $\Delta c = 1.5$ to $\Delta a = 2.59$; $\Delta b = 2.59$; $\Delta c = 2.59$ (Tables 5.86 and 5.87).

Table 5.86 2^3 ROVOP simulation.

PROJECT : S-6
PHASE : III
CYCLE : 1

ROVOP SIMULATION

I. CENTER OF EXPERIMENT : Ao = 24.09, Bo = 24.09, Co = 24.09 [Gen Random]

II. FACTOR'S INTERVAL OF VARIATION : Δa = 2.59, Δb = 2.59, Δc = 2.59

III. PROCESS NOISE : STDV = 5.0

EVOP WORK SHEET

TRIALS ORDER		Block	FACTORS			RESPONSE									
			A	B	C	n = 1	n = 2	n = 3	n = 4	n = 5	n = 6	n = 7	n = 8	n = 9	n = 10
0		Block I	**24.09**	**24.09**	**24.09**	57.38	55.27	53.91	48.55	64.51	61.88	59.97	61.96	55.26	66.59
1		Block I	21.50	21.50	21.50	59.36	45.69	47.00	57.04	51.61	44.72	52.04	59.02	45.26	47.71
2		Block I	26.68	26.68	21.50	62.52	61.52	64.85	59.07	50.36	67.68	64.11	51.66	64.33	58.77
3		Block I	26.68	21.50	26.68	62.36	65.51	51.70	60.34	61.46	64.26	60.95	51.16	57.94	65.61
4		Block I	21.50	26.68	26.68	57.46	61.07	73.00	51.23	62.50	60.73	58.09	74.25	58.13	64.19
Ø		Block II	**24.09**	**24.09**	**24.09**	51.87	50.17	55.73	60.48	63.45	76.50	57.61	67.53	63.30	59.10
5		Block II	21.50	21.50	26.68	54.36	58.96	60.24	60.89	49.88	59.11	47.65	56.19	56.83	58.24
6		Block II	26.68	26.68	26.68	76.39	70.40	69.65	67.74	61.85	67.93	76.80	63.60	72.88	63.67
7		Block II	26.68	21.50	21.50	62.44	66.88	54.97	61.38	63.00	61.83	56.76	56.42	59.61	66.77
8		Block II	21.50	26.68	21.50	50.79	57.90	56.75	53.03	59.55	59.33	58.64	63.49	61.27	56.48

Table 5.87 2^3 ROVOP information board.

2^3 INFORMATION BOARD AFTER PHASE NO. III

Last Replica Completed: 10 Last Cycle Completed: 1

		RESPONSE No. 1	RESPONSE No. 2	RESPONSE No. 3
REQUIREMENT :		MAXIMIZE RESPONSE		
PROCESS AVERAGES :		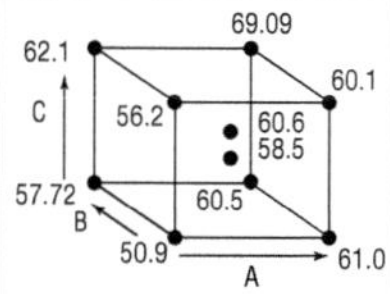		
2.S.E. LIMITS FOR AVERAGES		3.71		
EFFECTS WITH 2.S.E	PHASE MEAN	59.68		
	FACTOR A	5.94 ± 2.62		
	FACTOR B	5.26 ± 2.62		
	FACTOR C	4.34 ± 2.62		
	INTERACTION AB	-1.04 ± 2.62		
	INTERACTION AC	-0.48 ± 2.62		
	INTERACTION BC	2.13 ± 2.62		
	INTERACTION ABC	2.61 ± 2.62		
	CHANGE IN MEAN	0.13 ± 2.34		
ESTIMATED STANDARD DEVIATION S		5.87		
PRIOR ESTIMATE OF STANDARD DEV.		5.00		

NOTE:

1. The Center of Experiment-Reference Conditions (The Current Best-Known Process Conditions)
2. The Change in Mean provides a measure of whether the conditions being run the current phase yield an average result better or worse than the reference conditions.
3. The Phase Mean is the Mean Response Over All Conditions Being Run in the present Phase.
4*. The change in Mean is a measure of the over-all Curvature of the Response Surface. It will be positive near a MINIMUM and negative near a MAXIMUM.
5. Sequencing of the Runs-Randomization, if runs can be randomized without too much difficulty, than this should be done.

*A Diagonally oriented saddle represented by equal coefficients of opposite sign could occure and yield a zero change in mean effect, but such a phenomenon would rarely be met in practice.

Conclusions

- All three factors are significant with positive sign. Move the center of the experiment in positive direction.
- The *change in mean* effect is not significant.
- The estimated standard deviation is very close to the prior estimate of the standard deviation.

Recommendation for Phase IV

- Move the center of the experiment from $A_0 = 24.09$; $B_0 = 24.09$; $C_0 = 24.09$ to $A_0 = 30$; $B_0 = 30$; $C_0 = 30$ and start Phase IV by switching from REVOP to BEVOP.
- Keep the same factor's interval of variation from $\Delta a = 2.59$; $\Delta b = 2.59$; $\Delta c = 2.59$ (Tables 5.88 and 5.89).

Table 5.88 2^3 BEVOP simulation.

PROJECT : S-6
PHASE : IV
DATE :

BEVOP SIMULATION

I. CENTER OF EXPERIMENT : Ao = 30.00; Bo = 30.00; Co = 30.00 [Gen Random]

II. FACTOR'S INTERVAL OF VARIATION : Δa = 2.59; Δb = 2.59; Δc = 2.59

III. PROCESS NOISE : STDV = 5.0

EVOP WORK SHEET

TRIALS ORDER		Block	FACTORS			RESPONSE									
			A	B	C	n = 1	n = 2	n = 3	n = 4	n = 5	n = 6	n = 7	n = 8	n = 9	n = 10
0		Block I	**30.00**	**30.00**	**30.00**	73.66	71.55	70.19	64.83	80.79	78.16	76.25	78.24	71.54	82.87
1		Block I	27.41	27.41	27.41	77.48	63.81	65.12	75.15	69.72	62.84	70.15	77.14	63.38	65.82
2		Block I	32.59	32.59	27.41	78.19	77.19	80.52	74.73	66.03	83.35	79.77	67.33	80.00	74.44
3		Block I	32.59	27.41	32.59	78.03	81.18	67.36	76.01	77.13	79.93	76.62	66.83	73.61	81.28
4		Block I	27.41	32.59	32.59	73.13	76.73	88.67	66.90	78.16	76.40	73.75	89.92	73.79	79.86
Ø		Block II	**30.00**	**30.00**	**30.00**	68.15	66.45	72.01	76.76	79.73	92.78	73.89	83.81	79.58	75.38
5		Block II	27.41	27.41	32.59	71.26	75.86	77.13	77.78	66.77	76.00	64.54	73.09	73.72	75.13
6		Block II	32.59	32.59	32.59	90.83	84.84	84.10	82.18	76.29	82.37	91.24	78.05	87.32	78.11
7		Block II	32.59	27.41	27.41	79.33	83.77	71.87	78.27	79.90	73.72	73.65	73.31	76.50	83.66
8		Block II	27.41	32.59	27.41	67.68	74.79	73.65	69.92	76.44	76.22	75.53	80.38	78.16	73.37

Table 5.89 2^3 BEVOP information board.

2^3 INFORMATION BOARD AFTER PHASE NO. IV

Last Cycle Completed : n = 10

	RESPONSE No. 1	RESPONSE No. 2	RESPONSE No. 3
REQUIREMENT :	MAXIMIZE RESPONSE		
PROCESS AVERAGES :	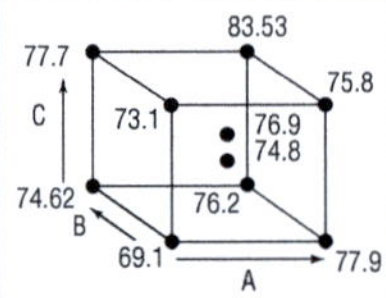		
2.S.E. LIMITS FOR AVERAGES	3.71		
EFFECTS WITH 2.S.E — PHASE MEAN	75.96		
FACTOR A	4.71 ± 2.62		
FACTOR B	4.04 ± 2.62		
FACTOR C	3.12 ± 2.62		
INTERACTION AB	-1.04 ± 2.62		
INTERACTION AC	-0.48 ± 2.62		
INTERACTION BC	2.13 ± 2.62		
INTERACTION ABC	2.61 ± 2.62		
CHANGE IN MEAN	0.13 ± 2.34		
ESTIMATED STANDARD DEVIATION S	5.87		
PRIOR ESTIMATE OF STANDARD DEV.	5.00		

NOTE:

1. The Center of Experiment-Reference Conditions (The Current Best-Known Process Conditions)
2. The Change in Mean provides a measure of whether the conditions being run the current phase yield an average result better or worse than the reference conditions.
3. The Phase Mean is the Mean Response Over All Conditions Being Run in the present Phase.
4*. The change in Mean is a measure of the over-all Curvature of the Response Surface. It will be positive near a MINIMUM and negative near a MAXIMUM.
5. Sequencing of the Runs-Randomization, if runs can be randomized without too much difficulty, than this should be done.

*A Diagonally oriented saddle represented by equal coefficients of opposite sign could occure and yield a zero change in mean effect, but such a phenomenon would rarely be met in practice.

Conclusions

- All three factors are significant with positive sign. Move the center of the experiment in positive direction.
- The *phase mean is* increased significantly.
- The *change in mean* effect is not significant.
- The estimated standard deviation is very close to the prior estimate of the standard deviation.

Recommendation for Phase V

- Move the center of the experiment from $A_0 = 30$; $B_0 = 30$; $C_0 = 30$ to $A_0 = 40$; $B_0 = 40$; $C_0 = 40$ and start a Phase V.
- Keep the same factor's interval of variation from $\Delta a = 2.59$; $\Delta b = 2.59$; $\Delta c = 2.59$ (Tables 5.90 and 5.91).

Table 5.90 2^3 BEVOP simulation.

PROJECT : S-6
PHASE : V
DATE :

BEVOP SIMULATION

I. CENTER OF EXPERIMENT : Ao = 40.00; Bo = 40.00; Co = 40.00 [Gen Random]

II. FACTOR'S INTERVAL OF VARIATION : Δa = 2.59; Δb = 2.59; Δc = 2.59

III. PROCESS NOISE : STDV = 5.0

EVOP WORK SHEET

TRIALS ORDER		Block	FACTORS A	B	C	RESPONSE n = 1	n = 2	n = 3	n = 4	n = 5	n = 6	n = 7	n = 8	n = 9	n = 10
0		Block I	**40.00**	**40.00**	**40.00**	91.66	89.55	88.19	82.83	98.79	96.16	94.25	96.24	89.54	100.87
1		Block I	37.41	37.41	37.41	98.58	84.92	86.22	96.26	90.83	83.94	91.26	98.25	84.49	86.93
2		Block I	42.59	42.59	37.41	95.15	94.15	97.48	91.70	82.99	100.31	96.74	84.29	96.97	91.40
3		Block I	42.59	37.41	42.59	94.99	98.14	84.33	92.97	94.09	96.89	93.58	83.79	90.57	98.24
4		Block I	37.41	42.59	42.59	90.09	93.70	105.63	83.86	95.13	93.36	90.72	106.88	90.76	96.82
Ø		Block II	**40.00**	**40.00**	**40.00**	86.15	84.45	90.01	94.76	97.73	110.78	91.89	101.81	97.58	93.38
5		Block II	37.41	37.41	42.59	90.29	94.89	96.17	96.82	85.81	95.04	83.58	92.12	92.76	94.17
6		Block II	42.59	42.59	42.59	105.72	99.74	98.99	97.07	91.18	97.27	106.14	92.94	102.21	93.00
7		Block II	42.59	37.41	37.41	98.37	102.81	90.90	97.30	98.93	97.76	92.68	92.35	95.54	102.70
8		Block II	37.41	42.59	37.41	86.72	93.83	92.68	88.95	95.48	95.26	94.56	99.42	97.20	92.40

Table 5.91 2^3 BEVOP information board.

2^3 INFORMATION BOARD AFTER PHASE NO. V Last Cycle Completed : n = 10			
	RESPONSE No. 1	RESPONSE No. 2	RESPONSE No. 3
REQUIREMENT :	MAXIMIZE RESPONSE		
PROCESS AVERAGES :	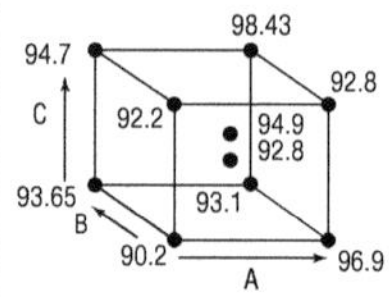		
2.S.E. LIMITS FOR AVERAGES	3.71		
EFFECTS WITH 2.S.E PHASE MEAN FACTOR A FACTOR B FACTOR C INTERACTION AB INTERACTION AC INTERACTION BC INTERACTION ABC CHANGE IN MEAN	93.96 2.64 ± 2.62 1.97 ± 2.62 1.04 ± 2.62 -1.04 ± 2.62 -0.48 ± 2.62 2.13 ± 2.62 2.61 ± 2.62 0.13 ± 2.34		
ESTIMATED STANDARD DEVIATION S	5.87		
PRIOR ESTIMATE OF STANDARD DEV.	5.00		

NOTE:

1. The Center of Experiment-Reference Conditions (The Current Best-Known Process Conditions)
2. The Change in Mean provides a measure of whether the conditions being run the current phase yield an average result better or worse than the reference conditions.
3. The Phase Mean is the Mean Response Over All Conditions Being Run in the present Phase.
4*. The change in Mean is a measure of the over-all Curvature of the Response Surface. It will be positive near a MINIMUM and negative near a MAXIMUM.
5. Sequencing of the Runs-Randomization, if runs can be randomized without too much difficulty, than this should be done.

*A Diagonally oriented saddle represented by equal coefficients of opposite sign could occure and yield a zero change in mean effect, but such a phenomenon would rarely be met in practice.

Conclusions

- All three factors are not significant. It seems we are very close to an optimum.
- The *phase mean is* increased significantly.
- The *change in mean* effect is not significant.
- The estimated standard deviation is very close to the prior estimate of the standard deviation.

Recommendation for Phase VI

- Move the center of the experiment from $A_0 = 40$; $B_0 = 40$; $C_0 = 40$ to $A_0 = 50$; $B_0 = 50$; $C_0 = 50$ and start a Phase VI.
- Keep the same factor's interval of variation from $\Delta a = 2.59$; $\Delta b = 2.59$; $\Delta c = 2.59$ (Tables 5.92 and 5.93).

Table 5.92 2^3 BEVOP simulation.

PROJECT : S-6
PHASE : VI
DATE :

BEVOP SIMULATION

I. CENTER OF EXPERIMENT : Ao = 50.00, Bo = 50.00, Co = 50.00 [Gen Random]

II. FACTOR'S INTERVAL OF VARIATION : Δa = 2.59, Δb = 2.59, Δc = 2.59

III. PROCESS NOISE : STDV = 5.0

EVOP WORK SHEET

TRIALS ORDER		Block	FACTORS			RESPONSE									
			A	B	C	n = 1	n = 2	n = 3	n = 4	n = 5	n = 6	n = 7	n = 8	n = 9	n = 10
0		Block I	**50.00**	**50.00**	**50.00**	97.66	95.55	94.19	88.83	104.79	102.16	100.25	102.24	95.54	106.87
1		Block I	47.41	47.41	47.41	107.69	94.02	95.33	105.37	99.94	93.05	100.37	107.35	93.60	96.04
2		Block I	52.59	52.59	47.41	100.12	99.12	102.45	96.66	87.96	105.28	101.70	89.26	101.93	96.37
3		Block I	52.59	47.41	52.59	99.95	103.11	89.29	97.94	99.05	101.86	98.55	88.75	95.54	103.20
4		Block I	47.41	52.59	52.59	95.06	98.66	110.60	88.83	100.09	98.33	95.68	111.85	95.72	101.79
Ø		Block II	**50.00**	**50.00**	**50.00**	92.15	90.45	96.01	100.76	103.73	116.78	97.89	107.81	103.58	99.38
5		Block II	47.41	47.41	52.59	97.33	101.93	103.20	103.85	92.84	102.07	90.61	99.16	99.80	101.20
6		Block II	52.59	52.59	52.59	108.62	102.63	101.88	99.97	94.07	100.16	109.03	95.83	105.10	95.90
7		Block II	52.59	47.41	47.41	105.41	109.85	97.94	104.34	105.97	104.79	99.72	99.38	102.57	109.73
8		Block II	47.41	52.59	47.41	93.76	100.87	99.72	95.99	102.51	102.30	101.60	106.46	104.24	99.44

Table 5.93 2^3 BEVOP information board.

2^3 INFORMATION BOARD AFTER PHASE NO. VI

Last Cycle Completed : n = 10

	RESPONSE No. 1	RESPONSE No. 2	RESPONSE No. 3
REQUIREMENT :	MAXIMIZE RESPONSE		
PROCESS AVERAGES :	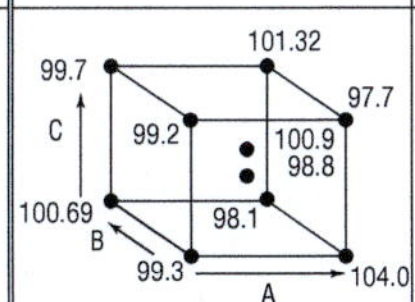		
2.S.E. LIMITS FOR AVERAGES	3.71		
EFFECTS WITH 2.S.E — PHASE MEAN	99.96		
FACTOR A	0.57 ± 2.62		
FACTOR B	-0.11 ± 2.62		
FACTOR C	-1.03 ± 2.62		
INTERACTION AB	-1.04 ± 2.62		
INTERACTION AC	-0.48 ± 2.62		
INTERACTION BC	2.13 ± 2.62		
INTERACTION ABC	2.61 ± 2.62		
CHANGE IN MEAN	0.13 ± 2.34		
ESTIMATED STANDARD DEVIATION S	5.87		
PRIOR ESTIMATE OF STANDARD DEV.	5.00		

NOTE:

1. The Center of Experiment-Reference Conditions (The Current Best-Known Process Conditions)
2. The Change in Mean provides a measure of whether the conditions being run the current phase yield an average result better or worse than the reference conditions.
3. The Phase Mean is the Mean Response Over All Conditions Being Run in the present Phase.
4*. The change in Mean is a measure of the over-all Curvature of the Response Surface. It will be positive near a MINIMUM and negative near a MAXIMUM.
5. Sequencing of the Runs-Randomization, if runs can be randomized without too much difficulty, than this should be done.

*A Diagonally oriented saddle represented by equal coefficients of opposite sign could occure and yield a zero change in mean effect, but such a phenomenon would rarely be met in practice.

Conclusions

- All three factors are not significant. Apparently, we are definitely sitting on the optimum.
- The *phase mean* reached maximum.
- The *change in mean* effect is not significant.
- The estimated standard deviation is very close to the prior estimate of the standard deviation.

Recommendation for Phase VII

- Move the center of the experiment far away from $A_0 = 50$; $B_0 = 50$; $C_0 = 50$ to $A_0 = 70$; $B_0 = 70$; $C_0 = 70$ to verify if we are moving away from optimum.
- Keep the same factor's interval of variation from $\Delta a = 2.59$; $\Delta b = 2.59$; $\Delta c = 2.59$ (Tables 5.94 and 5.95).

Table 5.94 2^3 BEVOP simulation.

PROJECT : S-6
PHASE : VII
DATE :

BEVOP SIMULATION

I. CENTER OF EXPERIMENT :	Ao = 70.00 Bo = 70.00 Co = 70.00	Gen Random
II. FACTOR'S INTERVAL OF VARIATION :	Δ a = 2.59 Δ b = 2.59 Δ c = 2.59	
III. PROCESS NOISE :	STDV = 5.0	

EVOP WORK SHEET

TRIALS ORDER		Block	FACTORS			RESPONSE									
			A	B	C	n = 1	n = 2	n = 3	n = 4	n = 5	n = 6	n = 7	n = 8	n = 9	n = 10
0		Block I	**70.00**	**70.00**	**70.00**	73.22	60.73	82.07	75.77	75.57	70.26	68.71	78.33	74.04	81.77
1		Block I	67.41	67.41	67.41	80.45	83.85	81.06	81.58	79.54	95.10	82.73	93.97	66.08	82.08
2		Block I	72.59	72.59	67.41	70.66	63.43	74.58	78.01	67.69	72.83	73.66	72.47	70.25	82.37
3		Block I	72.59	67.41	72.59	62.24	82.52	75.50	74.39	73.74	68.43	74.71	71.45	66.45	80.82
4		Block I	67.41	72.59	72.59	70.01	79.36	67.31	74.93	69.32	75.17	77.02	74.87	72.55	69.10
Ø		Block II	**70.00**	**70.00**	**70.00**	84.67	80.30	85.42	84.05	74.05	74.99	68.34	86.52	82.96	80.59
5		Block II	67.41	67.41	72.59	82.12	85.86	80.43	75.31	72.33	74.53	87.23	78.27	77.35	89.48
6		Block II	72.59	72.59	72.59	74.48	68.77	72.03	77.63	63.19	72.24	73.35	80.41	67.60	64.96
7		Block II	72.59	67.41	67.41	80.73	78.65	81.03	74.44	79.61	73.78	74.57	76.80	82.06	72.62
8		Block II	67.41	72.59	67.41	71.08	77.01	73.22	93.35	79.91	73.78	80.12	69.13	85.29	71.63

Table 5.95 2^3 BEVOP information board.

2^3 INFORMATION BOARD AFTER PHASE NO. VII

Last Cycle Completed : n = 10

	RESPONSE No. 1	RESPONSE No. 2	RESPONSE No. 3
REQUIREMENT :	MAXIMIZE RESPONSE		
PROCESS AVERAGES :	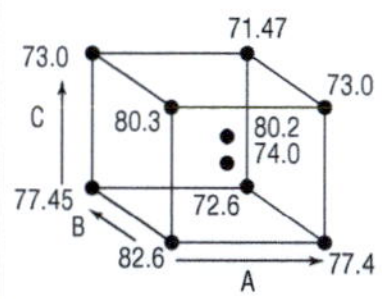		
2.S.E. LIMITS FOR AVERAGES	3.67		
EFFECTS WITH 2.S.E PHASE MEAN FACTOR A FACTOR B FACTOR C INTERACTION AB INTERACTION AC INTERACTION BC INTERACTION ABC CHANGE IN MEAN	 76.21 -4.71 ± 2.58 -4.73 ± 2.58 -3.09 ± 2.58 1.53 ± 2.58 0.33 ± 2.58 0.29 ± 2.58 1.35 ± 2.58 -0.91 ± 2.31		
ESTIMATED STANDARD DEVIATION S	5.80		
PRIOR ESTIMATE OF STANDARD DEV.	5.00		

NOTE:

1. The Center of Experiment-Reference Conditions (The Current Best-Known Process Conditions)
2. The Change in Mean provides a measure of whether the conditions being run the current phase yield an average result better or worse than the reference conditions.
3. The Phase Mean is the Mean Response Over All Conditions Being Run in the present Phase.
4*. The change in Mean is a measure of the over-all Curvature of the Response Surface. It will be positive near a MINIMUM and negative near a MAXIMUM.
5. Sequencing of the Runs-Randomization, if runs can be randomized without too much difficulty, than this should be done.

*A Diagonally oriented saddle represented by equal coefficients of opposite sign could occure and yield a zero change in mean effect, but such a phenomenon would rarely be met in practice.

Conclusions

- All three factors are significant with negative sign. It means we are far away from optimum.
- The *phase mean is* reduced significantly.
- The *change in mean* effect is not significant.
- The estimated standard deviation is very close to the prior estimate of the standard deviation.

Because of the complex geometry of the 2^3 ROVOP, we are going to summarize movement of the ROVOP from the best known conditions to the optimum:

ROVOP Phase I, Cycle 1–4	Phase mean	46.0	Moving toward optimum
ROVOP Phase II, Cycle 1	Phase mean	55.1	Moving toward optimum
ROVOP Phase III, Cycle 1	Phase mean	59.7	Moving toward optimum
BEVOP Phase IV, Cycle 10	Phase mean	76.0	Moving toward optimum
BEVOP Phase V, Cycle 10	Phase mean	94.0	Very close to optimum
BEVOP Phase VI, Cycle 10	Phase mean	100.0	Optimum
BEVOP Phase VII, Cycle 10	Phase mean	76.2	Moving away from optimum

5.9 Random Evolutionary Operation – REVOP

REVOP claims to be particularly useful in experimentation where there are a large number of relevant factors, the types of relationships between the variables are not known, and it is anticipated that they are quite complex. REVOP is effective in complex situations because it is essentially a non-parametric method. In that case, evaluation of specific parameters is avoided in assumed functional relationship and we work directly with desired result, rather than indirectly through assumed functions and parameters.

For REVOP, the number of experiments required to find a set of optimum conditions appears to be independent of the number of factors and of the complexity of the relationship. REVOP involved random choices of data points for the experimentation and generally, data derived by REVOP are unbalanced and very inefficient for evaluation of parameters, functional relationships, and their associated parameters.

As EVOP, REVOP also requires:

1. A starting point usually is the best known conditions.
2. A measure of goodness or response to measure the result of the experimentation.
3. A feedback of information to make the procedure truly sequential.

REVOP procedure is as follows:

1. A list is made of all controllable factors acting on the process, with the range for investigation of each factor. It is assumed that any uncontrollable factors acting on the process do so in an essentially random manner.
2. The lower limit of each factor is coded as zero, (0.0), and the upper limit of each factor is coded as 10.0. Then if these limits are symmetrical about the best known conditions, the starting points for each factor will be coded as 5.0. The code 2.0 represents a factor at 20% of the range from its lower limit. The first experiment is run at values 5.0 for each factor.
3. Obviously the next experiment must be run at different conditions and where no appropriate choice of direction is available a random direction is selected. Digits 0–9 are written at random for each factor (d_1) and the factor $D_1 = d_1^2/c$ is arranged so that the maximum step length for any one factor is about 20% of the range of investigation for that factor (corresponds to code 2.0). D_1 values are calculated for each factor are given + or − signs at random. Coefficient c is factor's steps expansion/contraction with range 1–40.
4. Step increments are added and experimentation is made as long as the response continues to improve. Where no improvement is found, another set of random numbers is established for d_2 and the procedure continues. Any D calculated from a new set of d is added to the previous best set of conditions.
5. If the newly calculated experimental conditions (A) show negative values, these are reflected to their equivalent positive values. If the newly calculated experimental value (in A) is greater than 10, and quite outside the predetermined range of experimentation, a decision must be taken whether to permit the calculated value or to restrict it to a value of 10·0.

The advantages of REVOP are:

1. Any number of factors can be handled.
2. Calculations are extremely simple.

The disadvantages of REVOP are:

1. Only quantitative factors can be handled.
2. Movement of experimentation is by random selection.
3. The calculated step length, for any value *d*, may not be measurable on the process control instruments.
4. No attempt can be made in separating the effects of any of the factors.
5. Calculation of interactions is not involved.
6. It is not possible to build up a response surface from REVOP experimentation.

Example 5.7 This is REVOP application example found in the literature for ten process factors with objective to maximize response value [35]. All steps following REVOP procedure are shown in Table 5.96. The maximum response value was achieved in step E_{17}.

Table 5.96 REVOP work sheet.

Coordinates	x_1	x_2	x_3	x_4	x_5	x_6	x_7	x_8	x_9	x_{10}	y
E_0	5	5	5	5	5	5	5	5	5	5	14.2
d_1	4	9	1	5	7	5	4	8	5	9	—
$D_1 = d_1{}^2/40$	+0.4	−2.0	+0	−0.6	−1.2	+0.6	−0.4	−1.6	−0.6	−2.0	—
$E_1 = E_0 + D_1$	5.4	3.0	5.0	4.4	3.8	5.6	4.6	3.4	4.4	3.0	22.3
$E_2 = E_1 + D_1$	5.8	1.0	5.0	3.8	2.6	6.2	4.2	1.8	3.8	1.0	16.0
d_2	0	5	5	4	5	5	5	0	4	3	—
$D_2 = d_2{}^2/12$	−0	+2.1	−2.1	−1.3	−2.1	+2.1	−2.1	−0.0	+1.3	+0.7	—
$E_3 = E_1 + D_2$	5.4	5.1	2.9	3.1	1.7	7.7	2.5	3.4	5.7	3.7	31.8
$A_4 = E_3 + D_2$	5.4	7.2	0.8	1.8	−0.4	9.8	0.4	3.4	7.0	4.4	—
$E_4=$	5.4	7.2	0.8	1.8	0.4	9.8	0.4	3.4	7.0	4.4	39.6
$A_5 = E_4 + D_2$	5.4	9.3	−1.3	0.5	−2.5	11.9	−1.7	3.4	8.3	5.1	—
$E_5=$	5.4	9.3	1.3	0.5	2.5	11.9	1.7	3.4	8.3	5.1	31.3
d_3	1	4	8	7	1	6	0	3	5	0	—
$D_3 = d_3{}^2/30$	−0.0	+0.5	+2.1	+1.8	+0.0	−1.2	+0.0	+0.3	+0.8	−0.0	—
$A_6 = E_4 + D_3$	5.4	7.7	2.9	3.6	−0.4	8.6	0.4	3.7	7.8	4.4	—
$E_6=$	5.4	7.7	2.9	3.6	0.4	8.6	0.4	3.7	7.8	4.4	30.8
$A_7 = A_4 - D_3$	5.4	6.7	−1.3	0.0	0.4	11.0	0.4	3.1	6.2	4.4	—
$E_7=$	5.4	6.7	1.3	0.0	0.4	11.0	0.4	3.1	6.2	4.4	38.0
d_4	3	8	9	7	6	7	4	9	5	1	—

Table 5.96 (Continued)

Coordinates	x_1	x_2	x_3	x_4	x_5	x_6	x_7	x_8	x_9	x_{10}	y
$D_4 = d_4{}^2/40 - 0.2$	−0.2	+1.5	+2.0	−1.2	−0.9	−1.2	−0.4	+2.0	−.06	−0.0	—
$A_8 = A_4 + D_4$	5.2	8.8	2.8	0.6	−1.3	8.6	0.0	5.4	6.4	4.4	—
$E_8 =$	5.2	8.8	2.8	0.6	1.3	8.6	0.0	5.4	6.4	4.4	23.5
$A_9 = A_4 - D_4$	5.6	5.6	−1.2	3.0	0.5	11.0	0.8	1.4	7.6	4.4	—
$E_9 =$	5.6	5.6	1.2	3.0	0.5	11.0	0.8	1.4	7.6	4.4	32.8
d_5	9	7	3	1	2	6	1	7	1	8	—
$D_5 = d_5{}^2/40$	−2.0	−1.2	−0.2	−0.0	−0.1	−0.0	+0.0	+1.2	+0.0	+1.6	—
$A_{10} = A_4 + D_5$	3.4	6.0	0.6	1.8	0.5	9.1	0.4	4.6	7.0	6.0	—
$E_{10} =$	3.4	6.0	0.6	1.8	0.5	9.1	0.4	4.6	7.0	6.0	55.1
$A_{11} = A_{10} + D_5$	1.4	4.8	0.4	1.8	0.4	8.2	0.4	5.8	7.0	6.0	—
$E_{11} =$	1.4	4.8	0.4	1.8	0.4	8.2	0.4	5.8	7.0	6.0	71.6
$A_{12} = A_{11} + D_5$	−0.6	3.6	0.2	1.8	0.3	7.3	0.4	7.0	7.0	9.2	—
$E_{12} =$	0.6	3.6	0.2	1.8	0.3	7.3	0.4	7.0	7.0	9.2	88.15
$A_{13} = A_{12} + D_5$	2.6	2.4	0.0	1.8	0.2	6.4	0.4	8.2	7.0	10.8	—
$E_{13} =$	2.6	2.4	0.0	1.8	0.2	6.4	0.4	8.2	7.0	10.8	82.3
d_6	1	1	7	4	2	6	9	3	8	1	—
$D_6 = d_6{}^2/40$	+0.0	+0.0	−1.2	−0.4	−0.1	−0.9	+2.0	+0.2	+1.6	+0.0	—
$A_{14} = A_{12} + D_6$	−0.6	3.6	−1.0	1.4	0.2	6.4	2.4	7.2	8.6	9.2	—
$E_{14} =$	0.6	3.6	1.0	1.4	0.2	6.4	2.4	7.2	8.6	9.2	90.42
$A_{15} = A_{14} + D_6$	−0.6	3.6	−2.2	1.0	0.1	5.5	4.4	7.4	10.2	9.2	—
$E_{15} =$	0.6	3.6	2.2	1.0	0.1	5.5	4.4	7.4	10.2	9.2	84.5
d_7	4	3	3	6	1	2	8	8	5	9	—
$D_7 = d_7{}^2/40$	−0.4	−0.2	+0.2	−0.9	+0.0	+0.1	−1.6	+1.6	−0.6	−2.0	—
$A_{16} = A_{14} + D_7$	−1.0	3.4	−0.8	0.5	0.2	6.5	0.8	8.8	9.2	7.2	—
$E_{16} =$	1.0	3.4	0.8	0.5	0.2	6.5	0.8	8.8	9.2	7.2	82.6
$A_{17} = A_{14} - D_7$	−0.2	3.8	−1.2	2.3	0.2	6.3	4.0	5.6	8.0	11.2	—
$E_{17} =$	0.2	3.8	−1.2	2.3	0.2	6.3	4.0	5.6	8.0	11.2	92.58

5.9.1 REVOP Simulation

5.9.1.1 Simulation No. 7

The same mathematical formula from simulation No.4 with three factors was used to test REVOP method. As mentioned in REVOP procedure, the lower limit of each factor is coded zero, (0.0), and the upper limit of each factor is coded 10.0. Then if these limits are symmetrical about the best known process, the starting points for each variable will be coded 5.0. Experimental conditions are defined based on random numbers for each factor (d_i) and the coded factor $D_i = d^2/c$ steps are calculated so that the maximum step length for any one factor is about 20% of the range of investigations for that factor. The values calculated for D_i for each factor are given + or – sign at random. Step increments are added and experimentation is made as long as the response continues to improve. Where no improvement is found, another set of random numbers is established for d and the procedure continues. Any D calculated from a new set of d is added to the previous best set of conditions. The appropriate mathematical function was used as before to generate process yield numbers plus noise (STDEV multiplied by Random Normal Deviates) to mask yield with process noise. The objective of this study is to maximize product yield using REVOP.

Phase I: The settings (Center Point, Factor's Interval, and Process Noise, coefficient of factor's steps expansion/contraction) and yield for different process conditions have been calculated using Microsoft Excel simulation in Table 5.97. This simulation started with higher process noise. The corresponding calculations for each phase are shown below. The simulator will stop if there is no improvement after ten (10) phases (Tables 5.98).

Table 5.97 REVOP simulation.

PROJECT : S-7 [Reset Button]
PHASE : I
DATE :

THE SIMULATOR WILL STOP IF THERE IS NO IMPROVEMENT AFTER 10-PHASES.

SIMULATION

I. NUMBER OF FACTORS[1-10]: [Set K] K = 3 [Go REVOP]

II. CENTER OF EXPERIMENT-CODED VALUE: Ao = 1.00; Bo = 1.00; Co = 1.00

III. FACTORS RANGE:

CODE	ACTUAL	CODE	ACTUAL
A[0.0] =	0.00	A[10.0]	100.00
B[0.0] =	0.00	B[10.0]	100.00
C[0.0] =	0.00	C[10.0]	100.00

IV. COEFFICIENT-c FACTOR'S STEPS EXPANSION/CONTRACTION:

c = 20.00 RANGE[1-40]

V. d_i-RANDOM NUMBERS IN RANGE [0-9]: dA = 7; dB = 8; dC = 9

Microsoft Excel: The Best Trial Order Found in Phase 1 Or Trial Order 1 [OK]

VI. D_i-CODED FACTOR'S STEPS: $Di=di^2/c$ Δa = 2.45; Δb = 3.20; Δc = 4.05

VII. PROCESS-RESPONSE NOISE: STDV = 5.0 [Gen Random]

VIII. SET NUMBER OF REPLICATES[n=1-10] 6 [Set N]

REVOP WORK SHEET

TRIALS ORDERS	FACTORS			RESPONSE						AVG
	A	B	C	n=1	n=2	n=3	n=4	n=5	n=6	
E0	1.00	1.00	1.00	-3.90	-0.79	3.39	5.71	9.12	9.95	3.91
E1	3.45	4.20	5.05	92.92	95.68	88.80	98.19	96.36	97.48	94.91
E2	5.90	7.40	9.10	57.29	60.69	52.03	55.10	52.11	56.74	55.66

The second REVOP test with the same simulation but with lower process noise (STDEV) has been shown below. The best trial order found in Phase 2 with response value: 91.32 (Table 5.99)

Table 5.98 REVOP simulation.

PROJECT : S-7 [Reset Button]
PHASE : I
DATE :

THE SIMULATOR WILL STOP IF THERE IS NO IMPROVEMENT AFTER 10-PHASES.

SIMULATION

I. NUMBER OF FACTORS[1-10]: [Set K]	K = 3		[Go REVOP]
II. CENTER OF EXPERIMENT-CODED VALUE:	Ao = 1.00 Bo = 1.00 Co = 1.00		
III. FACTORS RANGE:	CODE	ACTUAL	CODE ACTUAL
	A[0.0] =	0.00	A[10.0] 100.00
	B[0.0] =	0.00	B[10.0] 100.00
	C[0.0] =	0.00	C[10.0] 100.00
IV. COEFFICIENT-c FACTOR'S STEPS EXPANSION/CONTRACTION:	c = 20.00		RANGE[1-40]
V. d_i-RANDOM NUMBERS IN RANGE [0-9]:	dA = 3 dB = 4 dC = 0		
VI. D_i-CODED FACTOR'S STEPS: $Di=di^2/c$	Δa = 0.45 Δb = 0.80 Δc = 0.00		
VII. PROCESS-RESPONSE NOISE:	STDV = 1.0		[Gen Random]
VIII. SET NUMBER OF REPLICATES[n=1-10]	6		[Set N]

REVOP WORK SHEET

TRIALS ORDERS	FACTORS			RESPONSE						AVG
	A	B	C	n=1	n=2	n=3	n=4	n=5	n=6	
E0	1.00	1.00	1.00	2.42	3.04	3.88	4.34	5.02	5.19	3.98
E1	1.45	1.80	1.00	22.12	22.67	21.29	23.17	22.81	23.03	22.51
E2	1.90	2.60	1.00	38.07	38.75	37.02	37.63	37.03	37.96	37.74
E3	2.35	3.40	1.00	49.22	48.09	48.38	47.72	47.81	47.50	48.12
E4	2.80	4.20	1.00	57.67	57.03	57.20	57.02	56.42	58.19	57.26
E5	3.25	5.00	1.00	63.29	63.05	63.22	63.60	61.39	61.97	62.75
E6	3.70	5.80	1.00	63.70	63.83	63.89	63.47	63.64	62.31	63.47
E7	4.15	6.60	1.00	62.80	62.81	61.05	60.78	62.60	61.69	61.95

Table 5.99 REVOP simulation.

PROJECT : S-7
PHASE : 2
DATE :

THE SIMULATOR WILL STOP IF THERE IS NO IMPROVEMENT AFTER 10-PHASES.

SIMULATION

I. NUMBER OF FACTORS[1-10]:	K = 3		
II. CENTER OF EXPERIMENT-CODED VALUE:	Ao = 3.70 Bo = 5.80 Co = 1.00		
III. FACTORS RANGE:	CODE	ACTUAL	CODE ACTUAL
	A[0.0] =	0.00	A[10.0] = 100.00
	B[0.0] =	0.00	B[10.0] = 100.00
	C[0.0] =	0.00	C[10.0] = 100.00
IV. COEFFICIENT-c FACTOR'S STEPS EXPANSION/CONTRACTION:	c = 20.00		RANGE[1-40]
V. d_i-RANDOM NUMBERS IN RANGE [0-9]:	dA = 5 dB = 1 dC = 9		Microsoft Excel The Best Trial Order found in Phase 2 Or Trial Order 8 [OK]
VI. D_i-CODED FACTOR'S STEPS: $Di=di^2/c$	Δa = 1.25 Δb = 0.05 Δc = -4.05		
VII. PROCESS-RESPONSE NOISE:	STDV = 1.0		
VIII. SET NUMBER OF REPLICATES[n=1-10]	6		

REVOP WORK SHEET

TRIALS ORDERS	FACTORS			RESPONSE						AVG
	A	B	C	n=1	n=2	n=3	n=4	n=5	n=6	
A8	4.95	5.85	-3.05							
E8	4.95	5.85	3.05	90.67	92.70	89.24	92.29	93.26	89.77	91.32
A9	6.20	5.90	-1.00							
E9	6.20	5.90	1.00	63.63	63.06	63.40	64.37	63.87	64.15	63.75

The third REVOP test with the same simulation but with very high process noise (STDEV) has been shown below. The best trial order found in Phase 13 with response value: 102.11 (Tables 5.100 and 5.101).

Table 5.100 REVOP simulation.

PROJECT : S-7
PHASE : I
DATE :

Reset Button

THE SIMULATOR WILL STOP IF THERE IS NO IMPROVEMENT AFTER 10-PHASES.

SIMULATION

I. NUMBER OF FACTORS[1-10]: Set K K = 3 Go REVOP

II. CENTER OF EXPERIMENT-CODED VALUE: Ao = 1.00, Bo = 1.00, Co = 1.00

III. FACTORS RANGE:

CODE	ACTUAL	CODE	ACTUAL
A[0.0] =	0.00	A[10.0]	100.00
B[0.0] =	0.00	B[10.0]	100.00
C[0.0] =	0.00	C[10.0]	100.00

IV. COEFFICIENT-c FACTOR'S STEPS EXPANSION/CONTRACTION: c = 20.00 RANGE[1-40]

V. d_i-RANDOM NUMBERS IN RANGE [0-9]: dA = 5, dB = 2, dC = 7

Microsoft Excel
The Best Trial Order found in Phase 13
Or Trial Order 26
OK

VI. D_i-CODED FACTOR'S STEPS: $Di=di^2/c$ Δa = 1.25, Δb = -0.20, Δc = -2.45

VII. PROCESS-RESPONSE NOISE: STDV = 10.0 Gen Random

VIII. SET NUMBER OF REPLICATES[n=1-10] 6 Set N

REVOP WORK SHEET

TRIALS ORDERS	FACTORS			RESPONSE						AVG
	A	B	C	n=1	n=2	n=3	n=4	n=5	n=6	
E0	1.00	1.00	1.00	-11.81	-5.58	2.77	7.42	14.24	15.90	3.83
E1	2.25	0.80	-1.45	-35.59	-30.07	-43.84	-25.05	-28.71	-26.46	-31.62

Table 5.101 REVOP simulation.

PROJECT : S-7
PHASE : 13
DATE :

THE SIMULATOR WILL STOP IF THERE IS NO IMPROVEMENT AFTER 10-PHASES.

SIMULATION

I. NUMBER OF FACTORS[1-10]: K = 3

II. CENTER OF EXPERIMENT-CODED VALUE: Ao = 3.15, Bo = 5.75, Co = 5.10

III. FACTORS RANGE:

CODE	ACTUAL	CODE	ACTUAL
A[0.0] =	0.00	A[10.0] =	100.00
B[0.0] =	0.00	B[10.0] =	100.00
C[0.0] =	0.00	C[10.0] =	100.00

IV. COEFFICIENT-c FACTOR'S STEPS EXPANSION/CONTRACTION: c = 20.00 RANGE[1-40]

V. d_i-RANDOM NUMBERS IN RANGE [0-9]: dA = 4, dB = 2, dC = 3

Microsoft Excel
The Best Trial Order found in Phase 13
Or Trial Order 26
OK

VI. D_i-CODED FACTOR'S STEPS: $Di=di^2/c$ Δa = 0.80, Δb = -0.20, Δc = -0.45

VII. PROCESS-RESPONSE NOISE: STDV = 10.0

VIII. SET NUMBER OF REPLICATES[n=1-10] 6

REVOP WORK SHEET

TRIALS ORDERS	FACTORS			RESPONSE						AVG
	A	B	C	n=1	n=2	n=3	n=4	n=5	n=6	
A26	3.95	5.55	4.65							
E26	3.95	5.55	4.65	99.29	105.86	95.45	107.98	100.66	103.43	102.11
A27	4.75	5.35	4.20							
E27	4.75	5.35	4.20	99.40	119.36	90.79	101.69	80.18	95.56	97.83

5.10 Quick-Start EVOP – QSEVOP

Billed as a procedure for continuously improving a process, QSEVOP allows operators to identify factors settings, which result in progressively higher yield or throughput.

The goal of QSEVOP is to directly relate changes in process responses to modifications in operating conditions. This means that some kind of one-factor-at-a-time scheme should be used, because multi-factor techniques require completion of the entire investigation before the effects of each factor could be determined. Even then, it takes more effort to separate the effects of each factor.

QSEVOP is a simpler version of original EVOP procedure developed by Box and Draper to improve processes without generating off-specification products. EVOP uses such statistical methods as factorial designs of experiments to identify any interactions between factors. However, subsequent EVOP implementations have been directed toward rapid movement of the factor settings at an increased risk of specification violations. Also, no more than three-process factors can be considered in EVOP because of constraints imposed by simultaneously changing all factor's settings.

Although EVOP has been successfully applied to a wide variety of processes in chemical industries, extent of its usage is still less than expected. The quick-start version described here was developed with the assumption that EVOP would become more acceptable if operators could directly link changes in process responses to alterations in process factors.

QSEVOP uses small changes in operating conditions, which also results in small changes in process output. However, observations or measurements of the output are likely to include some error, or noise, typically caused by such factors like raw-material variations, equipment fluctuations, and instrument deviations. Therefore, repeated measurements are necessary to make sure that the small difference in performance before and after the adjustment of the factor settings is indeed real, and not due to a chance or an error.

QSEVOP prescribes a strategy for replicating the cycles of process runs involving sufficiently small perturbations in the factor settings. Each cycle is made up of a repeated series of measurements following a change in one of the independent factors of the process. After cycles have been run on all the factors, a new set of operating conditions is established, completing one "phase" of QSEVOP.

5.10.1 The way QSEVOP works

A process factor, such as pressure, temperature or flow rate, is slightly changed or perturbed from its current or base setting. The change can be in either direction, by increasing or decreasing the current value.

N-repeated observations or measurements (replicates) are made at the base and new settings. Responses of the process from N-pairs of replicates are compared (with or without the use of statistics) to determine if the alteration in the factor setting causes the process output to change in desirable or undesirable direction.

If a favorable effect, such as an increase in yield, is observed, the exploration is brought to a halt and that setting becomes the new base point for the factor.

The base point is not changed if an adverse effect occurs, such as a drop in the process throughput. The same procedure is employed to sequentially examine each process factor. When all factors were considered, many of the factor settings are likely to be changed from the original base point. In that way, the first phase of the study is completed.

The new factor settings provide a base point for the next phase, and the sequence of factor perturbations is repeated.

Successive phases will eventually lead the process conditions toward maximum throughput or yield with respect to the factors chosen. Unlike traditional EVOP procedure, QSEVOP permits an operator to simply add or delete process factors during any phase of the exploration. Also, any type and number of factors could be considered.

Generally, current operating conditions provide a reliable benchmark for evaluating progressive improvement of a process via QSEVOP. This approach also recognizes that most processes are being operated adequately. The anticipated improvement in what is being optimized is probably less than 10%. Therefore, it is unlikely that the factor settings will be altered drastically.

QSEVOP uses a modification of an algorithm called Hooke–Jeeves (HJ) method for perturbing factor settings. Because the HJ optimization technique does not reach an optimum point quickly enough, textbooks on process optimization seldom recommend it.

However, in many industrial applications, simplicity of the HJ method can more than compensate for its lack of mathematical elegance or theoretical efficiency. In QSEVOP, only searching part of the method, which uses only one factor at a time, is applied. The subsequent pattern-search part, meant to accelerate an approach to the optimum by simultaneously changing all factors at the next phase, was dropped from QSEVOP procedure. The reason: pattern-search part is simply not consistent with a requirement that no off-specification product should be generated during an investigation.

In QSEVOP, previous operating experience is used for determining the size of the perturbation. The objective is to make small changes and avoid specification violations. But the changes must be large enough to be detected in the presence of random process error.

Because random errors can sometimes mask the limited effects of the process generated by small factor changes, statistical reasoning can be used to select the minimum number of replications of observations or measurements. Of course, the accuracy of process response measurements will depend on the number of observations (also known as sample size). If the sample size increases, the mean sample value more closely represents the "true" response of the process.

A statistical concept known as confidence interval (CI) can be used to construct a range of interval in such a way that the "true" process response (which is always unknown) lies within this interval. A predetermined probability, which is called a confidence level, is also associated with the interval. For example, when mean $\overline{X}$ and standard deviation s (for the sample) are known, the CI is expressed as follows:

$$\overline{X} - t \times s < \mu < \overline{X} + t \times s$$

where t is a parameter of a statistical distribution called Student's distribution, which is analogous to the *Normal or Gaussian Distributions*. While the normal distribution is ideal for large samples (with at least 100 observations), Student's is more suited for the smaller samples of data typical obtained by carrying out QSEVOP.

When applied process response pairs from the QSEVOP, CI refers to the inherent variability that would have occurred by chance, had there not been a real difference between paired responses before and after the introduction of changes in variable settings. The CI's are calculated using tabulated values of t-statistic to determine if the average difference between a number of paired responses is statistically significant. A statistical significance level of approximately 90% means that after calculating the CI, the average experimental difference outside this range is likely to be significant.

The probability of it happening by chance is only about 10%. As the number of response pairs increases, the CI becomes smaller because the average difference can be better estimated.

The size of CI is directly proportional to the t-statistic, which decreases with the increments of the number of observations. *As a result, there is even a danger of relying too much on the statistical interpretation of the significance of changes in process response via CI because almost any difference will appear to be statistically significant if the number of pairs of responses is made large enough.* QSEVOP specifies the minimum number of observations necessary, to ensure that the difference in process performance is statistically significant.

For statistical significance level (or confidence level) of 90% on the difference between paired observations, statistical table shows the dependence of the CI size on the number of response pairs. A *Relative Confidence Interval* (*RCI*) is calculated to show consequences of increasing the number of replications on reducing the size of the CI. Clearly, a rapid initial reduction in the RCI is followed by its decline at much diminishing rate.

From an engineering viewpoint, it may not be useful to continue replications if the reduction of RCI is less than 5%. This level of reduction in RCI occurs when the number of replicates goes from six to seven. *Therefore, seven replicates were selected for process-improvement studies through QSEVOP.*

Two different criteria can be used to identify significant changes in factors in the presence of random error. The practical criterion is that *a real effect has occurred if at least five of seven replicates* displace the process in the same way (toward or away from the desired level of product yield or process throughput).

The direction of the effects is important, not their size. Although this approach lacks elegance, experience has shown it to be satisfactory. Strict examination of positive or negative effects would involve calculating a standard deviation calculation for both the baseline and the new setting. In other words, there is a statistically significant effect on the process due to factor changes. Meeting either criterion, one based on a "majority vote" and the other on statistical significance, is sufficient to identify the effect.

5.10.2 How to Recover From "Hang-ups"

Often, during execution of QSEVOP, the process will reach a state where neither criterion for identifying significant changes will be satisfied for any factor. Such misleading "hangups" occurs when none of the factor perturbation results in positive effects (leading toward the maximum yield). It is possible that the highest level of process performance has been reached. However, to make sure that this is indeed a valid interpretation, one should relax the criterion for verifying a positive effect and restart search for the maximum yield using larger changes in factor settings.

For example, the previous phase of QSEVOP can be examined to see if for any factor the difference between the average baseline responses and the new settings has positively changed the process response to the desired level of performance. As soon as such a factor is found, the search should be restarted *by doubling the magnitude of the perturbation of that factor in the appropriate direction*. From that point on, it is recommended to use normal criteria to identify any significant effects. Note that the relaxed criterion of comparing average effects from a previous phase is used only to initiate movement of the process response in the desired direction.

To date, no heuristic rule has been developed to recommend when a factor should be dropped. It is the best to leave that to the discretion of those most familiar with the process.

Ultimately, a condition will be found where even relaxation of the criteria or introduction of larger perturbations would fail to nudge the process to a higher yield. The process performance will either be adversely affected or no significant favorable effects can be realized by changing factor settings. Even under these circumstances, QSEVOP should not be terminated once and for all.

Continued use of QSEVOP will ensure timely detection and rapid recovery from possible declines in product yields, due to gradual process changes caused by such factors as equipment deterioration, instrument drift, and changes in raw material composition. Through QSEVOP's ongoing practice, operators develop the confidence and experience needed to quickly determine the appropriate perturbations sizes for each factor setting.

The advantages of QSEVOP are:

1. Any type of factors can be handled.
2. Any number of factors can be handled.

3. Permits the operator to simply add or delete process factors during any phase.
4. Calculations are extremely simple. Simplicity is the key.

The disadvantages of QSEVOP are:

1. The calculated step length may not be measurable on the process control instruments.
2. No attempt can be made in separating the effects of any of the factors.
3. Movement of experimentation can be in either direction, by increasing or decreasing the current value, which means one factor at a time experimentation.
4. Previous operating experience is used for determining the size of the perturbation.
5. Calculation of interactions is not involved.
6. It is not possible to build up a response surface from QSEVOP experimentation.

As far as the author is aware, no publications have been made in which this QSEVOP modification of EVOP has been used. The best way to test this idea is to run simulation, which was already verified by BEVOP method.

5.11 QSEVOP Simulation

5.11.1 Simulation No. 8

The same mathematical function from Simulation No.1 was used to generate process yield numbers plus noise (STDEV multiplied by Random Normal Deviates) to mask yield with process noise, to test QSEVOP (Tables 5.102–5.106).

Table 5.102 QSEVOP simulation.

PROJECT : S-8
PHASE : I
DATE :

Reset Button

OPTIMIZATION OBJECTIVE: To Maximize Response

SIMULATION

I. NUMBER OF FACTORS[1-10]: Set K K = 2 Go QSEVOP

II. THE CURRENT OPERATING CONDITIONS: Ao = 2.00, Bo = 2.00

III. FACTORS RANGE:

CODE	ACTUAL	CODE	ACTUAL
A[0.0] =	0.00	A[10.0] =	100.00
B[0.0] =	0.00	B[10.0] =	100.00
C[0.0] =	0.00	C[10.0] =	100.00

IV. Δ_i-CODED FACTOR'S STEPS: Δa = 1.00, Δb = 1.00

V. PROCESS NOISE: STDEV = 0.2

VI. THE FIXED NUMBER OF REPLICATES 7 Gen Random

QSEVOP WORK SHEET

CYCLES	FACTORS		RESPONSE														AVERAGE	CRITERIA	
	A	B	n=1	n=2	n=3	n=4	n=5	n=6	n=7	R=1	R=2	R=3	R=4	R=5	R=6	R=7		NUMBER OF SIGNES>5	t-TEST SIGNIFICANT
E0	2.00	2.00	63.94	63.74	64.05	64.26	64.24	64.35	63.56								64.02		
E1	3.00	2.00	73.86	73.66	73.63	73.80	73.85	73.58	73.89	9.92	9.92	9.58	9.55	9.61	9.23	10.32	73.75	Yes	Yes
E2	1.00	2.00	49.93	49.93	50.27	49.98	49.96	49.90	50.39	-23.93	-23.74	-23.36	-23.82	-23.88	-23.68	-23.49	50.05	No	Yes
E3	3.00	3.00	84.33	83.68	84.11	84.18	84.38	83.98	83.90	34.40	33.75	33.84	34.20	34.42	34.09	33.50	84.08	Yes	Yes
E4	3.00	1.00	59.71	59.83	59.70	59.93	59.99	60.01	59.94	-24.62	-23.85	-24.41	-24.25	-24.39	-23.98	-23.96	59.87	No	Yes

t-Test: Two-Sample Assuming Unequal Variances

	Variable 1	*Variable 2*
Mean	84.0800229	59.87134428
Variance	0.062051845	0.016397893
Observations	7	7
Hypothesized Mean Dif	0	
df	9	
t Stat	228.6779859	
P(T<=t) one-tail	1.4877E-18	
t Critical one-tail	1.833112933	
P(T<=t) two-tail	2.9754E-18	
t Critical two-tail	2.262157163	

Table 5.103 QSEVOP simulation.

PROJECT : S-8
PHASE : II
DATE :

Reset Button

OPTIMIZATION OBJECTIVE: To Maximize Response

SIMULATION

I. NUMBER OF FACTORS[1-10]: Set K K = 2 Do QSEVOP

II. THE CURRENT OPERATING CONDITIONS: Ao = 3.00 Bo = 3.00

III. FACTORS RANGE:

CODE	ACTUAL	CODE	ACTUAL
A[0.0] =	0.00	A[10.0] =	100.00
B[0.0] =	0.00	B[10.0] =	100.00
C[0.0] =	0.00	C[10.0] =	100.00

IV. Δ_i-CODED FACTOR'S STEPS: Δa = 1.00 Δb = 1.00

V. PROCESS NOISE: STDEV = 0.2

VI. THE FIXED NUMBER OF REPLICATES 7 Gen Random

QSEVOP WORK SHEET

CYCLES	FACTORS		RESPONSE														CRITERIA		
	A	B	n=1	n=2	n=3	n=4	n=5	n=6	n=7	R=1	R=2	R=3	R=4	R=5	R=6	R=7	AVERAGE	NUMBER OF SIGNES>5	t-TEST SIGNIFICANT
E0	3.00	3.00	83.94	83.74	84.05	84.26	84.24	84.35	83.56								84.02		
E1	4.00	3.00	89.86	89.66	89.63	89.80	89.85	89.58	89.89	5.92	5.92	5.58	5.55	5.61	5.23	6.32	89.75	Yes	Yes
E2	2.00	3.00	73.93	73.93	74.27	73.98	73.96	73.90	74.39	-15.93	-15.74	-15.36	-15.82	-15.88	-15.68	-15.49	74.05	No	Yes
E3	4.00	4.00	96.33	95.68	96.11	96.18	96.38	95.98	95.90	22.40	21.75	21.84	22.20	22.42	22.09	21.50	96.08	Yes	Yes
E4	4.00	2.00	79.71	79.83	79.70	79.93	79.99	80.01	79.94	-16.62	-15.85	-16.41	-16.25	-16.39	-15.98	-15.96	79.87	No	Yes

t-Test: Two-Sample Assuming Unequal Variances

	Variable 1	*Variable 2*
Mean	96.0800229	79.87134428
Variance	0.062051845	0.016397893
Observations	7	7
Hypothesized Mean Dif	0	
df	9	
t Stat	153.1090581	
P(T<=t) one-tail	5.49732E-17	
t Critical one-tail	1.833112933	
P(T<=t) two-tail	1.09946E-16	
t Critical two-tail	2.262157163	

Table 5.104 QSEVOP simulation.

PROJECT : S-8
PHASE : III
DATE :

Reset Button

OPTIMIZATION OBJECTIVE: To Maximize Response

SIMULATION

I. NUMBER OF FACTORS[1-10]: Set K K = 2 Do QSEVOP

II. THE CURRENT OPERATING CONDITIONS: Ao = 4.00
Bo = 4.00

III. FACTORS RANGE:	CODE	ACTUAL	CODE	ACTUAL
	A[0.0] =	0.00	A[10.0] =	100.00
	B[0.0] =	0.00	B[10.0] =	100.00
	C[0.0] =	0.00	C[10.0] =	100.00

IV. Δ_i-CODED FACTOR'S STEPS: Δa = 1.00
Δb = 1.00

V. PROCESS NOISE: STDEV = 0.2

VI. THE FIXED NUMBER OF REPLICATES 7 Gen Random

QSEVOP WORK SHEET

CYCLES	FACTORS		RESPONSE														CRITERIA		
	A	B	n=1	n=2	n=3	n=4	n=5	n=6	n=7	R=1	R=2	R=3	R=4	R=5	R=6	R=7	AVERAGE	NUMBER OF SIGNES>5	t-TEST SIGNIFICANT
E0	4.00	4.00	95.94	95.74	96.05	96.26	96.24	96.35	95.56								96.02		
E1	5.00	4.00	97.86	97.66	97.63	97.80	97.85	97.58	97.89	1.92	1.92	1.58	1.55	1.61	1.23	2.32	97.75	Yes	Yes
E2	3.00	4.00	89.93	89.93	90.27	89.98	89.96	89.90	90.39	-7.93	-7.74	-7.36	-7.82	-7.88	-7.68	-7.49	90.05	No	Yes
E3	5.00	5.00	100.33	99.68	100.11	100.18	100.38	99.98	99.90	10.40	9.75	9.84	10.20	10.42	10.09	9.50	100.08	Yes	Yes
E4	5.00	3.00	91.71	91.83	91.70	91.93	91.99	92.01	91.94	-8.62	-7.85	-8.41	-8.25	-8.39	-7.98	-7.96	91.87	No	Yes

t-Test: Two-Sample Assuming Unequal Variances

	Variable 1	*Variable 2*
Mean	100.0800229	91.87134428
Variance	0.062051845	0.016397893
Observations	7	7
Hypothesized Mean Dif	0	
df	9	
t Stat	77.54013029	
P(T<=t) one-tail	2.49715E-14	
t Critical one-tail	1.833112933	
P(T<=t) two-tail	4.9943E-14	
t Critical two-tail	2.262157163	

Table 5.105 QSEVOP simulation.

PROJECT : S-8
PHASE : IV
DATE :

Reset Button

OPTIMIZATION OBJECTIVE: To Maximize Response

SIMULATION

I. NUMBER OF FACTORS[1-10]: Set K — K = 2 — Go QSEVOP

II. THE CURRENT OPERATING CONDITIONS: Ao = 5.00; Bo = 5.00

III. FACTORS RANGE:

CODE	ACTUAL	CODE	ACTUAL
A[0.0] =	0.00	A[10.0] =	100.00
B[0.0] =	0.00	B[10.0] =	100.00
C[0.0] =	0.00	C[10.0] =	100.00

IV. Δ_i-CODED FACTOR'S STEPS: Δa = 1.00; Δb = 1.00

V. PROCESS NOISE: STDEV = 0.2

VI. THE FIXED NUMBER OF REPLICATES 7 — Gen Random

QSEVOP WORK SHEET

CYCLES	FACTORS		RESPONSE														CRITERIA		
	A	B	n=1	n=2	n=3	n=4	n=5	n=6	n=7	R=1	R=2	R=3	R=4	R=5	R=6	R=7	AVERAGE	NUMBER OF SIGNES>5	t-TEST SIGNIFICANT
E0	5.00	5.00	99.94	99.74	100.05	100.26	100.24	100.35	99.56								100.02		
E1	6.00	5.00	97.86	97.66	97.63	97.80	97.85	97.58	97.89	-2.08	-2.08	-2.42	-2.45	-2.39	-2.77	-1.68	97.75	No	Yes
E2	4.00	5.00	97.93	97.93	98.27	97.98	97.96	97.90	98.39	0.07	0.26	0.64	0.18	0.12	0.32	0.51	98.05	Yes	Yes
E3	5.00	6.00	98.33	97.68	98.11	98.18	98.38	97.98	97.90	0.40	-0.25	-0.16	0.20	0.42	0.09	-0.50	98.08	No	No
E4	5.00	4.00	97.71	97.83	97.70	97.93	97.99	98.01	97.94	-0.62	0.15	-0.41	-0.25	-0.39	0.02	0.04	97.87	No	No

t-Test: Two-Sample Assuming Unequal Variances

	Variable 1	*Variable 2*
Mean	98.05237065	97.87134423
Variance	0.038410982	0.016397893
Observations	7	7
Hypothesized Mean Dif	0	
df	10	
t Stat	2.045810665	
P(T<=t) one-tail	0.033991645	
t Critical one-tail	1.812461123	
P(T<=t) two-tail	0.06798329	
t Critical two-tail	2.228138852	

Table 5.106 QSEVOP simulation.

PROJECT : S-8
PHASE : V
DATE :

Reset Button

OPTIMIZATION OBJECTIVE: To Maximize Response

SIMULATION

I. NUMBER OF FACTORS[1-10]: Set K K = 2 Go QSEVOP

II. THE CURRENT OPERATING CONDITIONS: Ao = 7.00; Bo = 7.00

III. FACTORS RANGE:

CODE	ACTUAL	CODE	ACTUAL
A[0.0] =	0.00	A[10.0] =	100.00
B[0.0] =	0.00	B[10.0] =	100.00
C[0.0] =	0.00	C[10.0] =	100.00

IV. Δ_i-CODED FACTOR'S STEPS: Δa = 1.00; Δb = 1.00

V. PROCESS NOISE: STDEV = 0.2

VI. THE FIXED NUMBER OF REPLICATES 7 Gen Random

QSEVOP WORK SHEET

CYCLES	FACTORS		RESPONSE														CRITERIA		
	A	B	n=1	n=2	n=3	n=4	n=5	n=6	n=7	R=1	R=2	R=3	R=4	R=5	R=6	R=7	AVERAGE	NUMBER OF SIGNES>5	t-TEST SIGNIFICANT
E0	7.00	7.00	83.94	83.74	84.05	84.26	84.24	84.35	83.56								84.02		
E1	8.00	7.00	73.86	73.66	73.63	73.80	73.85	73.58	73.89	-10.08	-10.08	-10.42	-10.45	-10.39	-10.77	-9.68	73.75	No	Yes
E2	6.00	7.00	89.93	89.93	90.27	89.98	89.96	89.90	90.39	16.07	16.26	16.64	16.18	16.12	16.32	16.51	90.05	Yes	Yes
E3	6.00	8.00	80.33	79.68	80.11	80.18	80.38	79.98	79.90	-9.60	-10.25	-10.16	-9.80	-9.58	-9.91	-10.50	80.08	No	Yes
E4	6.00	6.00	95.71	95.83	95.70	95.93	95.99	96.01	95.94	15.38	16.15	15.59	15.75	15.61	16.02	16.04	95.87	Yes	Yes

t-Test: Two-Sample Assuming Unequal Variances

	Variable 1	*Variable 2*
Mean	80.0800229	95.87134428
Variance	0.062051845	0.016397893
Observations	7	7
Hypothesized Mean Dif	0	
df	9	
t Stat	-149.1666532	
P(T<=t) one-tail	6.95147E-17	
t Critical one-tail	1.833112933	
P(T<=t) two-tail	1.39029E-16	
t Critical two-tail	2.262157163	

The QSEVOP move to the optimum has been shown in summary below:

Phase/cycle	Center-coded value	The best cycle	Remark
Phase I	$A_0 = 3.0; B_0 = 3.0$	$E3 = 84.08$	Moving toward optimum
Phase II	$A_0 = 4.0; B_0 = 4.0$	$E3 = 96.08$	Moving toward optimum
Phase III	$A_0 = 5.0; B_0 = 5.0$	$E3 = 100.08$	Optimum
Phase IV	$A_0 = 5.0; B_0 = 5.0$	$E0 = 100.02$	Optimum confirmed
Phase V	$A_0 = 7.0; B_0 = 7.0$	$E0 = 84.02$	Moving away

5.12 Simplex Evolutionary Operation – SEVOP

5.12.1 The Basic Simplex Method

In 1962, an efficient sequential optimization method called the basic simplex method was presented by Spendley at al [37]. Mathematical theory of experiments belongs to a group of non-gradient optimization techniques in multi-dimensional factor space. As a difference to gradient methods, this method does not require a mathematical model of researched phenomenon or does not require derivative of a response.

Movement to optimum by simplex method is done step by step and by comparing obtained response values, whereby a single response value is determined in each step. These properties of simplex design are considered for its important advantage:

- Simplex designs are applied in extreme experiments (Min./Max. experiments) with a large process noise.
- Simplex designing is successfully applied for determining optimum of a research subject with more responses, multi-response optimization.
- Simplex optimization is used in extreme experiments where besides quantitative – continuous also qualitative – categorical factors are included.

Under simplex designing, we understand finding the optimum of the response function by moving *Simplex Figure* on the response surface. Simplex movement is done step by step, whereby in each new step-trial-run the simplex vertex with least favorable response value is rejected.

Simplex is a simple geometrical figure, which forms $(k+1)$ independent points-vertices in k-factor space. Number of vertices in the simplex figure is minimal and it is for k-factors $k+1$. A simplex is called *regular*, if distances between its vertices are identical. By corresponding mathematical transformations, however, any simplex may be made regular. In a plane (two-factorial experiment), regular simplex has the form of an equilateral triangle. In a three-dimensional or three-factorial space, a simplex is a tetrahedron, etc. From any simplex, if we add one point, we can switch to a new simplex by dropping one vertex from the initial one. This is exactly a property of simplex designing on which movement to the optimum around response surface is based.

Movement to the optimum is realized step by step in such a way that the vertex with the most undesirable response value is successively rejected and a new point or vertex, which is the physical mirror image of the rejected vertex, is constructed. In the next step, the experiment is done in the new vertex, and then again the vertex with the least favorable response value is rejected and the procedure repeated. Movement to the optimum is here realized after each step and not after a series of trials. Simplex movement to the optimum is geometrically a "zig-zag" line, while the center of those simplexes moves along a line close to a gradient. Geometrical interpretation is given in Figure 5.101.

Figure 5.101 Simplex movement to optimum.

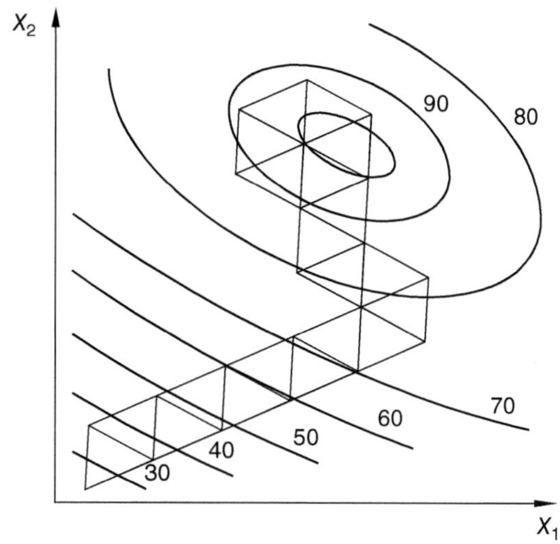

Strategy of simplex optimization consists of three steps:

- construction of an initial simplex.
- determination of a new vertex coordinates.
- timely stopping of a movement around response surface.

Figure 5.102 Simplex in coordinate system.

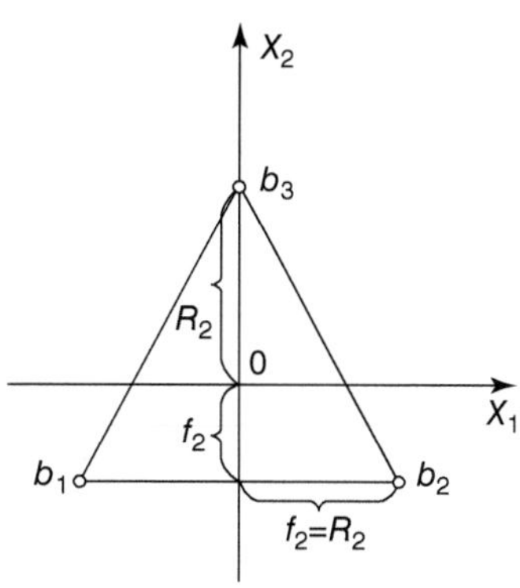

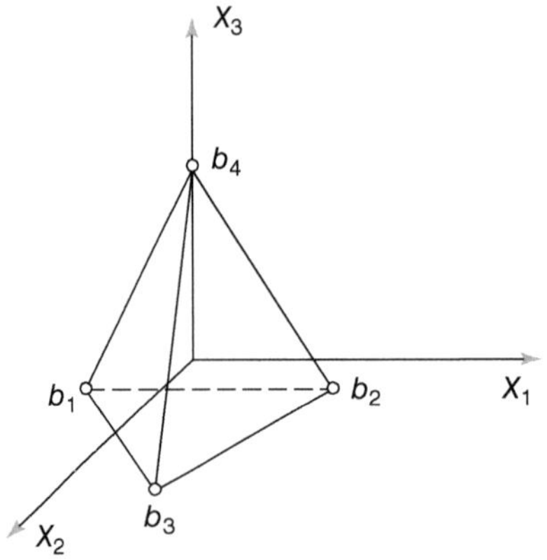

Figure 5.103 Tetrahedron in coordinate system.

Besides these three steps, one should also know how to switch from simplex to a second-order design, which may describe the optimum area [3].

The first step in simplex optimization consists of constructing matrix of a design of experiments for initial simplex where coordinates of experimental points-vertices are given. In solving this problem, different orientations of initial simplex to the coordinate system are possible. A simplex center is mostly set in coordinate beginning, while the distance between simplex vertices (simplex sides) has a coded value of one. Simplex is as a rule oriented in a factor space in such a way that vertex b_{k+1} lies on x_k axis, while other vertices are distributed symmetrically with respect to coordinate axes. Simplexes of such a construction are shown in Figure 5.102 and 5.103.

The figures show that coordinate vertices are equal to radii of inscribed (r_k) and described (R_k) sphere of a k-dimensional simplex. For a two-dimensional simplex, one should know r_2 and R_2 radii. For a chosen orientation of a simplex in factor space, it is possible to determine the vertex coordinates from this matrix:

$$\begin{Vmatrix} -r_1 & -r_2 & -r_3 & \dots & -r_{k-1} & -r_k \\ R_1 & -r_2 & -r_3 & \dots & -r_{k-1} & -r_k \\ 0 & R_2 & r_3 & \dots & -r_{k-1} & -r_k \\ . & . & . & \dots & . & . \\ 0 & 0 & 0 & \dots & R_{k-1} & -r_k \\ 0 & 0 & 0 & \dots & 0 & R_k \end{Vmatrix}$$

Radii of inscribed (r_k) and described (R_k) sphere for a k-dimensional simplex are equal to:

$$r_k = \frac{1}{\sqrt{2k(k+1)}}; \quad R_k = \sqrt{\frac{k}{2(k+1)}} \tag{5.1}$$

Null-dimensional simplex is brought down to a point, which is in the center of coordinate system. For $k=1$, a regular simplex is a straight line with two vertices-ends, which lies on x_1 axis and has these coordinates: +0.500 and–0.500. For $k=2$, coordinates of a simplex vertex may be either constructed graphically or calculated from formula (5.1). Three vertices of this simplex have these coordinates (Figure 5.104):

$$b_1(-0.500; -0.289);\ b_2(+0.500; -0.289);\ b_3(0; 0.578)$$

Figure 5.104 Coordinates of simplex vertices.

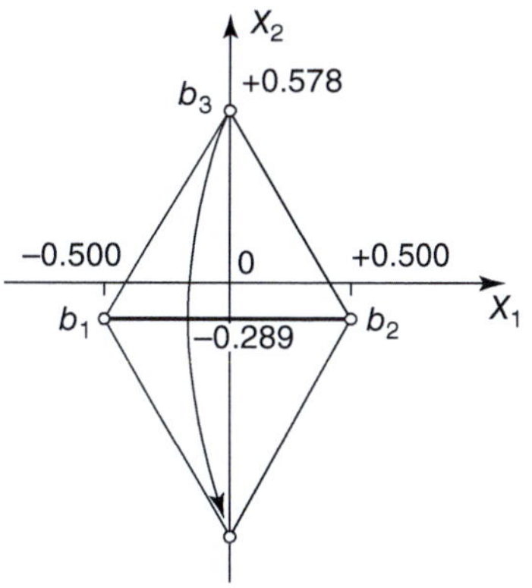

In accordance with formula (5.1), for $k = 7$ we got:

$$r_7 = \frac{1}{\sqrt{2 \cdot 7(7+1)}}; = 0.0945 \quad R_7 = \sqrt{\frac{7}{2(7+1)}} = 0.662$$

Table 5.107 is useful for the practical application.

Table 5.107 Coordinates of simplex vertices.

No	FACTOR VALUES									y_u
	X_1	X_2	X_3	X_4	X_5	X_6	X_7	...	X_k	
1	-0.500	-0.289	-0.204	-0.158	-0.129	-0.109	-0.0945	...		y_1
2	0.500	-0.289	-0.204	-0.158	-0.129	-0.109	-0.0945	...		y_2
3	0	0.578	-0.204	-0.158	-0.129	-0.109	-0.0945	...		y_3
4	0	0	0.612	-0.158	-0.129	-0.109	-0.0945	...	$\frac{1}{\sqrt{2k(k+1)}}$	y_4
5	0	0	0	0.632	-0.129	-0.109	-0.0945	...		y_5
6	0	0	0	0	0.645	-0.109	-0.0945	...		y_6
7	0	0	0	0	0	0.654	-0.0945	...		y_7
8	0	0	0	0	0	0	0.662	...		y_8
...	...	...	...	...	...	...	...	...	...	...
k+1	0	0	0	0	0	0	0	...	$\sqrt{\frac{k}{2(k+1)}}$	y_{k+1}

To solve a specific problem for a number of factors, we find the coordinates of the experimental points by using Table 5.107 and transforming the assumed values. To ensure the optimality of the used design of experiments, it is necessary to transform the values of the matrix.

It is well known that the design of experiments can be considered optimal if these three conditions are met:

- Matrix of design of experiments is orthogonal.
- Factors are symmetrical with respect to the center of the experiment.
- Sums of the squares of the elements from the matrix columns are equal.

The matrix of design of experiments in Table 5.107 does not meet only the third condition, and therefore all elements should be divided by the top value of element for the simplex. In case of $k = 4$, all elements should be divided by 0.632 to obtain the matrix in Table 5.108.

Table 5.108 Simplex matrix.

No. trials	Coded factor values			
	X_1	X_2	X_3	X_4
1	−0.790	−0.457	−0.323	−0.250
2	0.790	−0.457	−0.323	−0.250
3	0	0.914	−0.323	−0.250
4	0	0	0.969	−0.250
5	0	0	0	1.000

After choosing the simplex matrix of design of experiments with coded factors, we should switch to operational matrix with real factor values, taking into account factor variation intervals and coordinates of the center of the experiment. General formula for transfer from coded to real values (2.1) is also valid in this case:

$$X_k = \frac{x_k - x_{k_0}}{\Delta x_k}$$

Factor variation intervals in simplex designing are as a rule chosen so that they are five to ten times greater than the error square mean of their measurements (fixing error).

Example 5.8 Assume we have to do optimization of a phenomenon, which is defined by four factors ($k = 4$) [3]. We use Table 5.108 to define initial simplex. To determine operational matrix, we should know factor values in the center of the experiment and their variation intervals. These data are to be found in Table 5.109. The real factor values are obtained from relation (2.1):

$$\begin{aligned}
X_1 &= \frac{x_1 - 140}{20} \Rightarrow x_1 = 140 + 20X_1 \\
X_2 &= \frac{x_2 - 10}{5} \Rightarrow x_2 = 10 + 5X_2 \\
X_3 &= \frac{x_3 - 10}{4} \Rightarrow x_3 = 10 + 4X_3 \\
X_4 &= \frac{x_4 - 2}{0.5} \Rightarrow x_4 = 2 + 0.5X_4
\end{aligned} \tag{5.2}$$

Table 5.109 Factors and variation intervals.

Description	Factors	Dimensions	Center	Variation intervals
Temperature	X_1	(°C)	140	20
Quality of polymer	X_2	(%)	10	5
Time	X_3	(min)]	10	4
Quantity of catalyst	X_4	(g/l)	2	0,5

Table 5.110 Initial simplex.

No. trials	x_1	x_2	x_3	x_4
1	124.2	7.72	8.7	1.875
2	155.8	7.72	8.7	1.875
3	140.0	14.57	8.7	1.875
4	140.0	10.00	13.9	1.875
5	140.0	10.00	10.0	2.500

Real factors values in Table 5.110 have been calculated using Table 5.108, equations above, and Table 5.109.

By formation of initial simplex, we end up with the first problem of simplex optimization. One should now determine the direction of movement of initial simplex to optimum. As said before, at each step you move away from the least favorable conditions. By that simplex will move steadily toward more favorable results. This is not a problem if we analyze one response with quantitative values, since in searching for the response maximum, the least favorable is the trial where the response is the lowest, while in searching for response minimum, the most unfavorable trial-vertex is the one where the response has the highest value. When a response is qualitative, it is possible to determine the least favorable value by simple ranking response.

The simplex movement to optimum is not linked only with choice of the least favorable design point but also with limitations to factor changes. Simplex movement to optimum is realized by definite rule, which depends on the several situations:

- Simplex circling.
- Simplex swaying.
- Simplex movement in conditions with high process noise.
- Simplex movement at given limitations to responses and factors.

Simplex movement in factor space is realized by rejecting the vertex with the least favorable response value and by determining coordinates of new vertex, which is the mirror image of the rejected vertex:

$$X_{ij}^{(k+2)} = \frac{2}{k}\sum_{i=1}^{k} X_{ij} - X_{ij}^{*} \tag{5.3}$$

where

$X_{ij}^{(k+2)}$ are coordinates of new point-vertex.

X_{ij}^{*} are coordinates of vertex with the least favorable response value.

$\frac{2}{k}\sum_{i=1}^{k} X_{ij}$ is the coordinate average of all vertices except the rejected one.

k is number of factors; i current coordinate of factor.

j is current number of trials; $(k + 2)$ is the next number of trial.

Assume that by running trials from Table 5.110, we obtained the least favorable response value in the first vertex of trial No. 1. Then coordinates of the new sixth vertex are calculated in this way:

$$x_1^6 = \frac{2(140 \cdot 3 + 155.8)}{4} - 124.2 = 163.7;$$

$$x_2^6 = \frac{2(7.72 + 14.57 + 2 \cdot 10)}{4} - 7.72 = 13.43;$$

$$x_3^6 = \frac{2(8.7 \cdot 2 + 13.9 + 10.0)}{4} - 8.7 = 11.95;$$

$$x_4^6 = \frac{2(1.875 \cdot 3 + 2.5)}{4} - 1.875 = 2.187$$

Having done the trial according to obtained conditions, we compare once again the response values in vertices 2; 3; 4; 5; and 6. Trial with the least favorable response value is rejected and coordinates of the new seventh vertex are determined by the demonstrated rule. The procedure is continued until we reach the optimum. Successive rejection and addition of new vertices causes simplex center to move toward the optimum along a broken line. The shape of the line depends on sequence by which vertices are being rejected. A simplex center moves at maximal speed if the sequence of rejecting the vertices is periodically repeated. Geometric interpretation is given in Figure 5.105A. In case this sequence is not regular, simplex movement is slowed down, Figure 5.105B.

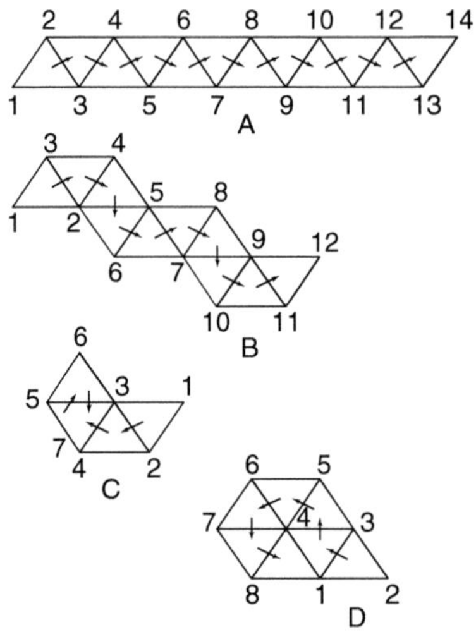

Figure 5.105 Simplex movement on response surface.

If the most inconvenient response value appears in the new vertex, which is then rejected, the simplex is returned to initial point, and we call this *Simplex Swaying*, Figure 5.105C. *We may avoid this simplex swaying by returning to initial vertex and rejecting the vertex, which is the second in a sequence of the least favorable response values*. If even then we do not reach a satisfactory result, we again return to initial vertex and reject the third one. In practice, we may be faced with false simplex movement to optimum as a consequence of the large process noise. In most cases, however, we should not pay great attention to process noise since it may prolong the procedure for only a few steps, but it cannot imperil a successful movement of simplex toward optimum.

Simplex optimization does not require a high response accuracy since the movement to optimum is based on comparing response values and not on their absolute values. Process noise does not importantly influence simplex orientation, since direction of simplex movement to optimum is in principle not repeated (for $k > 2$).

The effect of process noise (experimental error) on the efficiency of simplex optimization is reduced if the number of the factor increases. Caused by error of experiment, a simplex movement to optimum may turn into a circle, which is called *Simplex Circling*. In such cases we suggest *a replication of experiment*, and then the movement to optimum is either continued or terminated. The cause of simplex circling may be both the process noise and achieved optimum, Figure 5.105D. With a large number of factors, to get a more reliable estimate whether optimum has been reached, it is necessary to continue movement to optimum until the number of simplexes with the same vertex does not surpass this value:

$$N_{\max} = 1,65 \cdot k + 0,05 \cdot k^2 \quad (5.4)$$

Assuming that the optimum is in the vicinity of the circling simplex, it is possible to determine its position more precisely by reducing simplex dimensions and replicating the experiment. In case the simplex continues circling, *the procedure is repeated until factor variation intervals are closed to the value of the measurement error* (fixing error) [3].

Example 5.9 This is a typical cost optimization project. The objective was to minimize cost in the process of making a product, under the condition that the final product does not contain less than 40 [%] of active substance [3]. The process factors are:

x_1 is dosage of component A [g] ($x_{10} = 300$; $\Delta x_1 = 50$)
x_2 is ratio of components B/C [1] ($x_{20} = 0.45$; $\Delta x_2 = 0.1$)
x_3 is concentration of component B [%] ($x_{30} = 25$; $\Delta x_3 = 1.0$).

To form the initial simplex ($k = 3$), data from Table 5.107 have been used. All the data are divided by maximal value 0.612 to get a simplex design matrix. General formula for transfer from coded to real values (2.1) was used as before (Table 5.111):

$$x_1 = x_{10} + \Delta x_1 . X_1 = 300 + 50X_1 = 300 + 50 \cdot 0.8 = 340$$

$$x_2 = x_{20} + \Delta x_2 . X_2 = 0.45 + 0.1X_2 = 0.45 + 0.1 \cdot 0.5 = 0.50$$

$$x_3 = x_{30} + \Delta x_3 . X_3 = 25 + 1.0X_3 = 25 + 1.0 \cdot 0.3 = 25.3.$$

Table 5.111 Initial simplex.

No. of trials	Design matrix X_1	X_2	X_3	Operational matrix x_1	x_2	x_3	Response y
1	0.8	0.5	0.3	340	0.50	25.3	80
2	−0.8	0.5	0.3	260	0.50	25.3	95
3	0	−0.8	0.3	300	0.36	25.3	104
4	0	0	−1.0	300	0.45	24.0	105

The response values show that maximum cost is in trial No. 4, so that this vertex is rejected and the fifth one with these real coordinates is introduced (Table 5.112):

$$x_1^5 = \frac{2(340 + 260 + 300)}{3} - 300 = 300;$$

$$x_2^5 = \frac{2(0.5 + 0.5 + 0.36)}{3} - 0.45 = 0.46;$$

$$x_3^5 = \frac{2(25.3 + 25.3 + 25.3)}{3} - 24.0 = 27.0$$

Table 5.112 Simplex optimization.

Trials	Factors x_1	x_2	x_3	Response y_u
1	340	0.50	25	80
2	260	0.50	25	95
3	300	0.36	25	104
4	300	0.45	24	105
5	**300**	**0.46**	**27**	70
6	300	0.60	26	65
7	366	0.54	27	65
8	305	0.56	28	40
9	348	0.69	27	40
10	270	0.70	27	35
11	366	0.54	27	65
12	380	0.60	27	38
13	324	0.70	28	35
14	397	0.76	27	33
15	386	0.69	28	30
16	358	0.84	28	23
17	436	0.82	28	26
18	389	0.80	29	20
19	403	0.70	28	25

The outcomes undoubtedly confirmed that minimum cost was obtained in trial No. 18.

5.12.2 Simplex Evolutionary Operation – SEVOP

Simplex EVOP was proposed as an alternative to the original Box EVOP. The Simplex requires much less experimentation and reaches the optimum of a process much more quickly. Instead of the factorial experimentation, Simplex EVOP uses a succession of experimental designs in the form of a regular Simplex.

The regular Simplex is the first-order design, which requires the smallest number of experimental points. As mentioned before, for k factors (k-dimensional), $(k+1)$ experimental points are required. Thus for two factors the regular Simplex design is an equilateral triangle requiring three points; for three factors the design is a regular tetrahedron requiring four points. As in other forms of EVOP, more than three factors can be handled but designs cannot be shown graphically.

Rejecting one old point and introducing one new point in this most favorable direction of movement leads to the formation of a new simplex. Regularity is maintained at each stage and each new simplex after the first includes only one new point. The procedure can be repeated indefinitely.

1. The optimization begins with the initial trials. These initial trials form the first simplex. By a suitable linear transformation of co-ordinates, any simplex can be made "regular" (where "regular" has its usual geometric meaning), so that regularity in a specified co-ordinate system is not essential to the application of the technique.
2. In any simplex, determine the point which gives the least acceptable response, then move into the next simplex by rejecting this point and replacing it by its mirror image. The factor value (or coordinate) for the new point is given by:
   "*twice the average of the coordinates of the common points minus the co-ordinate of the rejected point*"

   $$2/n\,(A_1 + A_2 + A_3 + \ldots + A_{j-1} + A_{j+1} + \ldots + A_n + A_{n+1}) - Aj$$

   A succession of Simplex moves.
3. When this technique is used in error-prone observations, it is possible for the simplex system to revolve around a falsely high result, accepting this result as being the true optimum. To reduce this risk, the following rule applies:
   "*If in a simplex on k-factors an observations has occurred in (k + 1) successive simplexes and then is not eliminated under rule two above, do not move but discard this observation and repeat experimentation at that point*"
   If the original observation represents a true optimum, the repeated observation also will tend to be high. However, if the original observation is high because of an experimental error, it is unlikely that the repeated observation will give so high a result and the point is likely to be eliminated.
4. Never return to the factor levels that have just been rejected. In the event that such results are obtained, the following rule applies:

"Where the new point in a simplex has the lowest response, do not return to its mirror image, but move out of the simplex from the next lowest points."

According to the process observation, the second move made should send the simplex back to its original position, i.e. triangle 2 should move to triangle 1. The rules formulated for the simplex do not allow such swings, so instead of dropping the worst observation for the simplex 2, the next worst observation is dropped. This ensures that a completely new simplex is formed. This procedure will permit the system to circle continuously around an indicated optimum rather than oscillate between very narrow limits.

5. The calculated coordinates for trials outside the range of the control factors are not made. Instead a very unfavorable response is applied, forcing the simplex to move away from the factors limit.
6. Simplex can adjust its shape and size depending on the response in each step. This type of simplex is called the variable-size simplex method. Several new rules are added to the basic simplex rules. These new rules make the simplex:
   - Expand in a direction of more favorable conditions
     Or
   - Contract if a move was taken in a direction of less favorable conditions.

The procedure for expansion and contraction enables the modified simplex to accelerate along a successful track of improvement to the optimum conditions. Therefore, the modified simplex will usually reach the optimum region quicker than with the basic method and pinpoint the optimum levels more closely. The degree of contraction depends on how unfavorable the new response is.

The simplex method is especially appropriate when:

1. Process performance is changing over time.
2. More than three control factors are to be used.
3. The process requires a fresh optimization with each new lot of material.

Summary of SEVOP Operations process steps:

- Define process performance characteristics that need to be improved.
- Identify process factors whose small changes will lead the process improvement and record their current condition.
- Plan the incremental-interval changes steps (must be small steps; hence, it will not impact the production) for each identified process factor.
- For each factor, mark initial set of values as vertex of simplex (for two factors it's a triangle and use tetrahedron for three factors).
- Perform one run at current condition and also perform two runs with small incremental changes of one or both of the process factors.
- Record the results and identify the least favorable result from the runs, in other words identify the least favorable vertex.

- Then, perform a new run from the least favorable vertex (which is the reflection or mirror image of least favorable run).
- Now identify the new least favorable condition, this run will replace the existing one and also lead to another new run and this process will go on.
- The process starts at initial operating condition and moves on till more favorable results achieved.

The advantages of Simplex EVOP are:

1. In many processes, the optimum tends to move during time. Responses may indicate a moving optimum, even if the true optimum has not changed. It is unrealistic, and may be useless, to make process changes on the basis of out-of-date and irrelevant information; only the most recent observations should be used.
2. Simplex provides a rigorous definition of the frequency and extent of the changes to be made. Each move is from one Simplex to the adjacent Simplex. The least acceptable set of operating conditions is replaced by its mirror image in the plane (hyper-plane) of the remaining points.
3. When the real effects are small compared to the observational errors, they can be concealed and a false move can be withdrawn. As long as the changes are small compared to the changes in the basic design, there will be no major damage. In any case, since any decision made is constantly reviewed and corrected, the greater the negative effect, the faster it will be detected and eliminated.
4. Applying such a precise pattern of experimentation eliminates need for statistical analysis of data. The arithmetic involved is trivial and at no stage it is necessary to calculate a direction of steepest ascent. Although this procedure is ideal for control by means of a digital computer, plant supervisors optimizing a process with Simplex EVOP are under no disadvantage without a digital computer, an appropriate worksheet can be made.
5. The direction of progress depends only on the ranking of the responses and not on their scalar values. Therefore, Simplex EVOP is used when the response can be sorted in the order of preference for the response combination and the least preferred combination is discarded each time.
6. In a simplex experimentation new factors may be introduced at any stage without difficulty; if a factor has to be discarded that experiment needs to be restarted.

Summary of the advantages of Simplex EVOP:

1. The simplex method is especially appropriate when process performance is changing over time – new lot of raw material.
2. Any number of factors can be handled.
3. The qualitative responses are acceptable.
4. With high process noise, false move will be corrected quickly.
5. Calculations are extremely simple. Simplicity is the key.
6. SEVOP worksheet can be used, no disadvantage without a digital computer.

The disadvantages of Simplex EVOP are:

1. All factors must be quantitative.
2. If there is a high process noise, the simplex may behave erratically. Replicated runs could be made at each vertex to counteract the effect of noise, but this will increase a number of runs.
3. The simplex design does not provide information about the interaction effects.
4. The simplex is a resolution III design, main effects and two-factor interactions are confounded.
5. Simplex design will not provide information useful in building up process knowledge.

While some authors have advocated the sequential use of the simplex as a replacement for more conventional factorial-based EVOP (some suggested using the simplex as a replacement for factors screening and steepest ascent), the author of this book is not generally for this.

Example 5.10 "ABC Chocolate" production unit reported nearly 21.4% rejection, which leads to more rework, affecting the delivery schedules, and also customer satisfaction level [38].

Team realized that process will improve, if the two important parameters air pressure of machine (in psi) and also belt speed (rpm) settings are adjusted.

- Identify process performance characteristics. For example, reduction of scrap and the current rejection rate is 21.4%.
- Identify process factors whose small changes will lead the process improvement. Air pressure of machine (psi) and belt speed (rpm).
- The initial air pressure is 120 (psi) and also belt speed is 40 (rpm).
- Plan the incremental changes steps for each identified process factor. An increase in 10 (psi) as well as 5 (rpm) might lead to improve the process and reduces the rejection.
- For instance, the initial run is 120 (psi) and 40 (rpm) and 2nd and 3rd runs are 130 (psi), 40 (rpm) and 125 (psi), 45 (rpm), respectively, and reported the rejection rate.

The corresponding results are (Table 5.113):

Table 5.113 Initial simplex.

Run#	Air pressure (psi)	Belt speed (rpm)	Rejection (%)
1	120	40	21.4
2	130	40	20.2
3	125	45	20.4

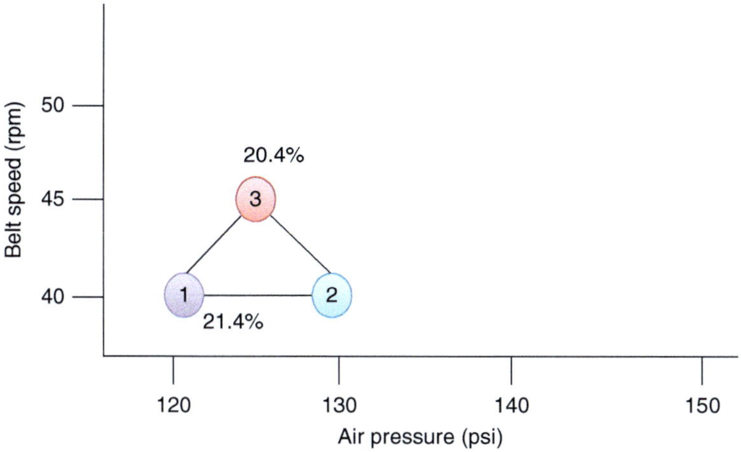

Figure 5.106 Initial simplex.

From the above picture and table it is clear that run 1 is the least favorable, in other words highest rejection rate, so initiate a new run reflection or mirror image of the run 1, new run 4 (Table 5.114).

Table 5.114 Simplex optimization.

Run#	Air pressure (psi)	Belt speed (rpm)	Rejection (%)
2	130	40	20.2
3	125	45	20.4
4	135	45	16.6

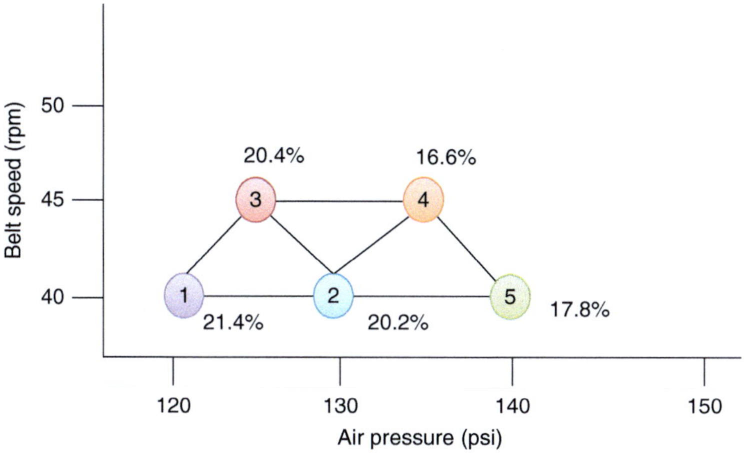

Figure 5.107 Simplex movement.

From the above picture and table it is clear that run 3 is least favorable, so run 5 mirror image of the run 3.

Procedure to calculate the new run value

New run value = (good value of process variable 1 + good value of process variable 2 – value of least favorable process variable)

To calculate the run 5 value (from 2,3,4)

- Run 5 Air pressure value = (value 4 + value 2 – value 3) = (135 + 130 – 125) = 140.
- Similarly, Run 5 Belt speed value = (value 4 + value 2 – value 3) = (40 + 45 – 45) = 40.

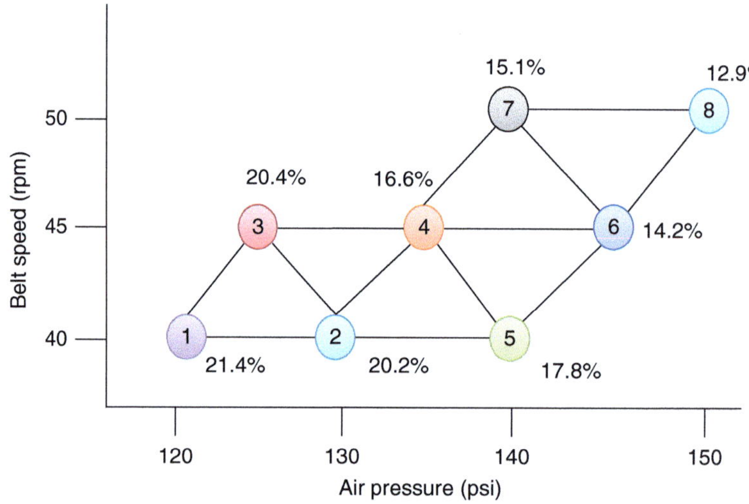

Figure 5.108 Simplex movement on response surface.

Similarly, follow the process and run the experiment till expected rejection rate achieved.

5.12.3 SEVOP Simulation

5.12.3.1 Simulation S-9

The mathematical formula from the simulation No. 1, Chapter 5.6, was used to demonstrate efficiency of the SEVOP method.

Phase I: The settings (Center Point, Factor's Interval, and Process Noise) and yield for different process conditions have been completed using Microsoft Excel simulation in Table 5.115. The initial simplex was created for two factors and codded factors were converted to the real values. The yield was calculated for all three vertices using simulation model. Those yield numbers were masked by a certain level of noise measured by standard deviation. By comparing responses in all three vertices, the vertex No. 1 with the lowest yield was rejected and the new vertex conditions were calculated as a mirror reflection of rejected conditions. Replicate runs were made at each vertex to counteract the effect of noise. This simulation started with the moderate factor's change (interval) and the large noise (STDEV).

In Phase XI, Table 5.116 and XII, Table 5.117, to avoid returning to factor levels that have just been rejected, following simplex procedure, the second least favorable condition was selected to move away from it (Table 5.118).

The maximum yield was achieved in Phases XI and XII.

Table 5.115 SEVOP simulation, Phase I.

PROJECT : S-9
PHASE : I
DATE :

SEVOP SIMULATION
S-9

I. CENTER OF EXPERIMENT : Ao = 20.00; Bo = 20.00 [Gen Random]

II. FACTOR'S INTERVAL OF VARIATION : Δa = 10.00; Δb = 10.00

III. PROCESS (PROPERTY-RESPONSE) NOISE : STDV = 10.00

EVOP WORK SHEET

TRIALS ORDER			FACTOR A	FACTOR B	RESPONSE n = 1	n = 2	n = 3	n = 4	n = 5	n = 6	n = 7	n = 8	n = 9	n = 10	Average
1		Block 1	15.00	17.11	59.67	67.01	49.81	64.18	71.61	74.55	55.97	40.88	45.27	58.22	58.72
2		Block 1	**25.00**	**17.11**	52.08	70.08	75.82	67.30	57.02	77.39	74.16	63.67	81.39	56.69	67.56
3		Block 1	**20.00**	**25.78**	76.14	67.59	73.53	70.82	70.32	76.51	71.44	49.66	54.08	71.05	68.11

INITIAL SIMPLEX

No	CODED VALUES A	CODED VALUES B	REAL VALUES a	REAL VALUES b
1	-0.500	-0.289	15.00	17.11
2	0.500	-0.289	25.00	17.11
3	0.000	0.578	20.00	25.78

Table 5.116 SEVOP simulation, Phase XI.

PROJECT : S-9
PHASE : XI
DATE :

SEVOP SIMULATION
S-9

I. CENTER OF EXPERIMENT : Ao = 20.00; Bo = 20.00 [Gen Random]

II. FACTOR'S INTERVAL OF VARIATION : Δa = 10.00; Δb = 10.00

III. PROCESS (PROPERTY-RESPONSE) NOISE : STDV = 10.00

EVOP WORK SHEET

TRIALS ORDER			FACTOR A	FACTOR B	RESPONSE n = 1	n = 2	n = 3	n = 4	n = 5	n = 6	n = 7	n = 8	n = 9	n = 10	Average
10		Block 1	**50.00**	**43.12**	104.34	92.39	111.77	97.83	103.29	111.19	104.39	85.16	86.95	84.82	**98.21**
12		Block 1	55.00	51.79	96.19	111.56	97.99	107.85	90.73	99.96	109.06	84.55	89.39	84.78	97.21
13		Block 1	60.00	43.12	105.02	92.97	75.08	102.39	108.48	82.98	97.59	108.64	101.78	95.03	96.99

Note:
The rule is never to return to factor levels that have just been rejected. The calculated reflection can also produce a unfavorable result. Without this rule the simplex would just oscillate between the two factor levels. This problem is nicely avoided by choosing the second least favorable condition and moving away from it.

Table 5.117 SEVOP simulation, Phase XII.

PROJECT : S-9
PHASE : XII
DATE :

SEVOP SIMULATION

S-9

I. CENTER OF EXPERIMENT : Ao = 20.00, Bo = 20.00

Gen Random

II. FACTOR'S INTERVAL OF VARIATION : Δ a = 10.00, Δ b = 10.00

III. PROCESS (PROPERTY-RESPONSE) NOISE : STDV = 10.00

EVOP WORK SHEET

TRIALS ORDER			FACTOR A	FACTOR B	RESPONSE n = 1	n = 2	n = 3	n = 4	n = 5	n = 6	n = 7	n = 8	n = 9	n = 10	Average
10		Block 1	**50.00**	**43.12**	104.34	92.39	111.77	97.83	103.29	111.19	104.39	85.16	86.95	84.82	**98.21**
13		Block 1	60.00	43.12	93.80	109.17	95.61	105.47	88.34	97.58	106.68	82.17	87.01	82.40	94.82
14		Block 1	55.00	34.45	102.63	90.58	72.69	100.00	106.09	80.59	95.20	106.25	99.39	92.64	94.61

Note:
The rule is never to return to factor levels that have just been rejected. The calculated reflection can also produce a unfavorable result. Without this rule the simplex would just oscillate between the two factor levels. This problem is nicely avoided by choosing the second least favorable condition and moving away from it.

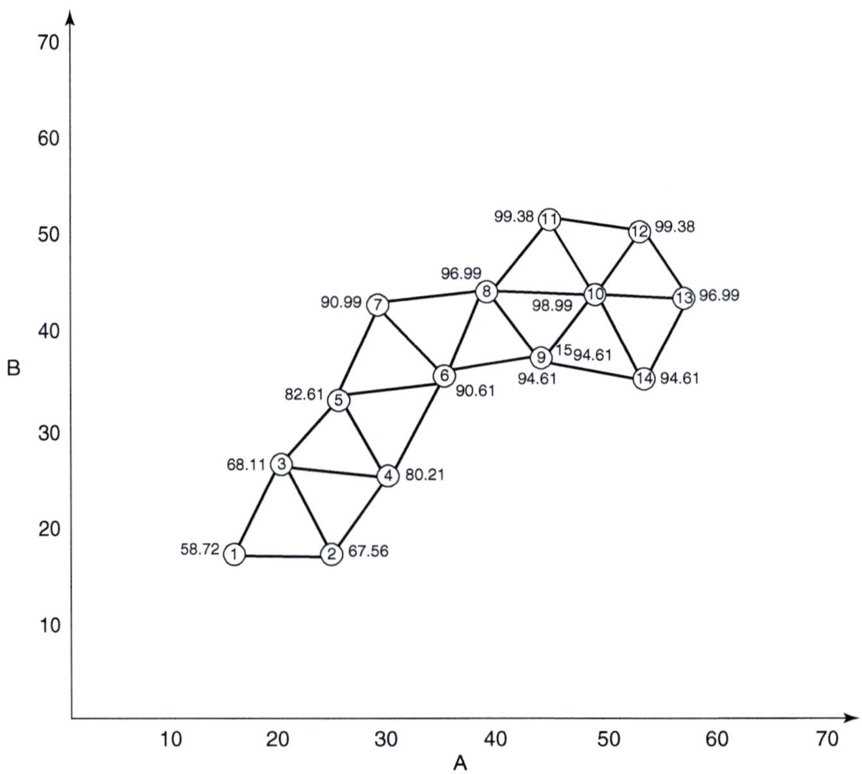

Figure 5.109 The simplex movement on the response surface.

Table 5.118 Summary of simplex movement.

.	FACTORS		RESPONSE
TRIALS	A	B	yu
1	15	17.11	58.72
2	25	17.11	67.56
3	20	25.78	68.11
4	25	22.89	80.21
5	30	31.56	82.61
6	35	34.45	90.61
7	30	43.12	90.99
8	40	43.12	96.99
9	45	34.45	94.61
10	50	43.12	98.99
11	45	51.79	99.38
12	55	51.79	99.38
13	60	43.12	96.99
14	55	34.45	94.61
15	45	34.45	94.61

5.12.3.2 Simulation S-10

The mathematical formula from simulation No. 4 was used to demonstrate efficiency of the SEVOP method.

Phase I: The settings (Center Point, Factor's Interval, and Process Noise) and yield for different process conditions have been completed using Microsoft Excel simulation in Table 5.119. The initial simplex was created for three factors and codded factors converted in real values. The yield was calculated for all four vertices using simulation model. Those yield numbers were masked by a certain level of noise measured by standard deviation. By comparing responses in all four vertices, the vertex No. 1 with the lowest yield was rejected and the new vertex conditions were calculated as a mirror reflection of rejected conditions. Replicate runs were made at each vertex to counteract the effect of noise. The corresponding calculations for phase IV have been shown in work sheets in Table 5.120. The maximum yield was achieved in vertex 7 (Table 5.121).

Table 5.119 SEVOP simulation, Phase I.

PROJECT : S-10
PHASE : I
DATE :

SEVOP SIMULATION

I. CENTER OF EXPERIMENT : Ao = 30.00, Bo = 30.00, Co = 30.00 [Gen Random]

II. FACTOR'S INTERVAL OF VARIATION : Δa = 10.00, Δb = 10.00, Δc = 10.00

III. PROCESS NOISE : STDV = 10.0

EVOP WORK SHEET

TRIALS ORDER		Block	FACTORS A	B	C	Response n = 1	n = 2	n = 3	n = 4	n = 5	n = 6	n = 7	n = 8	n = 9	n = 10	Average
1		Block I	21.83	25.28	26.67	60.50	58.61	57.95	61.58	59.21	63.63	53.17	41.61	77.78	65.64	59.97
2		Block I	38.17	25.28	26.67	51.26	82.22	62.59	101.48	76.05	69.14	101.05	84.96	67.92	66.46	76.31
3		Block I	**30.00**	**39.44**	26.67	83.37	90.20	77.61	80.38	67.08	46.08	78.50	84.53	72.80	78.71	**75.93**
4		Block I	30.00	30.00	40.00	91.73	91.76	82.67	99.98	92.15	98.18	97.46	99.92	87.32	69.78	91.10

INITIAL SIMPLEX

No	CODED VALUES A	B	C	REAL VALUES a	b	c
1	-0.817	-0.472	-0.333	21.83	25.28	26.67
2	0.817	-0.472	-0.333	38.17	25.28	26.67
3	0.000	0.944	-0.333	30.00	39.44	26.67
4	0.000	0.000	1.000	30.00	30.00	40.00

Table 5.120 SEVOP simulation, Phase I.

PROJECT : S-10
PHASE : IV
DATE :

SEVOP SIMULATION

I. CENTER OF EXPERIMENT : Ao = 30.00, Bo = 30.00, Co = 30.00 [Gen Random]

II. FACTOR'S INTERVAL OF VARIATION : Δa = 10.00, Δb = 10.00, Δc = 10.00

III. PROCESS NOISE : STDV = 10.0

EVOP WORK SHEET

TRIALS ORDER		Block	FACTORS A	B	C	Response n = 1	n = 2	n = 3	n = 4	n = 5	n = 6	n = 7	n = 8	n = 9	n = 10	Average
5		Block I	43.62	37.87	35.56	91.54	89.66	89.00	92.63	90.25	94.67	84.22	72.66	108.83	96.69	**91.02**
7		Block I	40.60	35.07	51.36	70.90	101.86	82.24	121.13	95.70	88.79	120.70	104.60	87.57	86.10	**95.96**
6		Block I	44.53	22.66	41.48	87.49	94.32	81.72	84.49	71.19	50.20	82.62	88.65	76.91	82.82	80.04
4		Block I	30.00	30.00	40.00	91.73	91.76	82.67	99.98	92.15	98.18	97.46	99.92	87.32	69.78	91.10

INITIAL SIMPLEX

No	CODED VALUES A	B	C	REAL VALUES a	b	c
1	-0.817	-0.472	-0.333	21.83	25.28	26.67
2	0.817	-0.472	-0.333	38.17	25.28	26.67
3	0.000	0.944	-0.333	30.00	39.44	26.67
4	0.000	0.000	1.000	30.00	30.00	40.00

Table 5.121 Summary of SEVOP movement .

.	FACTORS			RESPONSE
TRIALS	A	B	C	yu
1	21.83	25.28	26.67	59.57
2	38.17	25.28	26.67	76.31
3	30.00	39.44	26.67	75.93
4	30.00	30.00	40.00	91.10
5	43.62	37.87	35.56	91.02
6	44.53	22.66	41.48	80.04
7	40.00	35.07	51.36	95.96
8	31.62	45.97	43.13	89.02
9	44.53	22.66	41.48	80.04
10	33.13	20.62	53.00	75.81
11	24.62	34.47	54.76	78.88

5.13 Some Practical Advice About Using EVOP

Experience from different users of EVOP has led to several observations and suggestions about its practical implementation that may prove helpful. In this section, there are few ideas to share.

Usually, make sure that a reasonable list of the potential factors is available for EVOP analysis and experimentation. It is logical to start with two (or perhaps three) factors that operating personnel think are the most important. If several cycles occur (*five to eight, or so*) and no significant effects emerge, then new factors should be introduced into the design (or new levels of the old factors tried) and a new EVOP phase started. It is particularly important to keep an open mind when identifying potential factors. It was often that some factors were not considered as important, only because they had never been changed.

Sometimes a process experiences relative large run-to-run variation, and this is used as "excuse" for not using EVOP (or any statistically designed experiments for that matter). This is certainly not a valid argument, because each cycle of the EVOP design is a replicate, and replication is a very effective noise-reduction technique. By building up information over several cycles and working with effect estimates, that are based on averages of the responses at each design point, quite often large and important effects can be discovered even in noisy processes. It is also possible to discover that some factors settings result in less variability than the others (the range or standard deviation of the observation at each design point can be used as a second response). In noisy processes, sometimes more cycles may be required for the important effects to emerge, but the presence of process variability is not a problem to use designed experiments, even more it is the reason that statistically based designs have to be used.

Some are objected on the grounds that using EVOP is in violation to some principles of SPC. Specifically, SPC encourages operating personnel to leave the process alone so long as it is "statistical control," and this is directly opposed to EVOP procedure, which introduces changes in some process factors almost continuously. We must remember that the objectives of these two procedures are different. EVOP is concerned with optimization, including following the process that drifts over time. SPC, on the other hand, is concerned with the detection and elimination of external upsets in the process (called assignable causes) that may increase process variability or shift the process off-target. It may not be desirable to monitor and tightly control the process around the wrong target. Moreover, the process knowledge generated through any designed experiment may lead to improved operating performance far faster and more efficiently than the use of SPC methods alone.

Finally, training and education in EVOP methods should be offered to process engineers, quality engineers, development personnel, researchers, and other technical professionals. Usually such training lasts three to five days. Shorter version of this course (one to two days) should be given to key management personnel, so that they could understand the basics of designed experiments as well as the EVOP strategy. Plant operating personnel can usually be trained to run and evaluate EVOP schemes in about half a day. It was found that training and education are crucial for the successful implementation of both EVOP and experimental design methods in general.

6

EVOP Software

The software "EVOP Engine" is under construction and will include five modules, i.e. five different EVOP methods: BEVOP, ROVOP, REVOP, SEVOP, and QSEVOP. This special software for full-scale process optimization will provide different EVOP options for different types of the process. Based on the characteristics of each process like:

- Batch/Continuous
- Large-scale production/Small-scale production
- Running the same product for longer time/Short time
- Process is not running on optimum/On optimum
- Factors adjustment easy/Not easy
- Reaching steady conditions quickly after upset/Slowly
- Factors are quantitative/Qualitative
- Response measurements quantitative/Qualitative
- Response measurements available in short period of time/Long time
- High process noise/Low noise.

Users will be able to select the right EVOP method for the certain process. The demo version for the BEVOP is available on the request.

Full Scale Plant Optimization in Chemical Engineering: A Practical Guide, First Edition. Živorad R. Lazić.

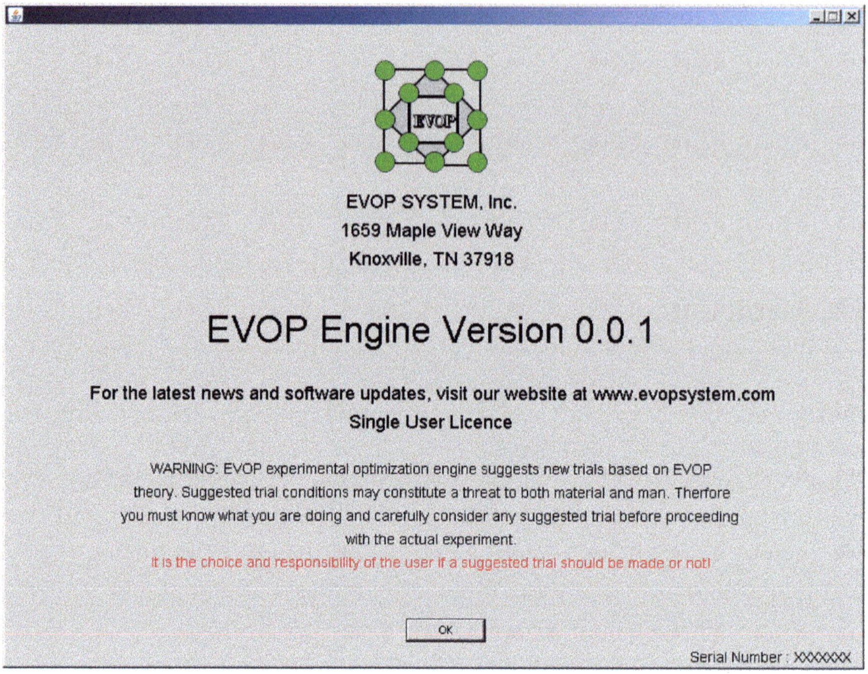
EVOP
EVOP SYSTEM, Inc.
1659 Maple View Way
Knoxville, TN 37918
EVOP Engine Version 0.0.1
For the latest news and software updates, visit our website at www.evopsystem.com
Single User Licence
WARNING: EVOP experimental optimization engine suggests new trials based on EVOP theory. Suggested trial conditions may constitute a threat to both material and man. Therfore you must know what you are doing and carefully consider any suggested trial before proceeding with the actual experiment.
It is the choice and responsibility of the user if a suggested trial should be made or not!
OK
Serial Number : XXXXXXX

Appendix A

The Approximate Method of Estimating the Standard Deviation in EVOP

Suppose there are k operating conditions and n cycles have been performed. The following data will then be available:

	Condition number				
Description	**1**	**2**	**3**	**...**	***k***
Average after $n-1$ cycles	$\bar{y}_{1,n-1}$	$\bar{y}_{2,n-1}$	$\bar{y}_{3,n-1}$		$\bar{y}_{k,n-1}$
Observations in nth cycle	$\bar{y}_{1,n}$	$\bar{y}_{2,n}$	$\bar{y}_{3,n}$		$\bar{y}_{\delta,n}$
Differences $d_{\text{in}} = y_{\text{in}} - \bar{y}_{i,n-1}$	d_{1n}	d_{2n}	d_{3n}		d_{kn}

Note: $\bar{y}$ average value.

Each difference as calculated on line (iv) of the EVOP form) is the difference between an individual observation (whose variances is σ^2) and an independently distributed average of $n-1$ observations (each with variance σ^2). Consequently, the variance of each of the differences is

$$\sigma^2(d_n) = \sigma^2 + \sigma^2/(n-1) = n\,\sigma^2/(n-1)$$

so that

$$\sigma = \sigma\,(d_n)\,[(n-1)/n]^{1/2}$$

For simplicity, we use an estimate of $\sigma(d_n)$ based on the range R_n of $d_{1n}, d_{2n, \ldots,} d_{kn}$. As we already know, an estimate of the standard deviation of a normal distribution can be obtained by multiplying the sample range by a factor w given in the next table.

N	**w[a]**
2	0.886
3	0.591
4	0.486
5	0.430
6	0.395
7	0.370

Full Scale Plant Optimization in Chemical Engineering: A Practical Guide, First Edition. Živorad R. Lazić.

N	w[a]
8	0.351
9	0.337
10	0.325
11	0.315
12	0.307

a) Note that, for $2 \leq n \leq 10$, w is approximately equal to $1/\sqrt{n}$. This is useful to remember, since it permits rough calculation to be made when a table of w is not available.

An estimate of σ (d_n) is provided therefore by $W_k R_n$, and on multiplying this quantity by $\sqrt{(n-1)/n}$ we obtain an estimate S_n of σ contributed by the nth cycle. We have

$$S_n = R_n[(n-1)/n]^{1/2}\, W_k$$

that is,

$$S_n = R_n f_{k,n}$$

where $f_{k,n} = [(n-1)/n]^{1/2}$ and w_k is the factor evaluated and recorded in the next table and also given on the EVOP forms; for example, when $k = 5$, $n = 2$, $w_k = 0.4299$, and $f_{k,n} = 0.4299/\sqrt{2}$=0.30 as in the table below.

After each cycle of the EVOP scheme except the first, an estimate of the σ can be calculated in this way. After n cycles, we have $n-1$ estimates, $S_2, S_3, \ldots, S_n$, and average $(S_1 + S_2 + S_3 + \ldots + S_n)/(n-1)$ provides the current estimate of σ.

No. of cycles	**Number of runs in the block (*k*)**								
n	**2**	**3**	**4**	**5**	**6**	**7**	**8**	**9**	**10**
2	0.63	0.42	0.34	0.30	0.28	0.26	0.25	0.24	0.23
3	0.72	0.48	0.40	0.35	0.32	0.30	0.29	0.27	0.26
4	0.77	0.51	0.42	0.37	0.34	0.32	0.30	0.29	0.28
5	0.79	0.53	0.43	0.38	0.35	0.33	0.31	0.30	0.29
6	0.81	0.54	0.44	0.39	0.36	0.34	0.32	0.31	0.30
7	0.82	0.55	0.45	0.40	0.37	0.34	0.33	0.31	0.30
8	0.83	0.55	0.45	0.40	0.37	0.35	0.33	0.31	0.30
9	0.84	0.56	0.46	0.40	0.37	0.35	0.33	0.32	0.31
10	0.84	0.56	0.46	0.41	0.37	0.35	0.33	0.32	0.31
15	0.86	0.57	0.47	0.42	0.38	0.36	0.34	0.33	0.31
20	0.86	0.58	0.47	0.42	0.38	0.36	0.34	0.33	0.32

Appendix B

2^2- and 2^3-Factor Box EVOP Calculations with Center Point

The completed work sheet after the first cycle is shown in Figure 5.2. Since no previous observations are available lines (i), (ii), and (iv) on this worksheet are left blank. The observations from each set of operating conditions (1), (2), (3), (4), and (5) are recorded in line (iii) labeled "New Observations." In lines (v) and (vi), labeled "New Sums" and "New Averages," the sum and average of all the observations are recorded at each of the operating conditions up to and including the cycle considered. At the end of the first cycle, these sums and averages are simply the observations themselves. The standard deviation σ cannot be estimated from the data of this phase since only one observation is available at each of the operating conditions. When, as in this Example 5.1, *a prior* estimate of the standard deviation is available, this can be used in calculating provisional error limits as indicated in the lower right-hand portion of the work sheet.

At cycle 2 we begin to obtain information from the data themselves concerning the value of σ relevant to this phase. The steps in filling in the worksheet for cycle 2, illustrated in Figure 5.3, and all subsequent cycles are as follows:

Step 1: The data from lines (v) and (vi) on the calculation worksheet of the previous cycle are copied in lines (i) and (ii) of the worksheet for the present cycle.

Step 2: The observations recorded during the present cycle are listed in line (iii).

Step 3: The difference between the quantities listed on lines (ii) and (iii) is written in line (iv). These are the differences with appropriate signs attached between the new observations and the averages over previous cycles taken at each set of operating conditions.

Step 4: The new sum and new average for each set of conditions are recorded on lines (v) and (vi).

Step 5: The effects are calculated as given on the worksheet using the new averages computed at each of the sets of operating conditions. Thus, the main effect for the factor "time" at the end of cycle 2 is given by: Time effect = 3.3.

After the first cycle, it is possible to obtain an estimate of the standard deviation of the individual observations (and hence of the error limits for the averages and effects) from the data themselves. These calculations are performed in the space provided on the right-hand side of the calculation worksheet.

Full Scale Plant Optimization in Chemical Engineering: A Practical Guide, First Edition. Živorad R. Lazić.

Step 6: The largest and smallest differences recorded in line (iv) are underlined. The range of the differences is the difference between these underlined values. This is entered at the right-hand end of line (iv).

Step 7: The range in line (iv) is multiplied by factor $f_{k,n}$ obtained from Table 4.4 to obtain an estimate s of σ contributed by this cycle. This result is entered after "Ne $f_{k,n}$ w s" in line (iii). To obtain factor $f_{k,n}$ enter Table 4.4 with the appropriate cycle number n and appropriate value of k (the number of sets of operating conditions in the program). At the end of cycle 2 we have, for $k = 5$, $f_{k,n} = 0.30$. The estimate of the standard deviation contributed by this cycle is then $s = \text{Range} \times f_{k,n} = 4.6 \times 0.30 = 1.38$.

All the entries in the calculation work sheet can be filled at the third cycle. Using the data given earlier in Table 5.1, the remaining items will be given for the work sheet shown in Figure 5.4, appropriate at the conclusion of cycle 3.

Step 8: The value "New s" obtained from Step 7 in line (iii) is added to the "Previous Sum s" in line (i) and the result recorded on line (v) as "New Sum s." This quantity is then divided by $(n-1)$ to give a "New Average s" in line (vi).

Step 9: The error limits for the k averages and the effects are obtained by direct substitution of "New Average s" for σ in the equations in the lower right portion of the calculation work sheet.

Step 10: The new averages, estimated effects, and their error limits are recorded on the Information Board. The Information Board at the end of cycle 4 is shown in Table 5.2.

B.1 2^3 Three-Factor Box EVOP Calculations with Center Point

Block II Summary of Calculation

1. Enter project number, phase, date, response, and cycle number.

Calculation of Averages

2. Copy lines (i) and (ii) from lines (v) and (vi) of block II worksheet for last cycle.
3. Enter new observations corresponding to conditions Φ, 5,6,7 and 8 in line (iii).
4. Calculate differences (ii) less (iii) and enter in (iv). Underline largest and smallest entries, taking full account of sign.
5. Add entries in lines (i) and (iii) and enter in line (v).
6. Multiply entries in line (v) by $1/n$ (see table of multiplying constants). Enter results in line (vi).

Calculation of Standard Deviations

7. Enter "Previous sum s" (all blocks) from block I worksheet in line (i).
8. Calculate range from difference of largest and smallest entries underlined on line (iv) (left of page) and enter in line (iv) on right of page.
9. Multiply entries in line (iv) by $f_{5,n}$ (see table of multiplying constants) to obtain "New s." Enter on line (iii).

10. Add entries in line (iii) and (i) to obtain "New sum *s*" all blocks. Multiply by constant $1/[2(n-1)]$ (see table of multiplying constants) to obtain "Average *s*." Enter in line (vi).

Calculation of Effects Using Yates Algorithm

11. After subtracting suitable constant (record at top of column) enter "New averages" in order shown. Sum squares.
12. Perform Yate's [3] additions and subtractions to obtain column (i). Repeat these operations with elements of column (i) to obtain column (ii) and with elements of column (ii) to obtain column (iii).
13. Check by computing sum of squares of column (iii). This should total exactly eight times the sum of squares calculated at Step 11.
14. Multiply entries in column (iii) by multipliers to obtain "Effects."
15. Enter observations for reference conditions 0 and Φ on right of worksheet (subtract suitable constant).
16. Compute "Reference mean," "Factorial mean," and "Phase mean" as shown (add back constant).
17. Obtain "Change in mean" from phase mean minus reference mean.

Calculation of 2 S.E Limits

18. Compute factors for "2 S.E. Limits" using multipliers in table.

Appendix C

Short Table of Random Normal Deviates

−0.26	−0.11	−0.42	−0.41	0.46	1.09	−0.58	−0.43	1.12	−0.06
0.58	−0.46	0.76	−1.89	0.03	0.05	0.18	−0.14	−0.74	−1.04
0.91	−0.74	0.13	−1.01	−0.49	−0.38	0.55	0.64	0.03	0.38
0.17	1.72	0.72	−1.19	1.83	1.22	−0.19	0.83	−0.23	−0.32
0.11	1.36	−0.22	0.17	1.15	−0.76	−0.16	−1.14	1.33	−1.18
−0.63	0.23	−0.43	1.36	2.40	1.66	0.16	0.20	−0.97	0.94
1.67	1.13	−1.44	0.27	0.99	0.56	−2.08	−0.07	−0.04	0.58
−0.90	0.47	−0.18	−0.51	1.70	−0.58	1.16	−0.13	0.42	−1.05
−1.29	0.62	−0.65	0.73	−0.23	−0.83	−0.96	1.07	2.37	1.44
−0.59	−2.43	−1.00	−1.10	−0.25	−0.30	0.86	−0.05	−0.70	1.28
−0.37	0.74	1.63	−0.52	−0.01	1.01	−0.61	2.28	0.89	−0.29
0.87	−0.60	−1.70	2.38	0.01	−1.66	0.03	0.53	0.98	−1.05
−0.01	−1.01	0.39	0.75	0.28	−0.40	2.31	−0.49	0.35	0.11
−1.39	−1.65	0.47	−2.26	0.33	1.52	0.55	2.32	0.62	0.24
−0.56	−0.60	−0.12	0.59	−0.29	−1.63	0.68	−0.17	−0.56	−0.67
1.26	−0.75	1.90	0.43	−1.75	−1.30	1.64	−0.80	−0.33	−1.98
−2.55	−0.03	−1.11	−0.38	0.23	0.41	−0.48	1.06	0.61	−0.20
1.55	0.45	0.24	0.58	1.56	0.25	−1.91	0.90	−0.58	0.11
−0.30	0.68	2.17	−0.90	0.88	1.39	−0.34	−0.05	0.94	−0.14
−0.56	−2.07	−1.25	0.18	1.68	1.32	3.39	1.21	−1.72	−0.29
−0.24	−0.47	−0.53	0.83	0.55	1.85	2.23	−1.16	0.12	−0.16
−0.16	−2.59	0.30	−1.22	−.079	−0.12	−0.25	−0.01	−0.25	0.09
−1.34	0.60	−0.27	−0.20	−0.17	1.83	−1.80	−0.04	−1.84	0.16
0.14	2.09	−0.30	0.26	−0.15	−0.95	−1.32	0.57	−0.13	0.32
−0.16	−2.51	−1.64	−0.88	0.92	1.54	−0.85	1.43	−1.57	−2.22
−0.90	−0.26	0.97	1.38	−0.02	1.41	−0.00	−0.79	0.14	−0.45
0.53	0.44	−2.50	−1.11	0.93	−0.18	−0.23	−0.37	0.38	1.39

Full Scale Plant Optimization in Chemical Engineering: A Practical Guide, First Edition. Živorad R. Lazić.

0.30	−1.43	0.32	0.95	1.35	1.24	1.42	−0.20	0.71	−0.32
−2.03	−0.52	0.07	−0.84	0.63	0.68	0.53	−0.65	0.54	0.01
0.44	−0.11	0.58	−0.25	1.92	−1.41	−0.56	0.19	−0.86	−0.31
−0.87	0.18	1.17	−1.68	0.97	−0.85	−0.22	0.75	1.20	0.01
−1.85	−0.33	0.98	−2.24	−0.76	1.42	1.16	−1.00	3.20	1.18
1.18	0.27	−2.44	2.29	0.01	−1.34	0.21	−0.48	−0.69	0.50
1.15	−1.22	−1.78	−0.01	−0.36	0.11	−1.03	2.08	0.85	0.58
−2.01	−0.07	−1.15	−0.97	−0.11	0.24	−0.32	−0.51	−0.51	−0.01
1.11	−0.71	−0.52	−1.27	−0.21	−1.60	0.11	0.52	−1.63	−0.78
−0.50	−0.07	0.03	1.24	−0.40	1.22	−0.91	0.33	−1.66	1.52
0.51	0.72	1.42	2.08	−0.48	−0.04	−0.27	−1.29	−0.24	−0.61
−0.17	0.36	0.32	−0.68	−0.68	−2.38	−0.56	−2.77	−1.71	−1.04
−0.73	−0.71	−0.51	0.70	1.12	0.17	−1.54	−0.94	0.31	−0.17

Appendix D

How Many Cycles Are Necessary to Detect Effects of Reasonable Size

The approximate number of cycles in EVOP should be enough to detect the difference Δ (difference between average response values) with STDEV – σ at the significance level α.

Example D1: The process STDEV is $\sigma = 822$, detect the difference $\Delta = 905$ with significance $\alpha = \beta = 0.1$.

Steps

- $1-\alpha = 0.9$; $\alpha = \beta = 0.1$.
- Calculate signal to noise ratio: $D = \Delta/\sigma = 905/822 = 1.1$.
- Define the number of observations from Table E.1 [3], using $D = 1.1$, and $\alpha = \beta = 0.1$: $n = 15$.
- Define the number of cycles in EVOP scheme for the averages of n observations at each factors level. It means for 2^2 EVOP, required number of cycles is: $n/2 = 15/2 = 7.5$ rounded 8.

Full Scale Plant Optimization in Chemical Engineering: A Practical Guide, First Edition. Živorad R. Lazič.

The same approach but using Microsoft Excel calculation:

The Approximate Method of Estimating the Number of Cycles in EVOP
2^2 FULL FACTORIAL DESIGN

SIGNAL TO NOISE RATIO	SIGNIFICANCE THRESHOLD FOR POWER	α	α	NUMBER OF CYCLES [#]
EFFECT SIZE-SIGNAL [Δ/σ]	0.1	0.05	0.01	2
0.5 STDEV	15.9[%]	0.0[%]	1.5[%]	
1.0 STDEV	32.3[%]	15.9[%]	5.1[%]	
2.0 STDEV	73.0[%]	61.7[%]	21.5[%]	
	SIGNIFICANCE THRESHOLD FOR POWER			NUMBER OF CYCLES [#]
EFFECT SIZE-SIGNAL [Δ/σ]	0.1	0.05	0.01	3
0.5 STDEV	25.5[%]	11.9[%]	3.1[%]	
1.0 STDEV	47.6[%]	32.9[%]	12[%]	
2.0 STDEV	93.4[%]	85.7[%]	56.9[%]	
	SIGNIFICANCE THRESHOLD FOR POWER			NUMBER OF CYCLES [#]
EFFECT SIZE-SIGNAL [Δ/σ]	0.1	0.05	0.01	4
0.5 STDEV	21.6[%]	16.2[%]	4.4[%]	
1.0 STDEV	59.6[%]	45.2[%]	20.0[%]	
2.0 STDEV	95.3[%]	95.6[%]	80.5[%]	
	SIGNIFICANCE THRESHOLD FOR POWER			NUMBER OF CYCLES [#]
EFFECT SIZE-SIGNAL [Δ/σ]	0.1	0.05	0.01	5
0.5 STDEV	28.8[%]	18.9[%]	5.5[%]	
1.0 STDEV	69.0[%]	55.9[%]	28.5[%]	
2.0 STDEV	99.6[%]	96.7[%]	82.2[%]	
	SIGNIFICANCE THRESHOLD FOR POWER			NUMBER OF CYCLES [#]
EFFECT SIZE-SIGNAL [Δ/σ]	0.1	0.05	0.01	6
0.5 STDEV	32.5[%]	21.5[%]	7.4[%]	
1.0 STDEV	75.4[%]	64.5[%]	37.2[%]	
2.0 STDEV	99.9[%]	95.6[%]	87.1[%]	
	SIGNIFICANCE THRESHOLD FOR POWER			NUMBER OF CYCLES [#]
EFFECT SIZE-SIGNAL [Δ/σ]	0.1	0.05	0.01	7
0.5 STDEV	35.1[%]	24.8[%]	9.0[%]	
1.0 STDEV	82.3[%]	71.8[%]	45.5[%]	
2.0 STDEV	99.9[%]	95.9[%]	99[%]	
	SIGNIFICANCE THRESHOLD FOR POWER			NUMBER OF CYCLES [#]
EFFECT SIZE-SIGNAL [Δ/σ]	0.1	0.05	0.01	8
0.5 STDEV	39.7[%]	27.7[%]	10.7[%]	
1.0 STDEV	90.7[%]	77.9[%]	50.4[%]	
2.0 STDEV	99.9[%]	95.9[%]	99.7[%]	
	SIGNIFICANCE THRESHOLD FOR POWER			NUMBER OF CYCLES [#]
EFFECT SIZE-SIGNAL [Δ/σ]	0.1	0.05	0.01	9
0.5 STDEV	43.1[%]	30.7[%]	12.5[%]	
1.0 STDEV	93.2[%]	82.9[%]	50.9[%]	
2.0 STDEV	99.9[%]	95.9[%]	99.9[%]	
	SIGNIFICANCE THRESHOLD FOR POWER			NUMBER OF CYCLES [#]
EFFECT SIZE-SIGNAL [Δ/σ]	0.1	0.05	0.01	10
0.5 STDEV	46.3[%]	32.7[%]	14.3[%]	
1.0 STDEV	92.7[%]	86.8[%]	57.0[%]	
2.0 STDEV	99.9[%]	96.9[%]	99.9[%]	

EVOP RECOMMENDS: α=β=0.1 & Δ/σ=1.0 : TO GET SIGNIFICANT EFFECT DESIRED PROBABILITY MUST BE MINIMUM 90[%]!
IT MEANS FOR 2^2 FULL FACTORIAL DESIGN MINIMUM NUMBER OF EVOP CYCLES IS: 8

The Estimated Number of CYCLES for 2^2 Full Factorial Design at α=β=0.1 alpha level for effect of

	0.5 Std. Dev.	1 Std. Dev.	2 Std. Dev.
n	25	8	3

Appendix E

Multiple Responses: The Desirability Approach

The desirability function approach is one of the most widely used methods in industry for the optimization of multiple response processes. It is based on the idea that the "quality" of a product or process that has multiple quality characteristics, with one of them outside of some "desired" limits, is completely unacceptable. The method finds operating conditions x that provide the "most desirable" response values.

For each response $Y_i(x)$, a desirability function $d_i(Y_i)$ assigns numbers between 0 and 1 to the possible values of Y_i, with $d_i(Y_i) = 0$ representing a completely undesirable value of Y_i and $d_i(Y_i) = 1$ representing a completely desirable or ideal response value. The individual desirabilities are then combined using the geometric mean, which gives the *overall desirability D*

$$D = (d_1(Y_1) \times d_2(Y_2) \times \cdots \times d_k(Y_k))^{1/k}$$

with k denoting the number of responses. Notice that if any response Y_i is completely undesirable ($d_i(Y_i) = 0$), then the overall desirability is zero. In practice, fitted response values $\widehat{Y}_i$ are used in place of the Y_i.

Depending on whether a particular response Y_i is to be maximized, minimized, or assigned a target value, different desirability functions $d_i(Y_i)$ can be used. A useful class of desirability functions was proposed by Derringer and Suich (1980). Let L_i, U_i , and T_i be the lower, upper, and target values, respectively, that are desired for response Y_i, with $L_i \leq T_i \leq U_i$.

If a response is of the "target is best" kind, then its individual desirability function is

$$d_i(\widehat{Y}_i) = \begin{bmatrix} 0 & \text{if} \quad \widehat{Y}_i(x)\langle L_i \\ \left(\dfrac{\widehat{Y}_i(x) - L_i}{T_i - L_i}\right)^s & \text{if} \quad L_i \leq \widehat{Y}_i(x) \leq T_i \\ \left(\dfrac{\widehat{Y}_i(x) - U_i}{T_i - U_i}\right)^t & \text{if} \quad T_i \leq \widehat{Y}_i(x) \leq U_i \\ 0 & \text{if} \quad \widehat{Y}_i(x)\rangle U_i \end{bmatrix}$$

with the exponents s and t determining how important it is to hit the target value. For $s = t = 1$, the desirability function increases linearly toward T_i; for $s < 1, t < 1$, the function is convex, and for $s > 1, t > 1$, the function is concave (see the example below for an illustration).

If a response is to be maximized instead, the individual desirability is defined as

$$d_i(\widehat{Y}_i) = \begin{bmatrix} 0 & \text{if} \quad \widehat{Y}_i(x) \langle L_i \\ \left(\dfrac{\widehat{Y}_i(x) - L_i}{T_i - L_i}\right)^s & \text{if} \quad L_i \leq \widehat{Y}_i(x) \leq T_i \\ 1 & \text{if} \quad \widehat{Y}_i(x) \rangle T_i \end{bmatrix}$$

with T_i in this case interpreted as a large enough value for the response.

Finally, if we want to minimize a response, we could use

$$d_i(\widehat{Y}_i) = \begin{bmatrix} 1.0 & \text{if} \quad \widehat{Y}_i(x) \langle T_i \\ \left(\dfrac{\widehat{Y}_i(x) - U_i}{T_i - U_i}\right)^s & \text{if} \quad T_i \leq \widehat{Y}_i(x) \leq U_i \\ 0 & \text{if} \quad \widehat{Y}_i(x) \rangle U_i \end{bmatrix}$$

with T_i denoting a small enough value for the response.

The desirability approach consists of the following steps:

1. Conduct experiments and fit response models for all k responses.
2. Define individual desirability functions for each response.
3. *Maximize* the overall desirability D with respect to the controllable factors.

Example E.1 Derringer and Suich (1980) present the following multiple response experiment arising in the development of a tire tread compound. The controllable factors are: x_1, hydrated silica level, x_2, silane-coupling agent level, and x_3, sulfur level. The four responses to be optimized and their desired ranges are:

Source	Desired range
PICO Abrasion index, Y_1	$120 < Y_1$
200% modulus, Y_2	$1000 < Y_2$
Elongation at break, Y_3	$400 < Y_3 < 600$
Hardness, Y_4	$60 < Y_4 < 75$

The first two responses are to be maximized, and the value $s = 1$ was chosen for their desirability functions. The last two responses are "target is best" with $T_3 = 500$ and $T_4 = 67.5$. The values $s = t = 1$ were chosen in both cases.

The following experiments were conducted using a central composite design.

Run number	x_1	x_2	x_3	Y_1	Y_2	Y_3	Y_4
1	−1.00	−1.00	−1.00	102	900	470	67.5
2	+1.00	−1.00	−1.00	120	860	410	65.0
3	−1.00	+1.00	−1.00	117	800	570	77.5
4	+1.00	+1.00	−1.00	198	2294	240	74.5
5	−1.00	−1.00	+1.00	103	490	640	62.5
6	+1.00	−1.00	+1.00	132	1289	270	67.0
7	−1.00	+1.00	+1.00	132	1270	410	78.0
8	+1.00	+1.00	+1.00	139	1090	380	70.0
9	−1.63	0.00	0.00	102	770	590	76.0
10	+1.63	0.00	0.00	154	1690	260	70.0
11	0.00	−1.63	0.00	96	700	520	63.0
12	0.00	+1.63	0.00	163	1540	380	75.0
13	0.00	0.00	−1.63	116	2184	520	65.0
14	0.00	0.00	+1.63	153	1784	290	71.0
15	0.00	0.00	0.00	133	1300	380	70.0
16	0.00	0.00	0.00	133	1300	380	68.5
17	0.00	0.00	0.00	140	1145	430	68.0
18	0.00	0.00	0.00	142	1090	430	68.0
19	0.00	0.00	0.00	145	1260	390	69.0
20	0.00	0.00	0.00	142	1344	390	70.0

Using ordinary least squares and standard diagnostics, the fitted responses are:

$$\hat{y}_1 = 139.12 + 16.49x_1 + 17.88x_2 + 2.21x_3 - 4.01x_1^2 - 3.45x_2^2 - 1.57x_3^2 + 5.12x_1x_2 - 7.88x_1x_3 - 7.13x_2x_3$$
$(R^2 = 0.8369$ and adjusted $R^2 = 0.6903)$;

$$\hat{y}_2 = 1261.13 + 268.15x_1 + 246.5x_2 - 102.63x_3 - 83.5x_1^2 - 124.82x_2^2 + 199.2x_3^2 + 69.3x_1x_2 - 104.38x_1x_3 - 94.13x_2x_3$$
$(R^2 = 0.7137$ and adjusted $R^2 = 0.4562)$;

$\hat{y}_3 = 417.5 - 99.67x_1 - 31.4x_2 - 27.42x_3$ $(R^2 = 0.682$ and adjusted $R^2 = 0.6224)$;

$$\hat{y}_4 = 68.91 - 1.41x_1 + 4.32x_2 + 0.21x_3 + 1.56x_1^2 + 0.058x_2^2 - 0.32x_3^2 - 1.62x_1x_2 + 0.25x_1x_3 - 0.12x_2x_3$$
$(R^2 = 0.8667$ and adjusted $R^2 = 0.7466)$.

Note that no interactions were significant for response 3 and that the fit for response 2 is quite poor.

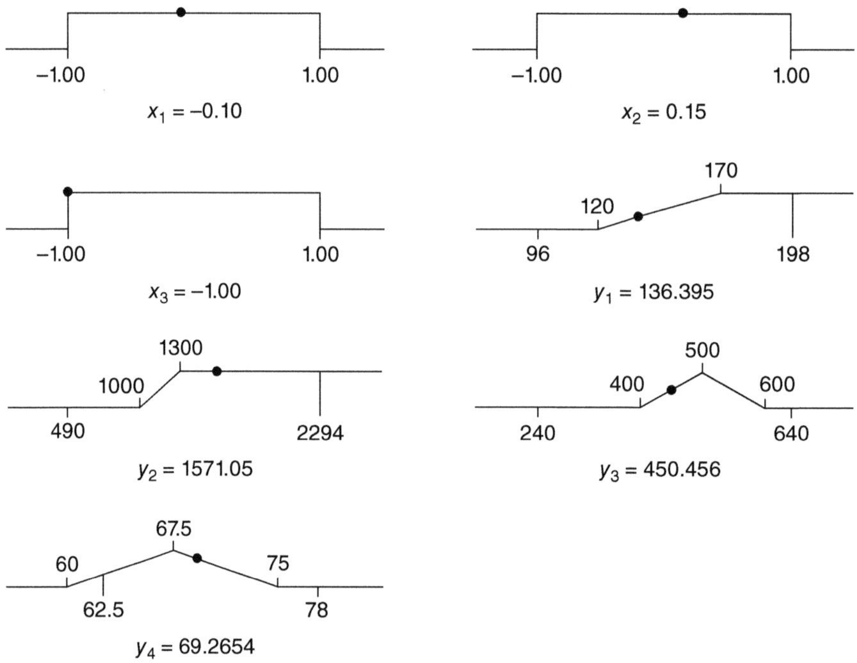

Figure E.1 Desirability functions and optimal solution for example problem.

Optimization of D with respect to x was carried out using the Design-Expert software. Figure E.1 shows the individual desirability functions $d_i(Y_i)$ for each of the four responses. The functions are linear since the values of s and t were set equal to one. A dot indicates the best solution found by the Design-Expert solver.

The best solution is $(x^*)' = (-0.10, 0.15, -1.0)$ and results in:

$$d_1(\widehat{Y}_1) = 0.34; \quad (\widehat{Y}_1(x^*) = 136.4)$$

$$d_2(\widehat{Y}_2) = 1.0; \quad (\widehat{Y}_2(x^*) = 157.1)$$

$$d_3(\widehat{Y}_3) = 0.49; \quad (\widehat{Y}_3(x^*) = 450.56)$$

$$d_4(\widehat{Y}_4) = 0.76; \quad (\widehat{Y}_4(x^*) = 69.26)$$

The overall desirability for this solution is 0.596. All responses are predicted to be within the desired limits.

Figure E.2 shows a 3D plot of the overall desirability function $D(x)$ for the (x_2, x_3) plane when x_1 is fixed at −0.10. The function $D(x)$ is quite "flat" in the vicinity of the optimal solution, indicating that small variations around x^* are predicted to not change the overall desirability drastically. However, the importance of performing confirmatory runs at the estimated optimal operating conditions should be emphasized. This is particularly true in this example given the poor fit of the response models (e.g. $\widehat{Y}_2$).

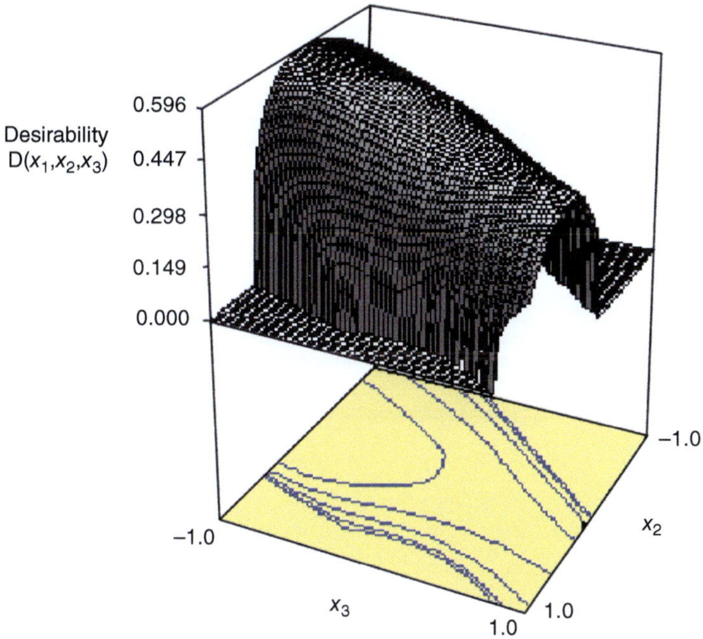

Figure E.2 Overall desirability function for example problem.

References

1 Box, G.E.P. and Draper, N.R. (1969). *Evolutionary Operation A Statistical Method for Process Improvement*. New York: Wiley.
2 Hockman, K.K. and Jenkins, M.W. (1994). *Design of Experiments: Neglected Key to Quality Reveals How Variables Interact To Optimize Process*. Managing Technology Today.
3 Lazic, R.Ž. (2004). *Design of Experiments in Chemical Engineering: A Practical Guide*. Wiley-VCH.
4 Lazic, R. Ž. (2013). *Kaolin ozonation and bleaching optimization Personal Communication*, BASF Catalysts LLC, Gordon, GA, USA.
5 Lazic, R. Ž. (2011). *This invention is related to processing kaolin by controlled cavitation (CFC), directed to kaolin ozonation and/or bleaching processing*. Patent Application US 2011/0297046A1.
6 McCulloch, W.S. and Pitts, W.H. (1943). A logical calculus of the ideas immanent in nervous activity. *Bulletin of Mathematical Biophysics* 5: 115–133.
7 Hebb, D.O. (1949). *The Organization of Behavior*. NY: Wiley.
8 Rumelhart, D.E., Hinton, G.E., and Williams, R.J. (1986). Learning internal representations by error propagation. *Nature* 323 (6088): 533–536.
9 Lazic, R. Ž. (2020). *Run Design of Experiment (DOE) on Neural-Network (NN) Model*. Research Gate.
10 Box, G.E.P. and Hunter, J.S. (1959). Condensed calculations for evolutionary operation programs. *Technometrics* 1: 77–95.
11 Lowe, C.W. (1964). Some techniques of evolutionary operations. *Transactions of the Institute of Chemical Engineering*. 42: T332–T344.
12 Anonymous (1961). EVOP in plant scale experiments. *Industrial and Engineering Chemistry* 53 (12): 40A–46A.
13 Anonymous (1961). Evolutionary operation, a tool to optimize a process without disturbing it. *Petroleum Week* 7: 48–49.
14 Hehner, H. O. (1963). Personal communication, December 10.
15 Anonymous (1960). In big computer test, plant raises NH_4 output. *Chemical Engineering* 67 (14): 110–114.
16 Cestoni, D.J., Ringelman, R.E., Olson, L.R. et al. (1960). *Chemical Engineering Progress* 56 (5): 73–77.

Full Scale Plant Optimization in Chemical Engineering: A Practical Guide, First Edition. Živorad R. Lazič.

17 Coutie, G.A. (1958). Use statistics for optimization. *Chemical Engineering* 66 (16): 190–191.

18 Coutie, G.A. (1959). *Statistical considerations in process optimization.* In: *Proc. of the Joint Symposium on Instrumentation and Computation in Process Development and Plant Design*, vol. 1. London.

19 Pursglove, S.D. (1961). Here's how you can use EVOP. *Chemical Processing* 23–24: 95–97.

20 DeBusk, R.E. (1962). Experience in EVOP at Tennessee Eastman Co. *Industrial Quality Control* 19 (4): 15–21.

21 Wilson, C.L. (1960). A designed experiment. *Industrial and Engineering Chemistry* 52 (6): 504–506.

22 The Chemstrand Corporation (1959). *Annual Report*, 13.

23 Davidson, I.D. (1961). EVOP a new strategy in experimental design. *The B.H.P. Technical Bulletin, Australia* 5: 6–11.

24 Thomas, M.D. and Webster, G.L. (1960). *Evolutionary Operation Applied to Resistance Welding of Automotive Sheet Metal.* Flint, Michigan: General Motors Institute (unpublished).

25 Thomas, M. D. (1965). Personal communication, December 13.

26 Weaver, O.R. (1963). EVOP in operation. In: *A Paper Read to Seventh Technical Conference of the Chem. Division of A.S.Q.C.* St. Paul, Minnesota.

27 Klingel, A.R. and McIntyre, R.G. (1962). An experimental strategy for investigating commercial process. *Applied Statistics* 11 (2): 79–92.

28 Samuel, O.C. (1962). Experimenting on the plant process without interfering with production. *Food Processing* 23 (2): 31–33, 49.

29 Koleff, A. M. (1963). *Application of EVOP to corn grinding operation at A.E. Stanley Manufacturing Co.* Personal communication.

30 Remmers, E.G. and Dunn, C.G. (1961). Process improvement of a fermentation product. *Industrial and Engineering Chemistry* 53 (9): 743–745.

31 Anonymous (1963). Maumee moves closer to optimum process to make saccharin. *Chemical and Engineering News* (December 9): 76–78.

32 Richard S. Bingham, Jr. (1963). Try EVOP For Systematic Process Improvement September 1963. *Industrial Quality Control* 20 (3): 17–23.

33 Lowe, C.W. (1974). Evolutionary operation in action. *Applied Statistics* 23 (2): 218–226.

34 Peter, T.K. (2005). Thyregod using evolutionary operation to improve yield in biotechnological processes. *Quality and Reliability Engineering International* 21 (5): 457–463.

35 Bruse, R. G. (1967). *Some Optimization Procedures Used in Response Surface Methodology. Master thesis of Science in Statistics.* Virginia Polytechnic Institute.

36 Myers, R.H. and Montgomery, D.C. (2016). *Response Surface Methodology: Process and Product Optimization Using Designed Experiments.* Wiley.

37 Spendley, W., Hext, G.R., and Himsworth, F.R. (1962). Sequential Application of Simplex Designs in Optimisation and Evolutionary Operation. *Technometrics* 4 (4): 441–461.

38 Ramana, P.V. (2020). *Evolutionary Operation Methodology (EVOP)*. Six Sigma Study Guide.

39 Nelder, J.A. and Mead, R.A. (1965). A simplex method for function minimization. *Computer Journal* 7: 308–313.

40 Aberg, E.R. and Gustavsson, A.G.T. (1982). Design and evaluation of modified simplex methods. *Analytica Chimica Acta* 144: 39–53.

41 Betteridge, D., Wade, A.P., and Howard, A.G. (1985). Reflections on the modified simplex-II. *Talanta* 32: 723–734.

42 Zadeh, L.A. (1965). Fuzzy sets. *Information and Control* 8: 338–353.

43 Otto, M. (1988). Fuzzy theory explained. *Chemometrics and Intelligent Laboratory Systems* 4: 101–120.

44 Derringer, G. and Suich, R. (1980). Simultaneous optimization of several response variables. *Journal of Quality Technology* 12: 214–219.

45 Banerjee, R. and Bhattacharyya, B.C. (1992). Evolutionary optimization (EVOP) to optimize protease biosynthesis by Rhizopus oryzae. *Bioprocess Engineering* 8: 151–155.

46 Banerjee, R. and Bhattacharyya, B.C. (1993). Evolutionary optimization (EVOP) to optimize three-dimensional biological experiments. *Biotechnology and Bioengineering* 13 (2): 149–155.

47 Chatto, K.A. and Kennard, R.W. (1961). Evolutionary operation in plant-scale experiment. *Industrial and Engineering Chemistry* 53 (12): A47.

48 Tunga, R., Banerjee, R., and Bhattacharyya, B.C. (1999). Optimization of n variable biological experiments by evolutionary operation-factorial design technique. *Journal of Bioscience and Bioengineering* 87 (2): 224–230.

Index

n

o

p

q

r

s

t